# British *Longcase* Clocks

Derek Roberts

1469 Morstein Road, West Chester, Pennsylvania 19380

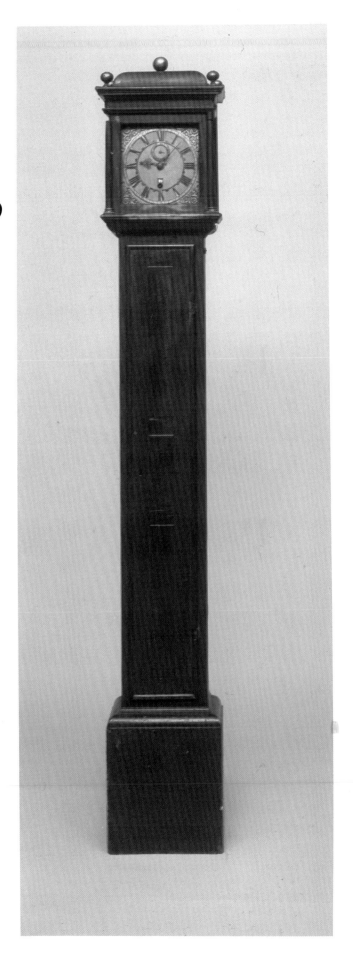

## Kings and Queens of England

Elizabeth I . . . . . . . . . . . . . . . . . . . . . . . . . . 1558-1603

James I . . . . . . . . . . . . . . . . . . . . . . . . . . . . . 1603-1625

Charles I . . . . . . . . . . . . . . . . . . . . . . . . . . . . 1625-1649

Commonwealth . . . . . . . . . . . . . . . . . . . . . 1649-1660

Charles II . . . . . . . . . . . . . . . . . . . . . . . . . . . 1660-1685

James II . . . . . . . . . . . . . . . . . . . . . . . . . . . . 1685-1689

William III and Mary II . . . . . . . . . . . . . . 1689-1694

William III . . . . . . . . . . . . . . . . . . . . . . . . . 1694-1702

Anne . . . . . . . . . . . . . . . . . . . . . . . . . . . . . . 1702-1714

George I . . . . . . . . . . . . . . . . . . . . . . . . . . . 1714-1727

George II . . . . . . . . . . . . . . . . . . . . . . . . . . 1727-1760

George III . . . . . . . . . . . . . . . . . . . . . . . . . 1760-1820

George IV . . . . . . . . . . . . . . . . . . . . . . . . . 1820-1830

William IV . . . . . . . . . . . . . . . . . . . . . . . . 1830-1837

Victoria . . . . . . . . . . . . . . . . . . . . . . . . . . . 1837-1901

## Furniture Periods in England

Thomas Chippendale . . . . . . . . . . c. 1760 (1718-1779)

Thomas Chippendale, younger . . c. 1789 (1749-1822)

Hepplewhite . . . . . . . . . . . . . . . . . . c. 1775 (died 1786)

Adam . . . . . . . . . . . . . . . . . . . . . . . . c. 1775 (1728-1792)

Sheraton . . . . . . . . . . . . . . . . . . . . . c. 1790 (1751-1806)

Regency . . . . . . . . . . . . . . . . . . . . . . . . . . . 1800-1830

---

Published by Schiffer Publishing, Ltd.
1469 Morstein Road
West Chester, Pennsylvania 19380
Please write for a free catalog
This book may be purchased from the publisher.
Please include $2.00 postage.
Try your bookstore first.

---

Copyright © 1990 by Derek Roberts.
Library of Congress Catalog Number: 89-64089.

All rights reserved. No part of this work may be reproduced or used in any forms or by any means—graphic, electronic or mechanical, including photocopying or information storage and retrieval systems—without written permission from the copyright holder.

Printed in the United States of America.
ISBN: 0-88740-230-5

**Title page photo:**
Figure 1. *John Wise*. An interesting and rare longcase clock full of originality, as is often the case in this early period. The panelled ebony veneered case standing only 5' 7½" high employs a shallow caddy with three ball finials. See Figure 32 for dial. Photo courtesy Sotheby's, Bond Street.

# Contents

The Kings and Queens of England .......... ii

Acknowledgments ........................ 6

Preface ................................ 7

Chapter 1.
The Longcase Clock, its Origins
    and Evolution ...................... 9

Chapter 2.
Early Square-Dial Longcase Clocks ......... 23
    Ebonized
    Walnut and Olivewood
    Parquetry
    Floral marquetry
    Arabesque marquetry
    Seaweed marquetry

Chapter 3.
Rare Clocks ............................ 65
    Long duration
    Grande Sonnerie Striking
    Night Clocks
    Those showing seconds in the arch

Chapter 4.
Astronomical and Equation Clocks .......... 79

Chapter 5.
Walnut Breakarch Longcase Clocks ........ 113
    (including various other woods)

Chapter 6.
Lacquer or Chinoiserie Decoration ......... 131

Chapter 7.
"London" Mahogany Longcase Clocks ..... 147

Chapter 8.
Quarter Chiming and Musical Clocks ....... 165

Chapter 9.
Brass-Faced Oak Eight-Day
    Longcase Clocks ................... 183

Chapter 10.
Country Mahogany Longcase Clocks ....... 201

Chapter 11.
Silvered Dials ......................... 213

Chapter 12.
White Dials ........................... 219

Chapter 13.
Thirty-hour Longcase Clocks ............. 241

Chapter 14.
The Victorian and Edwardian Eras ......... 253

Chapter 15.
Regional Characteristics ................. 263

Chapter 16.
Precision Timekeeping .................. 283

Chapter 17.
The Makers ........................... 304

Chapter 18.
Hands and Spandrels .................... 327

Chapter 19.
Care of a Longcase Clock ................ 337

Appendix: Maps and List of Place Names ... 346

Glossary of Terms ...................... 362
    including diagrams showing the components of
    a longcase clock

Makers Index ......................... 369

General Index ......................... 371

Figure 2. *Edward East, Londini.* Other views are seen in Figure 34 A and B. A fine early walnut longcase clock by Edward East shown just as it was discovered some 15-20 years ago. It has delicate Cherub spandrels; narrow chapter and seconds rings which would originally have been silvered and fine steel hands. The walnut case, a little distressed where it has been kicked on its' base over the years, is of excellent proportions and has walnut veneers on it anywhere up to 1/8″ thick, circa 1680.

Figure 3. *John Andrews, London, circa 1695.* A highly figured walnut longcase with an 11-inch square dial with ringed winding holes.

Figure 4 A. (left) *John Wise, London.* A small walnut longcase clock, the nine-inch dial signed John Wise. The center is engraved with tulips in an urn, the corners with unusual dragon and leaf engraved spandrels. The associated 30-hour plated movement with bell striking has an outside count wheel and anchor escapement. The case has a flat-topped hood flanked by Corinthian pillars, a trunk door with a lenticle and well-figured wood veneer, and a plinth with a recessed panel, the sides crossbanded and veneered. 6′ 2″ (188 cm) high.

A 30-hour clock by John Wise is illustrated in the journal of the Antiquarian Horological Society, March 1973, page 173. Photos courtesy Sotheby's, Bond St., London.

Figures 4 B and 35. *John Ebsworth, London.* A walnut longcase clock, circa 1685 the ten-inch dial signed *John Ebsworth Londini Fecit* with finely matted center, skeletonized chapter ring, seconds dial, calendar aperture and cherub spandrels. The five-pillar movement with bolt and shutter maintaining power, bell striking and outside count wheel. The case with a carved cresting, rising hood flanked by spiral columns and long trunk door with a lenticle, 6′ 9½″ (207 cm) high.

John Ebsworth was apprenticed in 1657, free of the Clockmakers Company in 1665, and Master in 1697; he died in 1703. See also Figure 35. Photo courtesy Sotheby's, London.

# Acknowledgments

Although the majority of the illustrations contained in this book are of clocks Derek Roberts Antiques has owned at one time or another, these by no means cover the entire spectrum of longcase clocks. Therefore, I am very grateful for the assistance given to me, mainly in specialized areas, of various individuals and authorities: Dr. Vaughan of the Science Museum in London supplied many of the photographs and descriptions for the introduction; the British Museum supplied photographs of the East Night clock, Mostyn Tompion, and the year equation clock by Quare; and the Fitzwilliam Museum of Cambridge supplied photographs of two of their fine clocks. We also are grateful to Her Majesty the Queen for granting us permission to illustrate various clocks in the Royal Collection.

Richard Garnier and Ben Wright at Christie's and Michael Turner and Tina Miller at Sotheby's kindly supplied photographs and details of several fine longcase clocks. Phillips, Edinburgh, supplied pictures of a variety of examples made in Scotland.

Mr. Ronald Lee spent much time and trouble in finding us suitable pictures of clocks by the Knibb Family and inevitably when writing we leaned heavily on his excellently researched book on this subject. Indeed, it would be appropriate at this stage to acknowledge all those who in the past have made contributions to our overall knowledge of longcase clocks and their makers. Examples that spring to mind include Ronald A. Lee's *The First Twelve Years of the Pendulum Clock*, which accompanied his exhibition on this subject in 1969. Cedric Jagger's *Royal Clocks*, which was published by Robert Hale in London in 1983. R.W. Symonds' fine book *Thomas Tompion, His Life and Work*, published by B.T. Batsford Ltd. of London. Percy G. Dawson, C.B. Drover, and D.W. Parkes' *Early English Clocks*, which was published by the Antique Collectors Club in 1982 and is a superbly researched and written account of English clockmaking prior to 1710. Tom Robinson's *The Longcase Clock*, also published by the Antique Collectors Club. Brian Loomes's research in many areas, including articles and books on the regional characteristics of various counties such as Lancashire, Yorkshire, and Westmorland; his two books giving a general list of clockmakers and details of the early clockmakers of Great Britain; and his research, articles, and book *The White Dial Clock*, which ventured into a previously uncharted field. The Antiquarian Horological Society, which has proved to be an invaluable medium for the publication and dissemination of much useful information by a large number of knowledgeable contributors over the past 30 years.

I am very grateful to the gifted horological illustrator David Penney for agreeing to supply all the line drawings of escapements included in this book and also those showing the components of the longcase clock. It should be noted that he retains all rights regarding any future publication of these drawings. John Martin kindly produced the most helpful series of pictures on how to set up and take care of your longcase clock.

During our first ten to twelve years in Tonbridge, many of the clocks that passed through our hands were photographed by Kenneth Clark. He is responsible for a considerable number of the photographs featured here, which is greatly appreciated. Subsequently, for convenience, we undertook all our own photography. Hence, the majority of the remaining photographs were produced by the author.

Throughout the preparation of this book my family and the entire staff of Derek Roberts Antiques have given me unfailing support and help in every way possible. However, virtually the entire production of the manuscript has been carried out by Julia Vella with untiring energy and unbounded enthusiasm, and it is this which made the whole thing possible. I also am greatly indebted to Nancy Schiffer for all the time, care and professional skills that she has brought to bear on the layout of this book and the numerous ways in which she has assisted me during its production.

Finally, it would be remiss of us to close without taking the opportunity to thank the people of Schiffer Publishing Ltd. for giving us the chance to publish this book which documents so many of the longcase clocks which have passed through our hands in the last 20 years and have given us so much pleasure.

# Preface

The longcase clock has been a part of the Englishman's home for some 300 years and has gained a place in his heart seldom if ever achieved by any other domestic item. An example of its importance in family life is that it, with the bed and the house, was nearly always mentioned specifically in any will.

Part of the attraction is in the fact that it is a living thing, striking out the hours, and soothing us with its gentle tick which at 60 to the minute is just a little slower than our hearts. Even the size and shape of a small country clock is not dissimilar from that of man, with its hood with a face the same height as ours, a trunk and a base.

Whereas with most furniture it is impossible to say where and by whom it was made, with a longcase clock, most of which are signed, we know not just who made it but in which town or village and also with a little research we can often even locate the street and the actual house where it was produced. Moreover, with a little more trouble we may find out something about the man who made it.

The purpose of this book is to enable all those who own or would like to own a longcase clock to learn a little more about the way in which they work, the maintenance they require, their origins and evolution, and the vast range of longcase clocks which have been produced over the last 300 years. By knowing more about their own longcase clock it is hoped that those who possess one will appreciate it more, and thus gain greater pleasure and take more care of it so that it can be passed on to future generations in at least as good a condition as when they received it.

Because of the enormity of the subject it is impossible to go into many areas in the depth some collectors might desire. Examples of this are the evolution of the longcase clock prior to 1700; precision timekeeping; astronomical clocks and technical details of many of the movements. However, it was felt to be more important to give readers an overall grasp of the subject; and for those who wish to know more, suitable references are provided.

For the foregoing reasons, rather than concentrating on the finest London pieces produced by a handful of makers (most of which are now in private collections or museums) this book devotes almost as much space to country clocks, including those of 30-hours duration which are often full of character and can give the average owner just as much pleasure.

It is to be hoped that this book enables us to share with the readers the great pleasure we have gained over the last twenty years from owning, often admittedly only for a short time, such an enormous variety of fine longcase clocks.

Figure 5. *Cleopatra's Needle*. This was built on the Heliopolis in Egypt by the Pharaoh Thothmes III about 1500 B.C. and would have been in effect one of several giant sundials predating our public clocks by nearly 3000 years. It was brought to England in 1877 and now stands on the Thames Embankment.

Figure 6. *Portable Shadow Clock*. A reproduction of an early (8th-10th century B.C.) shadow clock. To use the instrument it was placed in an east-west direction with the crossbar facing east in the morning and west in the afternoon. The shadow from the cross bar fell on the base which has a scale of hours marked upon it. Photo courtesy Science Museum, London.

Figure 7. *An Ivory Tablet Sundial*. The majority of these were made in the 16th and 17th centuries, frequently in Germany and were at that time thought of as pocket watches. They were turned due south by means of a compass in their base and the time was then read using the shadow cast by a thread stretched between the two halves of the sundial. They frequently incorporated correction tables for several major cities. This example is signed *Paulus Reinman Normbergae Faciebat*.

Figure 8. *An 18th century French ring sundial*. This indicates the time by light passing through a pinhole onto a graduated ring parallel to the earth's equator. This pinhole is in a piece which slides along an axial bar parallel to the earth's axis and is adjusted each day to correspond with the changes in the declination of the sun. The equatorial ring and axis bar are mounted inside a ring which is set in the meridian and which itself slides within an outer meridian ring which is fixed to the base. The inner meridian ring is calibrated so that it can be set for any latitude.

To tell the time the pinhole is set to the correct date and turned to face the sun. The whole instrument is then rotated about a vertical axis until the spot of light from the pinhole falls upon the hour circle. Photo courtesy Science Museum, London.

# Chapter 1.
# The Longcase Clock, its Origins and Evolution

Since his origins man has tried to measure time, probably by the built-in sense of time which all the animal species possess to a greater or lesser degree. The elevation of the sun in the sky and the movement of various heavenly bodies would also have given him guidance as would the length of the shadows cast by various natural bodies such as mountains and trees.

As civilization gradually progressed, a more precise method of measuring time was needed to regulate the life of the communities and decree at what hour public meetings or prayers, for instance, should take place. Thus *shadow clocks* such as Cleopatra's Needle (Figure 5), which now resides on the Thames Embankment, were erected in public places and, at a slightly later date, portable examples (Figure 6), were evolved which finally resulted in the beautiful little ivory tablet sundials (Figure 7) and ring sundials (Figure 8) which, in the 16th and 17th century, were the equivalent of the gentleman's pocket watch. Numerous different and fascinating forms of sundial were produced over the centuries (Figure 9); but undoubtedly the commonest was what might be referred to as the garden sundial (Figure 10). In the early days of the longcase clock this was used with the aid of equation tables to set the clock to its correct time and regulate it. Other early methods used to measure time were *sandglasses* (Figures 11 and 14), *clepsydra* (water clocks, see Figure 12) most of which were based on the speed at which water ran out of a container with a hole in the bottom of it, and *fire clocks*. These last ranged from a single candle marked down its length with the hours, and lamp timekeepers where the glass oil reservoir is graduated in hours (Figure 15), to complex clocks in which a trail of powder gradually burnt down a channel in a container, and even alarm clocks in which, when a flame reached threads and burnt through them, the balls they were supporting fell with a loud clang onto a metal tray.

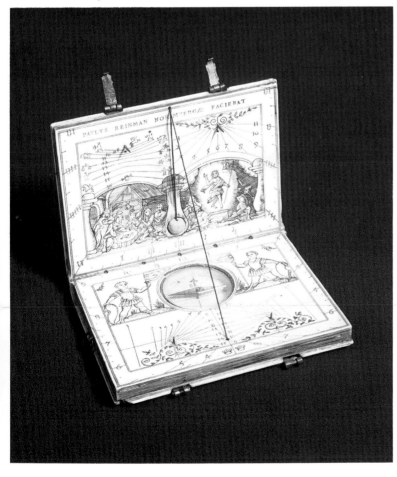

## 10 Longcase Clock

Figure 9. *Seventeenth Century Wooden Column Sundial.* This type of sundial, often portable, which in effect measures the altitude of the sun, has been in use since Roman times. The top of the sundial is rotated until the gnomon fixed to it is immediately above the correct date. The sundial is then suspended by a cord attached to a ring on top and turned until the gnomon faces the sun and the shadow from it is exactly vertical. The tip of the shadow now indicates the time.

In this particular sundial the top is removable and thus the gnomon can be folded up and stored inside. The base of the sundial can be rotated to indicate the time in various parts of the world provided the time difference is known. Photo courtesy Science Museum, London.

The first mechanical clocks to appear were probably made at the end of the 13th or early in the 14th century. Their production became possible because of the invention of a device known as "an escapement" which releases a train of wheels which are used to indicate time at set intervals.

On the earlier mechanical clocks which were all made for public use, there were often no hands, just a bell which sounded the hours and was used for instance to call people to prayers. Initially, these clocks made use of a foliot or rotating arm (Figure 13) on which weights could be moved in or out to vary its speed of oscillation and thus its time keeping ability.

The foliot was mounted on an arbor to which pallets (flanges) were attached which alternately released and arrested the "escape" wheel and thus the train of wheels to which the clock's hands were attached. The arm was kept swinging by the power transmitted to it from the train of wheels which were driven by a weight. One of the earliest turret clocks still in existence is that which was made for Wells Cathedral in 1392 (Figure 16).

As time progressed the foliot tended to be replaced by a balance wheel, particularly on domestic clocks and in the late 15th century Leonardo da Vinci made drawings illustrating the possible application of a pendulum, regulated by gravity, to a clock.

Between 1637 and 1639 Galileo wrote about his observations on the pendulum, inspired by the even swinging to and fro of a chandelier in a cathedral but he does not appear to have applied the principle to clocks although a drawing of a mechanism devised by him (Figure 17), which is possibly a pendulum clock, is still in existence. There is also the possibility that Johann Philipp Treffler made a pendulum clock in 1656.

Despite these earlier references the credit for the introduction of the pendulum clock is given to Christian Huygens, the Dutch mathematician and Solomon Coster who, in 1657, respectively designed and made the oldest pendulum clock known to exist. The news of their discovery rapidly spread through Europe and particularly to London and Paris where the design was readily adopted. Probably the first person to use it in London was Ahasuerus Fromanteel.

The use of the pendulum, initially in conjunction with the verge escapement (Figure 19), led to dramatic changes in case styles and the invention of the anchor escapement (Figure 20), probably by Hooke in collaboration with Clement around 1670, accelerated that process as it made possible the use of the "Royal" or seconds beating pendulum some 39.24 inches long.

To protect this long pendulum and also the driving weights of the clock, a case known as a "longcase" was devised which was also designed to harmonize with the furnishings of the period. Initially these clock cases were ebonized or ebony veneered to fit into the Puritan era into which they were born, but by the 1670s this austerity started to give way to harmonize with more decorative furnishings. Walnut and occasionally other veneers such as olive wood started to be used instead of ebony and within ten years inlay began to appear on clocks, initially in the form of parquetry, and by around 1685 floral marquetry was adopted. In the beginning this was laid in panels on the trunk door and base, but within a few years it had extended to the hood and the mould beneath it also. By the end of the century Arabesque marquetry started to appear as bold strapwork interspersed with figures of birds, animals and human beings. At its best this form of marquetry can be superb. Such was the speed of change of clock case design at this period that it was only to be another five years before seaweed marquetry was being produced. This is somewhat similar to, but much finer than, Arabesque and does not include any figures.

As the design of the case continued to change quite rapidly from 1660-1710, so did that of the movement and dial. The earliest dials were ten inches or less square and occasionally just engraved in the four corners (Figures 30 and 33). However, in the majority of cases, raised cast brass and frequently gilt decorations known as spandrels were used. The earliest of these was the winged cherub, the design of which gradually changed over the next five to ten years and finally gave way to the twin cherubs upholding the crown which is believed to have evolved when William and Mary took the throne. From that time on various other designs of spandrel were adopted. As the design of the dial and spandrels gradually changed so did that of the hands. The earliest longcases were small, probably only averaging six feet three inches or less, of excellent proportions and usually had dials of ten inches or a little less square. However as the affluence and the size of the houses in England and particularly London increased, so did the height of the clocks to match them. Thus by 1700-1710 the average height of a clock was well over seven feet and the dial size had been increased, first to 11 inches around 1685 and later to 12 inches to keep it in proportion.

The increase in height of the clock cases also brought about a change in case design. Whereas the short early clocks employed a "lift-up" hood (Figure 21) which runs up on runners and was locked there while the clock was being wound, this was obviously impractical with the taller clocks when the hood was raised. To overcome this problem the hood was modified so that it could be removed by sliding it forward. This had the added advantage that a "mask" could then be provided which fitted accurately up to the edges of the dial. Previously, this could not be used as it would have fouled the spandrels and hands when the hood was raised.

At approximately the same time as the hood was modified to slide forward the use of a front opening door was adopted to simplify winding by avoiding having to remove the hood.

Figure 10. *Garden Sundial by William Deane, London 1718.* This sundial which was made for the latitude of London (51½°) is divided to read minutes. Inside the hour circle is a ring engraved to give the "equation of time" for each day of the year so that mean time may be calculated and thus the sundial used to set a clock by. Photo courtesy Science Museum, London.

Opposite page:
Figure 12. *Cast taken from early Egyptian Water Clock.* The original of this was found in Karnak Temple, Upper Egypt and dates from 1415-1380 B.C. The vessel was filled with water which ran out slowly from a hole near the bottom. To compensate for the more rapid flow when the vessel is full, it has been tapered. The hours are read from a series of scales on the inner aspect of the vessel. As at that time the lengths of the hours varied each month, largely depending on the length of the day and night, each of which was divided into 12, 12 different scales were required, one for each month of the year. Photo courtesy Science Museum, London.

Figure 11. *Set of Four Sandglasses, Probably Italian, circa 1720.* Studs on the ebony and ivory frame indicate how long each sandglass takes to empty; one stud one-quarter hour; two studs one-half hour; 3 studs three-quarters hour; 4 studs one hour. Each of the glasses is joined at the center and is filled with fine emery powder. Photo courtesy Science Museum, London.

These changes of design were not just applied to the new clocks being made. At the same time many of the earlier clocks were modified and converted to forward sliding hoods, generally with, but occasionally without a door. Telltale signs of this are the vertical grooves at the back of each side of the hood and the absence of a mask around the dial.

The caddy top was a feature of a lot of the clock cases made before 1710 although in many instances these were subsequently removed; possibly to lower the height of the clock, and evidence of this may often be seen. The caddy top probably came to its ultimate development in the magnificent Arabesque and all over floral marquetry clocks that were produced around 1700-1710 (Figure 85) which often rose to a height of around eight feet three inches to eight feet six inches and were surmounted by five finials.

By around 1715 the use of marquetry had largely given way to the adoption of walnut veneers, often beautifully figured, with which to decorate the clock cases and at the same time the breakarch dial, in which a semicircle was added to the top of the square dial, was adopted, but this feature never completely supplanted the square dial.

Thomas Tompion was one of the first to break away from the square dial as early as 1695 using the additional space at the top for such features as showing the equation of time. Other uses to which it was rapidly put was the provision of a name plate and also sometimes to indicate the time of high water at different ports. Various other uses for this space may be seen elsewhere in this book.

A disadvantage of walnut had always been that it was, largely because of the beautiful figuring it often possessed, a relatively unstable wood and it also was susceptible to rotting due to damp floors and worm attacks. These problems were heightened by a shortage of good walnut brought about by a very severe winter in 1709 which killed many of the walnut trees in Europe and particularly in France, where, by 1720, this wood was banned as an export.

Figure 13. *The Dover Castle Clock circa 1600.* This clock is of particular interest as it still retains its original escapement and foliot, the movable weights for its regulation being seen suspended from it. The balance swings to and fro in about eight seconds and this allows the great wheel to rotate once an hour. It was removed from Dover Castle in 1872. Photo courtesy Science Museum, London.

A substitute thus had to be found for walnut and this turned out to be mahogany which was available in abundance in some of the colonies such as Cuba. The advantages of this material were that it was stronger than walnut; available in larger sizes and was not normally attacked by worm. Both straight grain and highly figured varieties were available and thus it could be used either both for building up the carcase of the case or providing beautiful highly figured veneers for the front.

Although walnut cases continued to be produced in London up until the 1750s or even later, often in conjunction with the square dial as favored by certain fine makers such as Graham, the vast majority of clock cases made in London between 1750 and 1830, after which time longcase clock production had largely ceased in the capital, were of mahogany. However in the country oak, which was more readily available there, was often used as a simpler and cheaper alternative.

### LACQUER CLOCK CASES

The term "lacquer work" is applied to furniture which is decorated with gilded gesso, usually raised, which is laid on a Japanned background. Although lacquer work is thought to have been employed in Japan and China as early as the 3rd century B.C., its importation into England only started in the early 17th century following the formation of the East India Company which brought back from the Orient some examples. These were usually small items such as boxes, furniture very rarely being shipped at this stage.

Lacquer work started being produced in Holland in the 1680s and within 20 years had reached quite a high standard, but very few lacquer longcase clocks were produced in England prior to 1700. The majority were made between 1720-1750 and production continued on until at least the 1770s, there being a considerable revival in Oriental work and designs in the mid-18th century. A typical example of this is the Chinese Chippendale designs contained in *The Director*. Contrary to popular belief very few clock cases were ever sent out to the Far East for lacquering.

Figure 14. *Four-Hour Ship's Sandglass*. This large sandglass, which was hung by its ropes and inverted every four hours, was used aboard ship where the 24-hour day was divided into six "watches" of four hours. Photo courtesy Science Museum, London.

## COUNTRY CLOCKMAKING

Until about 1690, the vast majority of longcase clocks were produced in London but after this time they started to be made in ever-increasing numbers throughout the country. At first production spread to the other major cities such as York, Salisbury, Exeter and Bristol to name but a few, but quite rapidly it involved the whole country. However, because the spread only radiated out from London slowly, the styles being used elsewhere often lagged 20-30 years behind that of the capital. Moreover, simpler designs and different woods were often employed, partly in the interests of economy but also because of the comparative lack of skill and training of some of the country craftsmen, particularly in the more remote areas.

## THIRTY-HOUR CLOCKS

A further simplification in the country was the widespread use of clocks of only 30-hours duration, which had to be wound once a day as opposed to once a week or even a month on the fine clocks being produced in London. These 30-hour clocks were usually small, i.e., less than six feet six inches tall, to fit into the country cottages, and were generally made of solid oak. The earlier 30-hour clocks tended to only have one hand, that being for the hours, but as the 18th century progressed it was increasingly common to incorporate a minute hand in the design.

## PAINTED DIALS

By the early 1770s painted dials started to appear on longcase clocks. Although this fashion was fairly rapidly adopted throughout most of the country, relatively few London-made clocks are to be seen with painted dials. By then longcase clock production was already beginning to decline in London, and where a plainer dial was preferred, London makers tended to use a flat brass one which was engraved and then black-waxed to accentuate the engraving of the numerals, and silvered all over, omitting the raised chapter ring and spandrels. However, some London-made clocks with painted dials are to be seen, usually with completely plain white dials and black numerals, and occasionally with the corners delicately decorated with floral designs, sometimes just gilt or occasionally in colors. These were produced by fine makers such as Eardley Norton, whose corner decoration, often little swags of flowers such as roses, were beautifully executed.

In the country the first painted dials were possibly plain white all over with black numerals, but this quickly gave way to more attractive designs, often incorporating flowers with an accompanying picture in the arch. Quite early these dials started to be produced in large numbers by specialist manufacturers such as Osborne and Wilson of Birmingham.

As the 19th century progressed the amount of decoration applied to painted dials continued to increase, particularly in the North of England, until virtually the whole dial except the chapter ring was decorated.

## 19th CENTURY LONGCASE CLOCKS

A limited number of longcase clocks were made in London in the Regency Period (1800-1830). These were usually well made, of simple pleasing lines, generally under seven feet tall, and often displayed a circular dial, sometimes engraved and silvered and sometimes painted.

By 1830 the era of the longcase clock had almost passed in London as imported clocks, first from France, then from Germany and America, filled the market. Production had largely ceased, even in the north of England, by 1860. Nevertheless, some longcase clocks were still produced until the advent of the first World War; their production staging a brief revival in the early 1900s when some magnificent and very expensive chiming longcase clocks, usually over eight feet tall, were produced.

Figure 15. *18th Century North German Oil Clock*. This consists of a pewter stand with a glass reservoir to the lamp, alongside which is a scale graduated from 8 P.M. to 7 A.M. which indicates the time by the level of the unburned oil. However, as the reservoir in this lamp is pear shaped and the scale is evenly divided, this particular example must have been very inaccurate. Photo courtesy Science Museum, London.

Figure 16. *Wells Cathedral Clock circa 1392*. This is thought to be, after the Salisbury Cathedral clock, the oldest to have survived in England. It was originally controlled by a foliot balance but was fitted with a pendulum and anchor escapement some time after 1670. It was in Wells until 1835 and came to London for exhibition in 1871. Photo courtesy Science Museum, London.

16 *Longcase Clock*

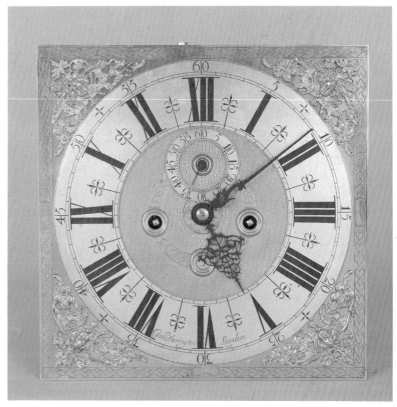

Figure 17. *Model of Galileo's Escapement.* This model, made in Florence in the 19th century, is based on the drawing showing the incomplete clock made by Galileo's son Vincenzio in 1649 to his father's design. Photo courtesy Science Museum, London.

Figure 18 A and B. *External Count Wheel Strike*

It is seen here on a month-duration movement by Etherington. Note the fine dial with wheatear engraving to the border and an additional engraved ring in the dial center, a feature also seen on some clocks by East.

## THE MECHANICAL DEVELOPMENT OF THE LONGCASE CLOCK

Only brief mention is made here of the mechanical development of the longcase clock as by 1710 its design had virtually become finalized and was to change little over the next 200 years other than in details such as the shape of the pillars and various other components. However a few points are worth noting.
— A small number of very early longcase clocks were made incorporating a verge escapement with short bob pendulum (see Figures 27 and 28).

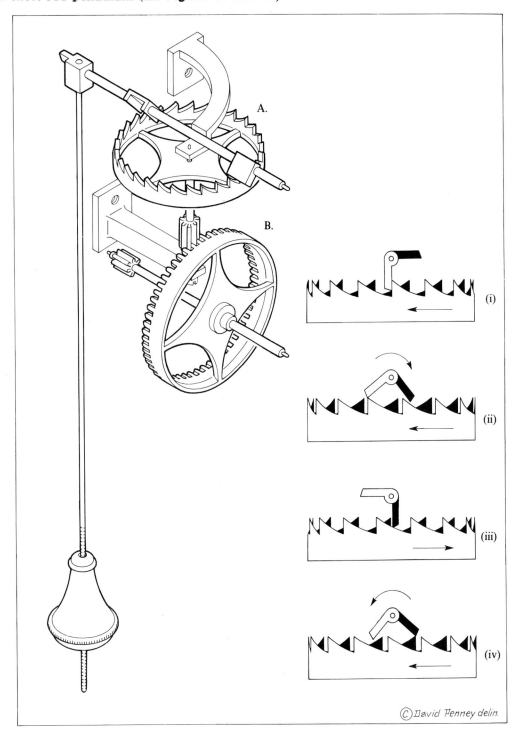

Figure 19. *The Verge Escapement.* In figure (i) the pendulum is at the extreme right of its swing and the front pallet is engaged with the vertical face of the 'scape wheel tooth. This tooth imparts impulse to the pendulum via the front verge pallet, as the pendulum swings from right to left until the unlocking occurs (ii), the tooth giving impulse drops, and the rear verge pallet engages another tooth at the 'scape wheel; the momentum of the pendulum giving recoil to the 'scape wheel until the pendulum comes to rest (iii) at the end of its swing. The same sequence of events then occurs with the pendulum swinging from left to right. Drawn by David Penney.

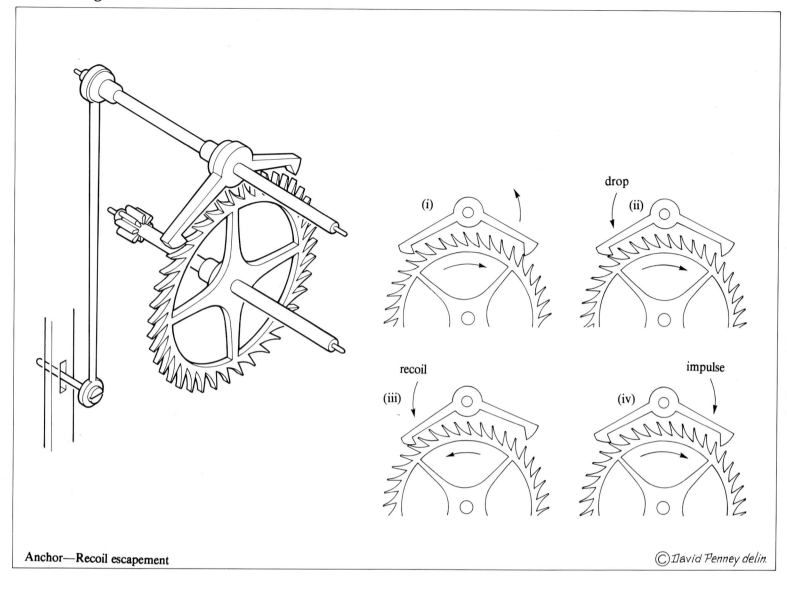

Anchor—Recoil escapement © David Penney delin.

Figure 20. *The Anchor Escapement.* This is basically simpler than the verge escapement in that the 'scape wheel is in the same plane as the other teeth in the train and thus no contrate wheel is required. The pallets are also not directly connected to the pendulum.

In figure (i) the pendulum is swinging to the right and the 'scape wheel is about to be released by the right hand (exit) pallet; as the pendulum continues its swing the wheel is released but is almost immediately arrested again by the left hand (entry) pallet which causes it to recoil (ii) as the pendulum completes its swing (iii). As the pendulum swings back to the left the 'scape wheel is released by the left hand pallet and the right-hand pallet, and thus the pendulum, receives impulse from the engaging tooth of the 'scape wheel. Drawn by David Penney.

Figure 21. *The "lift-up" hood.* The early longcase clocks had hoods which lifted up vertically by means of slots which engaged on either side of the backboard. They were held in that position by a catch and also had a pivoted spoon lock which prevented the hood moving when the trunk door was closed. With a lift up hood it is impossible to provide a mask for the dial.

Longcase Clock 19

— The countwheel was used as the method of controlling the strike on most clocks until nearly 1700 (Figures 22, 18, and 26). Initially the countwheel was placed external to the backplate of the movement, then just inside behind the greatwheel, and finally in front of the greatwheel. The disadvantage of this method was that it was unsynchronized with the time train, and thus, if the clock failed to strike for any reason, possibly because the weight had run down, then the strike would go out of phase with the time, i.e. it might strike 7 at 9 o'clock. This could only be corrected by either moving the hour hand forward to the correct hour if it was on a friction, or releasing the strike train one or more times by lifting up the detent locking the countwheel until the right hour had been struck.

To overcome this problem a system known as "rack striking" (Figure 23) was developed which was in fairly general use in London by 1700 and by 1720 in most other parts of the country. The use of the countwheel was to persist on some Northern clocks to as late as 1760 and was used on nearly all 30-hour clocks no matter when they were made.

Figure 22. (left) *Count Wheel Strike*. This is the earliest form of strikework. The number of hours struck are determined by a notched wheel either mounted on the backplate (Figure 18A) or between the plates either in front of (Figure 26) or behind the great wheel. It will be seen that the distance between the notches (in which the detent engages and locks the wheel and thus the strike) varies. The shortest distance will allow only one strike and the longest 12. On some early bracket clocks the wheel is numbered around its periphery to indicate the hour to be struck. This form of striking is not automatically synchronized with the time train.

Figure 23. (right) *Rack Striking*. With this method the striking is permanently synchronized with the time train and thus permits the use of repeat work and does not require resetting.

As the hour wheel (A) rotates it turns the minute wheel (B) to which a pin is attached. As the hour approaches this pin acts on the lifting piece (E) which raises the rack hook (F), thus releasing the rack (J) which drops to the left, aided by the spring (L). The amount which it falls back is determined by a pin (M) fixed to the end of the rack tail (K). This pin rests on the edge of the snail (fixed to the hour wheel) which has 12 steps in it. The deeper the step the farther in the pin and thus the rack tail moves and the farther to the left the rack falls. As each tooth on the rack corresponds to one strike, the farther it falls the greater the number of strikes. The rack is moved back to the right by a "gathering" pallet (G) which gathers in one tooth for each revolution it makes, which corresponds to one strike. On its final revolution the tail of the gathering pallet comes up against the locking pin (H) and then that particular hour strike is complete. Drawn by David Penney.

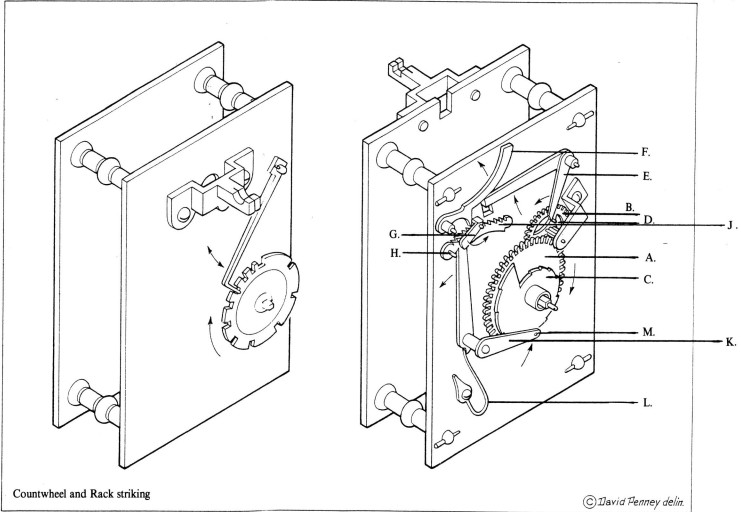

Countwheel and Rack striking

© David Penney delin.

## BOLT AND SHUTTER MAINTAINING POWER

While most eight-day clocks are being wound, they will automatically stop and thus lose time because the driving power is being removed from the train of wheels. To overcome this problem a system of maintaining power was devised that is known as "bolt and shutter" (Figure 454). In this a cord has to be pulled or a lever depressed before shutters are removed from the back of the winding holes and the key can be inserted to wind the clock. As the shutters are removed, a spring-loaded bolt engages on one of the teeth of a wheel in the going train and applies enough pressure to keep the clock going while it is being wound.

The use of bolt and shutter maintaining power was only applied to some of the clocks being produced in early times and its use had largely been discontinued by 1700. Because it was easily damaged, frequently when the key was inserted, it was often removed from clocks, and in most cases in which it is now encountered it has been reinstated.

An improved form of maintaining power was evolved by John Harrison, known as his "Going Ratchet," in the 1720s. However its use was largely confined to longcase regulators or "Master Clocks" as opposed to ordinary longcase clocks.

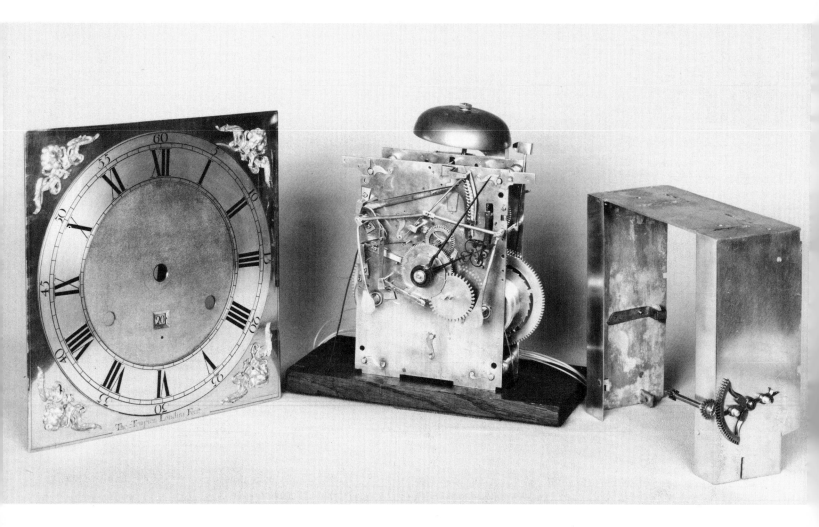

Figure 24. *Bolt and Shutter Maintaining Power.* A fine month-duration movement by Thomas Tompion with bolt and shutter maintaining power and a long horizontally mounted rack for the strike. It is protected by a brass dust cover which has pendulum rise and fall mounted on one side.

## THE DEAD BEAT ESCAPEMENT

An escapement called the "dead beat," because it stops the 'scape wheel "dead" and does not cause it or the other wheels to recoil as with the conventional anchor escapement, was introduced in the early 1700s (Figure 25). In the past its invention has been attributed to George Graham. However, it is now thought probable that it was Thomas Tompion who originated it. Its main application was in precision pendulum clocks or regulators where extreme accuracy was required and thus it was only occasionally applied to longcase clocks, mainly in the 19th century.

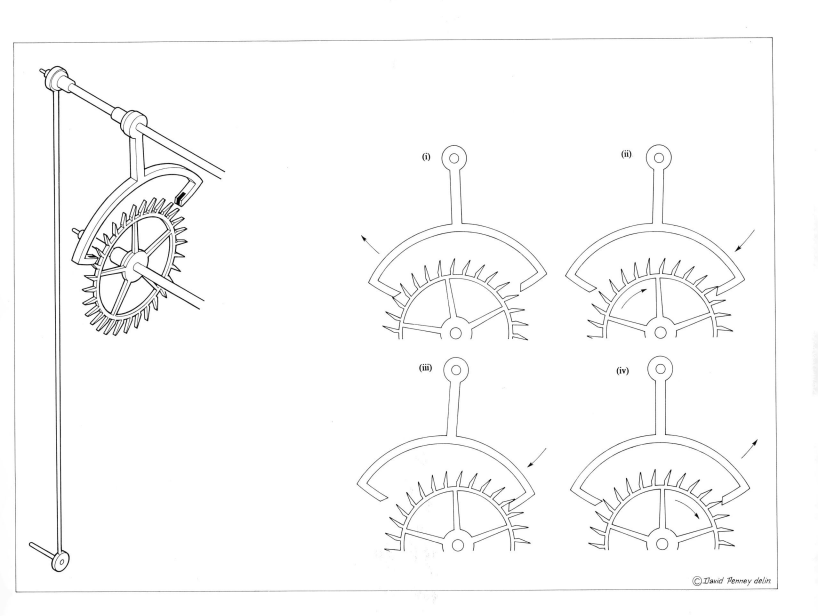

Figure 25. *The Dead Beat Escapement.* The advantage of this escapement over the anchor is that it does not force the 'scape wheel and thus the other wheels in the train to recoil and thus, in effect vary the power being transmitted. This is achieved by having a short impulse face and a long locking face which is curved so that it is, in effect, on a radius centered on the pallet arbor.

In (i) the pendulum has started its swing to the left and is gaining impulse from the scapewheel tooth as it slides down the impulse face prior to escaping, before being arrested again almost immediately (ii) by the exit pallet. The pendulum continues its swing to the left, increasing the depth of engagement of the pallet, but because of its form the wheel stays stationary and does not recoil (iii). As the pendulum swings back to the right the tooth of the 'scape wheel slides off the locking onto the impulse face of the pallet and provides impulse to the pendulum.

N.B. In the drawing the acting faces of the pallets are shown jewelled, a common practice on high-grade regulators to reduce both friction and wear. Drawn by David Penney.

Figure 26. *Internal Count Wheel Strike*

## COMPENSATED PENDULUMS

With the dramatic improvement in timekeeping achieved by the end of the 17th century, it was found that changes in the length of the pendulum which occurred with changes in temperature had an appreciable effect on the accuracy of the clocks. One of the first methods used to overcome this problem was the use of a pendulum rod made of wood instead of iron as this expands and contracts far less provided it is used in long grain and that grain is relatively straight. Moreover, even this expansion can be largely negated or compensated for by the upward expansion of the pendulum bob from the rating nut on which it rests.

George Graham was one of the first clockmakers in England to give serious attention to producing compensated pendulums. He measured the coefficient of expansion of various different metals to see how they could be used to produce a pendulum of invariable effective length. His hope was that the difference in coefficients of two metals would be sufficient for one to be used as the bob and the other the rod of the pendulum. However, this did not prove possible with conventional metals and thus his final solution was to use a column of mercury contained in a jar as the bob. As the temperature rises the mercury expands, rises up and compensates for the downwards expansion of the rod.

Harrison overcame the problem of temperature compensation in an entirely different way than Graham. He designed a pendulum with multiple rods to support the bob, some of which expanded up and others down. By choosing the correct effective overall length of the rods which expanded up and those which expanded down relative to their coefficients of expansion, a pendulum of constant length could be achieved. The usual design of this pendulum, known as the gridiron, was to use nine rods, a single central one and alternating rods of iron and brass on either side.

The problem of compensated pendulums was finally almost laid to rest when in 1897 Dr. Guillaume produced a metal called Invar which has an almost insignificant coefficient of expansion.

To produce the ultimate in precision timekeeping in the first 30-40 years of the 20th century, pendulums were placed in rooms or containers held by thermostats at a constant temperature, thus completely overcoming this problem.

This relatively brief account will, it is hoped, give the reader a reasonable idea of how the design of the longcase clock evolved in the 17th, 18th, and 19th centuries. The rest of the book will be devoted to illustrating those changes which occurred and the regional characteristics which developed, for instance in such areas as the West Country, Hull, Liverpool and Scotland to name but a few.

# Chapter 2.
# Early Square-Dial Longcase Clocks

The early history of the English longcase clock is a large and complex subject, being a period of rapid change and the introduction of numerous innovations. Many of these were of a minor nature but some, such as the adoption of anchor escapement and the long pendulum, had a profound affect on horology.

Not only did the design of the movement show many changes but so also did their dials and the cases in which they were housed. In a period of only 40-45 years at the end of the 17th century architectural cases were introduced and innovations included carved crested mouldings, caddy tops, the use of ebony, walnut, olivewood, laburnum and many other woods and several different forms of marquetry.

By the start of the breakarch dial, circa 1710-1715, a period of relative stability had been reached; some would say complacency. The design of longcase movements changed little in the following 150 years, except in the field of precision timekeeping (regulators) and case styles altered only slowly.

From the foregoing it well be realized that there is as much to be written about the first 40 years of the longcase clock as there is about everything which occurred after this and that therefore it is quite beyond the scope of this book to cover it in any depth. All we can do here is to pick out some of the highlights and illustrate these with pictures. However, for those who do find this early history of English horology fascinating, which it undoubtedly is, an excellent book on the subject *Early English Clocks* by Dawson, Drover and Parkes was published by The Antique Collectors Club in 1982.

The production of the first longcase clocks came shortly after the introduction of the pendulum to the verge escapement circa 1660 (Figure 19). These clocks (Figures 27 and 28) usually had brass dials around eight-and-one-half-inch square with raised chapter rings. In some instances raised decoration, normally small cherub spandrels, were placed in the four corners. Fromanteel, for instance, favored this style right from the start, whereas East, Hilderson and some other makers left them flat and engraved (Figure 30). The matting of the dial center, i.e. the production of a roughened center, commenced quite early and at this stage also solid silver chapter rings were sometimes used. On occasions they were fully fretted out, a most attractive feature. An appreciable number of the very early clocks were of 30-hours duration, but the movements were always plated and they were sometimes key-wound.

Before the introduction of the long pendulum with anchor escapement no seconds hand was provided but this was adopted almost immediately afterwards. Alarms were sometimes fitted virtually from the start of longcase clock production.

The earliest cherub spandrels were small and delicate and have never been bettered, but these gradually changed and got larger as the century progressed to suite the ever-increasing size of the dials.

Many of the early movements have shaped plates, which were sometimes rounded on top or in other instances cut out at the top corners (Figures 27 and 28). Instead of sitting straight onto the seatboard as on later clocks they were raised up on blocks.

In the early or Architectural period the cases were of simple pleasing lines such as that seen in Figures 27 and 29 and the first of these had very slender trunks. However these subsequently had to be increased in width a little to accommodate the introduction of long pendulums. Gilt or sometimes silver mounts were usually used on the cases. The very first of these had no mould around the top but this was introduced at a comparatively early period. They were usually ebony veneered but in some cases ebonized, generally using fruit wood.

Figure 27A and 27B. *An early architectural longcase clock attributed to Fromanteel.*

An unsigned ebony veneered longcase clock with panelled sides and a three-panelled door hung directly in the unframed trunk. The backboard extends up above the rising hood which is provided with a spoon lock and clip, fine gilt brass Corinthian capped columns, a gilt brass swag above the 12 o'clock position, and a cartouche in the tympanum above which are three foliate urn finials.

The latched six-pillar movement has circular cutouts to the top of the plates as favored by some early makers and rests on blocks. It has verge escapement with bob pendulum and external count wheel strike with a vertically pivoted hammer. Photos courtesy Christie's, London.

*Square-Dial* 25

Figure 28. *Ahasuerus Fromanteel*. An early longcase movement by Ahasuerus Fromanteel with verge escapement and short bob pendulum. It has blocks to either side lifting the movement up from the seatboard. Photos courtesy Ronald Lee, Fine Arts, London.

Figure 29. A small ebony veneered architectural longcase in a style popular for the first ten years of the pendulum clock.

Figure 30. *John Hilderson.* A beautiful florally engraved dial signed *John Hilderson, Londini Fecit* circa 1670. The four holes in the corners are signs of raised spandrels which were fixed by screws to the dial at a later date and subsequently removed.

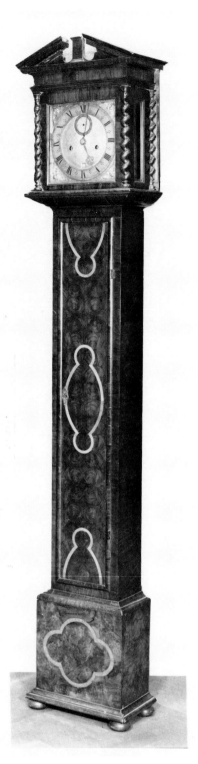

Figure 31. *Joseph Knibb, London.* An eight-day longcase clock with a 9½-inch square dial, large seconds ring, date aperture and cherub spandrels. The hood has a broken pediment and glazed sides. The sides of the case are veneered in walnut and the front with oyster veneer divided into lobed compartments in beechwood. 6′ 8″ (203 cm) high. Photo courtesy Christie's, London.

Figure 32. *John Wise.* An interesting and rare longcase clock full of originality, as is often the case in this early period. The panelled ebony veneered case stands only 5 feet 7½ inches high and supports a shallow caddy with three ball finials.

The eight-day five-pillar timepiece movement has latched plates and repeats the quarters on three bells. It has a hinged crutch with fine regulation above the backcock. The dial is just seven inches square and has a finely skeletonized chapter ring, a well matted center, a calendar aperture just above the winding hole, a seconds dial, and floral spandrels, circa 1680-1685. Another view is seen in Figure 1. Photo courtesy Sotheby's, Bond St., London.

*Square-Dial* 27

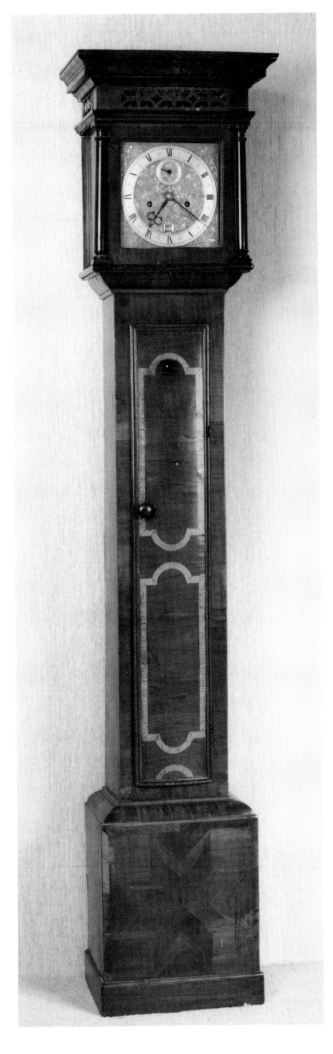

Figure 33 A and 33 B. An unsigned, country-made longcase clock veneered in laburnum with a ten-inch dial in the early style with a narrow chapter ring; the omission of spandrels and fine all over floral engraving. The case, originally with lift-up hood, now has a front opening door. The trunk door is inlaid with stringing and there is a simple parquetry design to the base.

28 *Square-Dial*

Figures 2, 34 A and 34 B. *Edward East, Londini.* A fine early walnut longcase clock by Edward East shown just as it was discovered some 15 to 20 years ago. It has delicate cherub spandrels, narrow chapter and seconds rings which would originally have been silvered, and fine steel hands. The walnut case, a little distressed where it has been kicked on its base over the years, is of excellent proportions and has walnut veneers on it ranging up to ⅛ of an inch thick, circa 1680.

Figure 35. *John Ebsworth, London, circa 1685.* See Figure 4 B for case.

A carved cresting (Figure 36) applied to a flat top started to appear as the architectural style went out of favor and this was subsequently gradually replaced by the caddy top.

With the introduction of the long pendulum many makers wished to display this feature as prominently as possible and thus placed glass lenticles in the trunk door or, in the case of 1¼ seconds pendulums, the base so that they could be seen swinging to and fro. However, this was by no means universally adopted. Knibb used the feature quite frequently but Tompion generally only seems to have employed it on some of his earlier clocks.

By the late 1670s the use of ebony had started to give way to woods such as walnut and olive and the use, particularly of the former, was to continue through to the end of the century along with the various forms of marquetry which were introduced. Indeed walnut went on being employed on clock cases long after marquetry had gone out of favor and was only finally displaced by mahogany.

30 *Square-Dial*

Figure 36. *Joseph Knibb, London.* A fine ebony veneered longcase clock with an elaborate carved cresting with a cherub's head to the front and side crestings also. It is surmounted by a gilt ball finial and has finely gilt mounts to the top of the front and sides of the hood. Note the beautifully executed capitals to the tops of the hood pillars. These were cast in many different pieces and soldered together prior to chasing and fire-gilding.

The dial has a solid silver skeletonized chapter ring and since the clock has Roman striking, four is denoted by IV instead of the usual IIII. There are cherub spandrels and wheatear engraving to the border, interrupted at the center of the bottom to leave space for Knibb's signature. Photos courtesy Ronald Lee, Fine Arts, London.

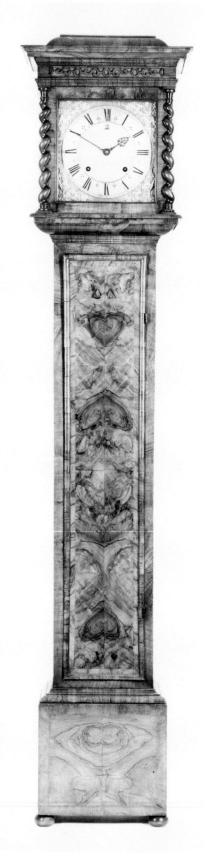

Figure 37. *Roman Striking Longcase Clock by Joseph Knibb, London.* A month going, walnut veneered longcase clock with shallow caddy; rising hood with spoon latch, and spiral twist columns. The five-pillar movement which is secured by spikes to the seatboard has latched plates and dial feet and is provided with Roman striking on two bells with large pin wheel and small outside count wheel. The pendulum is provided with butterfly suspension. Note the IV for the four which usually indicates Roman striking. 6′ 7″ (201 cm) high. Photo courtesy Christie's, London.

Figure 38. *Joseph Knibb, London, circa 1680.* This clock is the type most people think about when you mention a Knibb longcase clock. It has a lovely slim walnut case with a lenticle in the long door, a well-proportioned base and a carved crested moulding.

Most of these veneered cases had carcasses of oak, the few exceptions, approximately ten in number, were made of wood closely resembling mahogany. This is not like the heavy close-grained variety which is seen in the early mahogany furniture of George I's reign, but is of open grain and rather soft texture, similar in some respects to the wood used in modern cigar boxes.

This has been identified as a species of Cariniana, or Jequitiba rosa, which is referred to in Ronald Lee's book *The Knibb Family, Clockmakers* 1964, p.55, and is grown today in Brazil. As far as is known all carcasses made of this wood are associated with clocks of either John or Joseph Knibb which seem to have been made between 1676-1682.

The month-duration movement has bolt and shutter maintaining power, latched plates and a ten-inch dial with delicate cherub spandrels and a large seconds dial.

Figure 39 A and B. *Thomas Pare, London, circa 1690-95.* A slim walnut longcase clock surmounted by a caddy top and finials. It has a six-pillar eight-day movement with an 11-inch dial which has wheatear engraving to the border, late cherub spandrels, ringed winding holes, and an engraved center within the matting.

Thomas Pare (also spelled Parre or Parr) is a fine but not well known maker about whom comparatively little is recorded.

Figure 40. *Thomas Pare, London, circa 1705.* A clock by the same maker as that in Figure 39 A and B but made some ten years later. It has a very similar dial, 12 inches not 11 inches; there is a lenticle in the door and a concave mould beneath the hood.

Figure 41. *Thomas Trigg, London, circa 1695.* A late 17th century walnut longcase clock with unusually strong figuring. The 11-inch dial is attractively laid out with a tudor rose in the center, the name plaque above this, and an engraved cartouche around the date aperture. 7′ 2″ (218 cm) high.

Figure 42. A late 17th century walnut longcase clock having an eleven-inch dial, long trunk door with a lenticle, narrow spiral twist columns and a caddy top probably originally surmounted by three finials. 7′ 1″ (216 cm) high.

Figure 43. *Charles Goode, London.* An early 18th century, eight-day ebonized fruitwood longcase clock with 12-inch dial by Charles Goode. It will be seen that the moulds beneath the hood have by this time changed from convex to concave and the caddy top has become more pronounced. Other makers, including Tompion, also used this style of case. 7′ 10″ (239 cm) high.

Figure 44 A and B. *Gabriel Smith, Barthomley, circa 1700-1710.* Although the vast majority of longcase clocks made prior to 1700 were of London origin there was also some fine work being done outside the capital, particularly in and around the old cities such as Chester, York and Exeter.

Shown here is an eight-day clock by Gabriel Smith of Barthomley, a fine and quite prolific maker, born 1656 and died 1743. His son Joseph also worked initially in Barthomley but later moved to Chester the county town.

The movement is beautifully executed, and has six pillars as opposed to the five generally used in London. Interestingly all of these are screwed, not riveted, into the backplate, through which the domed heads may be seen protruding. The 12-inch dial is of equal quality to the movement with finely finished twin cherub spandrels, well-balanced hands, and an attractive cartouche around the date aperture.

## 36 Square-Dial

Figure 45 A, B, and C. *Claudius Du Chesne, London, circa 1705.* A fine clock by Du Chesne in what is almost the fully developed Queen Anne style of case with concave moulds below the hood and a tall caddy surmounted by three finials.

Mulberry, a comparatively rare wood whose leaves are loved by the silkworm, has a very warm color and is highly figured. Thomas Tompion used it on a few of his cases including some in exactly the same style as this but normally without a lenticle in the door. Note the extensive use of ebony moulds which produce a most dramatic appearance and highlight the mulberry veneers. The movement is of similar quality to the case, has chamfered corners and a 12-inch dial with wheatear engraving to the border, delicate blued steel hands and subsidiary dials in the top corners. On the left is a fast/slow indicator and the mechanism for this may be seen above the movement, a long pivoted bar which carries the pendulum at its midpoint being raised or lowered when the hand is turned by means of an eccentric cam on which the end rests. The strike/silent regulation may be seen in the opposite corner.

## Square-Dial 37

Figure 46. *Daniel Quare, London, circa 1710.* A fine walnut, month-duration longcase clock with 12-inch dial signed *Dan. Quare London* above the calendar aperture, with seconds dial, well modelled hands, and mask spandrels. The movement has five ringed pillars, the going train behind IX with five wheels, the striking with four wheels and controlled by a rack. The case has an ogee cresting to the hood flanked by turned finials, a brass border framing the dial and similar capitals and bases to the plain pilasters. The crossand feather-banded trunk door and plinth are veneered in richly figured burr-wood of warm golden color and the sides have inlaid panels. 8′ 4″ (254 cm) high. Top finial lacking. Photo courtesy Sotheby's, London.

Figure 47. *George Graham, London No 699, circa 1730-1735.* A good walnut longcase clock signed on both the bottom of the dial and also on a plaque and numbered on the backplate. It has a seconds ring, date aperture and mask, and leaf spandrels. All the five plate pillars are latched and it is fitted with maintaining power.

The case has a caddy top above a finely pierced fret. The hood has freestanding pillars and the long trunk door and base have finely figured veneers.

"Honest George," as he has been called, had a very conservative streak and produced clocks similar to this with a square dial from the time he took over Tompions workshops until he died some 40 years later.

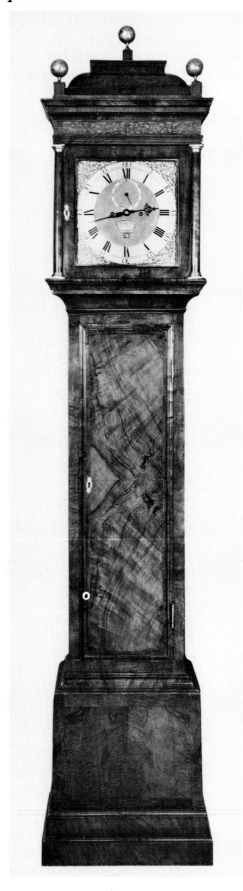

Figure 48. *Vulliamy and Gray, circa 1740.* Justin Vulliamy emigrated from Switzerland circa 1730 and took into partnership Benjamin Gray, his son in law, who was appointed clockmaker to George II in 1744.

As can be seen, the design of this clock is very similar to that of the Graham shown in Figure 47, although it has applied pillars to the hood.

Figure 49. *Francis Robinson, Northampton, circa 1720.* A Francis Robinson is recorded in London as well as Northampton and one wonders whether they are the same man who moved from one place to the other. The case is very similar to those being made in London but if it is compared with Figures 47 and 48 it will be seen that it is not quite so tall and elegant and has a shorter trunk door. It would seem possible that the bottom plinth is a replacement. Note that the late cherub spandrels are still being employed.

## Square-Dial 39

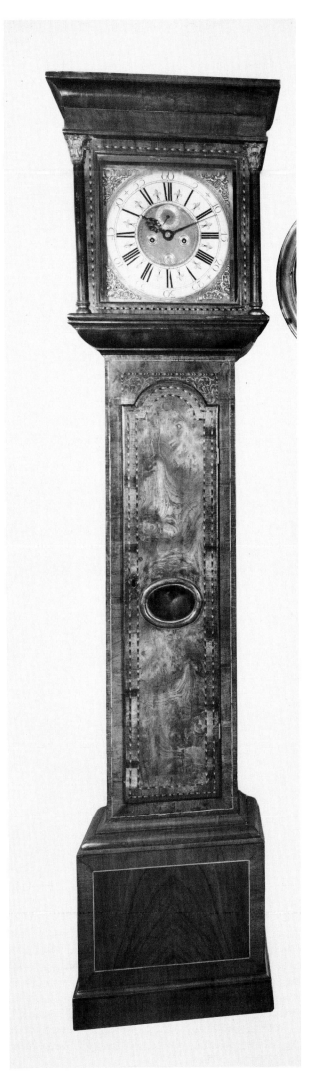

Figure 50. *Marshall, Dublin, circa 1725-1730.* This clock is included here as it typifies those being made in Dublin at this period. The dials which are usually thirteen inches and sometimes on the later clocks 14 inches square, often have a band of engraving around the inner aspect of the chapter ring. The earlier cases are generally veneered in walnut or, as in this instance, elm, and have decorative inlay. The base on this clock is a replacement.

The fashion for square dials persisted quite late in Dublin, still being used with mahogany cases with a swan-neck pediment.

Figure 51 A. *Tompion, No 64, A Walnut, Quarter-Repeating, Month-Duration Clock.* The fine ten-inch dial signed *Tho: Tompion Londini Fecit* within the engraved leaf border and with calendar aperture and cherub spandrels.

40 Square-Dial

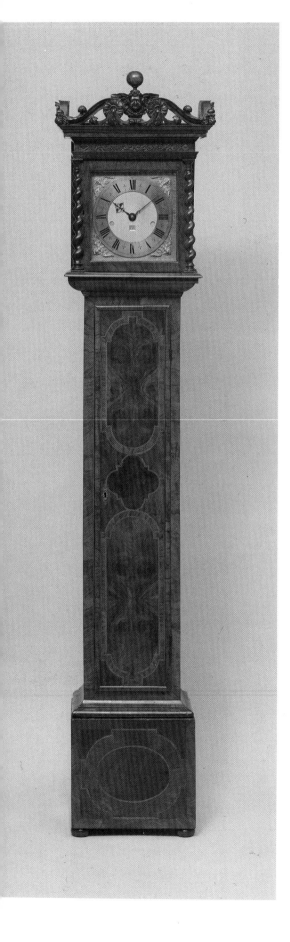

Figure 51 B and C. The fully latched movement has seven ringed pillars, unusual calendar work with an externally toothed wheel, and bolt and shutter maintaining power. The repeating work has a strike/silent control allowing repetition while on silence, the going train on the left with five wheels, the striking train with four, the front plate cut for repeating work, the backplate for the anchor and both for the hour bell. The rising hood has later carved cresting and spiral columns. The waist door has two quarter-veneered panels of richly figured wood, cross banded and outlined with stringing, and a central quartrefoil of darker grain. The plinth has an oval within a broader banding and the sides are also crossbanded. 6' 11" (241 cm) high. Photo courtesy Sotheby's, London.

## Parquetry And Marquetry

Marquetry is decorative work in which pieces of cut wood, shell or ivory are inlayed into wood veneer to produce patterns such as flowers and birds. A near relative of marquetry is parquetry in which the inlay is in geometric patterns, usually involving straight lines, one of the simplest examples of this being star shaped parquetry which popularly was used on furniture from circa 1680-1720.

It was in the early 1670s that the ebonised case started to give way to one veneered in walnut or sometimes olivewood. One of the earliest ways of making this more decorative was to use "oyster veneers." These are obtained by cutting transverse sections from a bough or trunk of a tree, usually walnut or olive, thus displaying its annular rings which give a most striking effect (Figure 52) particularly when a large number of these are used together (Figure 68). The name derives from the exterior of the oyster shell which usually has a similar design on its surface.

The first decorations applied to walnut cases were of parquetry, sometimes in the form of a single star but on other occasions more elaborate designs were used (Figure 64). By 1680-1685 parquetry decoration started to be replaced by floral marquetry (Figure 53) usually inhabited by birds (Figure 55) and sometimes butterflies. Mythological figures are also occasionally seen, mainly in Dutch marquetry (Figure 56 A and B).

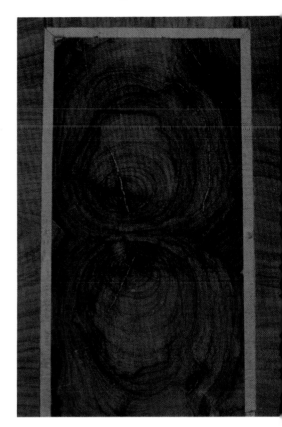

Figure 52. *Oyster veneers*

Figure 53. *Floral marquetry laid in panels.*

Figure 54. *A basket of flowers.*

Figure 55. *Birds were a common feature of floral marquetry.*

Initially, this marquetry was laid in panels and its application was confined to the trunk and base, but within a short while it spread to involve the hood also and even the mould beneath it and the caddy, if there was one, above. The earlier floral marquetry tended to be fairly bold in concept but later designs were often smaller and finer. Certain makers, such as Gretton in particular favored these more detailed designs (Figure 82). Although most marquetry only involved the use of wood, stained bone was also occasionally used (Figure 58). This has the advantage that the bone holds its color and thus still gives a most pleasing and bright appearance whereas any stained wood has by now lost nearly all its color.

By the 1690s instead of laying the marquetry in panels it was starting to be applied to the whole of the front of the case (Figure 61) and occasionally the sides also. It is thus often referred to as "all over" marquetry. This style of marquetry was to continue being used, although in ever decreasing amounts, until around 1715 and was even employed on some very large 12-inch square dial clocks which with substantial caddy and finials, often went up to eight feet six inches in height. However, by the turn of the century a new form of marquetry known as "arabesque" was starting to appear, inspired no doubt by the delicate tracery and designs of the Moorish castles such as the Alhamba in Granada which travellers were starting to visit and report on.

The basic principal of arabesque marquetry (Figure 57 A and B) is the employment of bold contrasting strapwork often with butterflies, figures or occasionally birds between. At its best it can be beautifully executed and have quite a dramatic appearance. Because of the period in which it was employed it is usually seen on the larger clocks with 12-inch dials but occasionally an example may be seen with an 11-inch dial and although it is usually applied "all over" it may sometimes be seen panelled (Figure 60).

Within a further five years the final development of marquetry, "seaweed," made its appearance on clock cases. This consists of delicate tracery, with only two different colors of wood being used and it virtually never employs any figures or animals (Figure 59). With the advent of the 12-inch breakarch dial and larger cases, usually seven feet six inches to eight feet six inches, which this necessitated, there was a strong swing back to the unadorned walnut case. However, the practice of decorating them with inlay did reoccur to some extent in the Sheraton and Edwardian periods and was also occasionally used in Victorian times.

Figure 56 A and B. *Dutch style floral marquetry inhabited by mythological figures.*

Square-Dial 43

Figure 57 A and B. *Different forms of Arabesque marquetry.*

## 44  Square-Dial

Figure 58. *The colorful use of stained bone in "bird and flower" marquetry.*

Figure 59. *Seaweed marquetry.*

Figure 60. *Panelled arabesque marquetry.*

Figure 61. *"All over" floral marquetry.*

46 *Square-Dial*

Figure 62. *Oyster Veneers*. An olivewood oyster veneered and parquetry longcase clock with rising hood and a 10 ½-inch dial which has a narrow chapter ring with cherub spandrels, fully latched plate pillars sand dial feet and would originally have had maintaining power. Photo courtesy of Sotheby's, Bond Street.

Figure 63. *A Walnut and Star parquetry miniature longcase clock*. This clock has a 7½-inch square dial with winged cherub spandrels inscribed *Edwardus East Londini Fecit* and stands only 5' 7½" (174 cm) high.

Figure 64. *William Clement, London*. A fine walnut longcase clock with oyster veneers, decorated with wavy panels of star parquetry to the trunk and base and a lenticle set into the trunk door. It has a convex mould beneath the hood which was at one time cut to form a door, a fairly common occurrence, and then resealed. The six-pillar movement (five latched) has external count wheel strike and the dial feet are also latched. The 10¼-inch square dial has a matted center which is gilded and winged cherub spandrels. 6' 7" (201 cm) high.

Figure 65. *Johannes Knibb Oxoniae Fecit*. A fine ten-inch dial with beautifully executed border and hands, circa 1680.

Photo courtesy Sotheby's, London.

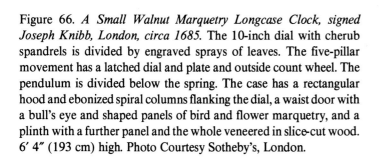

Figure 66. *A Small Walnut Marquetry Longcase Clock, signed Joseph Knibb, London, circa 1685.* The 10-inch dial with cherub spandrels is divided by engraved sprays of leaves. The five-pillar movement has a latched dial and plate and outside count wheel. The pendulum is divided below the spring. The case has a rectangular hood and ebonized spiral columns flanking the dial, a waist door with a bull's eye and shaped panels of bird and flower marquetry, and a plinth with a further panel and the whole veneered in slice-cut wood. 6' 4" (193 cm) high. Photo Courtesy Sotheby's, London.

48 *Square-Dial*

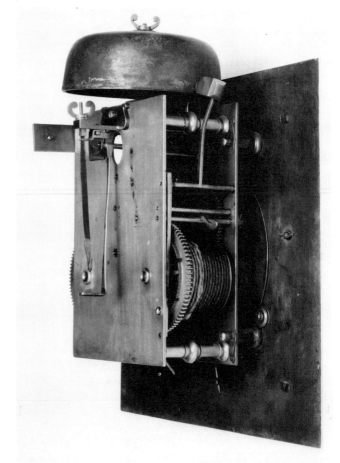

Figure 67 A, B, and C. *Joseph Knibb, Londini Fecit.* An olivewood and parquetry longcase clock, the ten-inch dial signed *Joseph Knibb Londini Fecit,* without seconds and with a calendar aperture below XII and cherub spandrels. The movement with latched dial feet, six ringed plate pillars and inside count wheel. The pendulum with butterfly nut suspension. The case veneered in well-figured oyster-cut olivewood, crossbanded and inlaid with parquetry stars and fan medallions, the low domed rising hood with securing bracket and spiral columns and the waist door with a bull's eye. 6′ 6″ (198 cm) high. Photo courtesy of Sotheby's, London.

Square-Dial 49

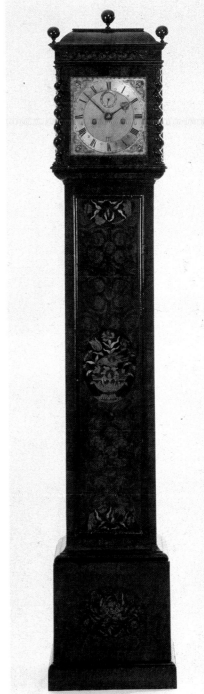

**Figure 68 A, B, and C.** *Joseph Knibb, A Very Small Olivewood Marquetry Longcase Clock.* The 8-inch dial signed *Joseph Knibb London* within the formal border and with narrow seconds ring, calendar aperture and cherub spandrels, the applied silver chapter ring with the minutes numbered in full 1-60 and *fleur-del-lis* half-hour marks. The eight-day movement with six ringed pillars, latched dial feet and plate pillars, bolt and shutter maintaining power and outside count wheel. The case crossbanded and veneered in oyster-cut olivewood of rich coloring, the waist door inlaid with an urn of tulips and carnations, a bird pecking at the blossoms, and two sprays of flowers, the plinth with a peony in a ring of summer flowers, all on an ebony ground and partly in green-stained bone, the rising hood with turned finials and ebonized spiral columns. 5′ 6″ (168 cm) high.

Both Lee and Symonds comment on the rarity the very small early longcase clock. Lee (Plate 55) illustrates another by Knibb in a panelled ebony case and Symonds records four by Christopher Gould. Some are of short duration and many are fitted with repeating work, leading Symonds to suggest they were made for bedroom use.

Seconds rings are seldom found in such clocks, and the present example is also notable for the silver chapter ring with fully numbered minutes, the early marquetry panels, and the outstanding quality of the oyster veneering. Photos courtesy Sotheby's, London.

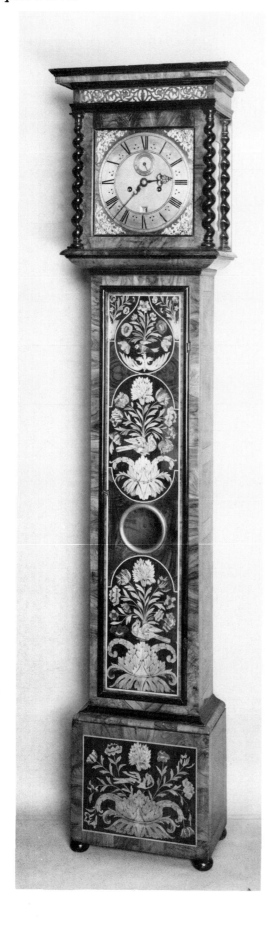

Figure 68-i. *Panelled Floral Marquetry, circa 1685-1690.* Delicate panels of floral marquetry having a bird incorporated in each. Note that on this clock the marquetry has not yet spread to involve the hood, and the moulds above and below the trunk are comparatively flat.

Figure 68-ii A and B. *James Wightman, London, circa 1685-1690.* James Wightman was apprenticed to Edward Eyston on 18th January 1663, became free of the clockmakers company on 16th January 1670, and was elected a steward in 1682. This clock like that seen in Figure 66 has the marquetry confined to the door and base and the moulds above and below the trunk are shallow. Note the careful balancing of the panels on either side of the lenticle. It has unusual decoration outside the ringed winding holes and inside the seconds ring. 6' 6" (197 cm) high.

Figure 68-iii A and B. *Edward Bird, circa 1690.* This month movement has external count wheel strike and bolt and shutter maintaining power. The shutters may be seen obsuring the winding holes and the lever for moving them and actuating the maintaining power is visible to the left of the movement. Besides showing the day of the month additional apertures above this give the day of the week and their ruling deity. Edward Bird is not a famous maker but his work is always of a high standard.

## 52 Square-Dial

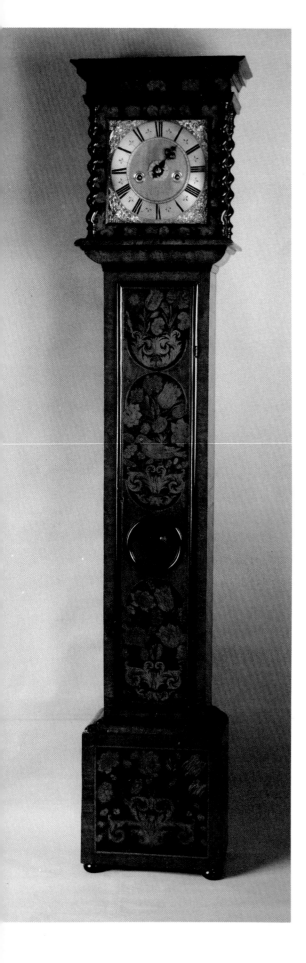

Figure 69 A, B, and C. *Francis Clement, Londini, circa 1690-1695.* A most interesting clock by Francis Clement, believed to be the brother of William Clement, this opinion being reinforced by the fact that they were both apprenticed to Thomas Chapman. Very little work is recorded by Francis which may mean that he worked for his brother for much of his life.

This clock is fitted with an alarm, the mechanism for which can be seen to one side of the movement and there is only a single hand as is also the case with some of William Clement's clocks. 6' 9" (206 cm) high.

Figure 70. *Robert Gregory, London, circa 1690-1695.* Note the poker work border to the door and base and the ebony moulds and pillars, quite often used at this period to highlight the marquetry. 6' 9" (206 cm) high.

Figure 71. *Bird and Flower Marquetry, circa 1695.* The base of this clock is particularly attractively laid out with a bird among foliage in the center and flowers in the four corners.

Figure 72. *Eleven-Inch Marquetry Longcase Clock, circa 1695.* Finely detailed bird and flower marquetry with birds and somewhat unusual unicorns being incorporated in the central door panel and the base. 6' 10" (208 cm) high.

## 54 Square-Dial

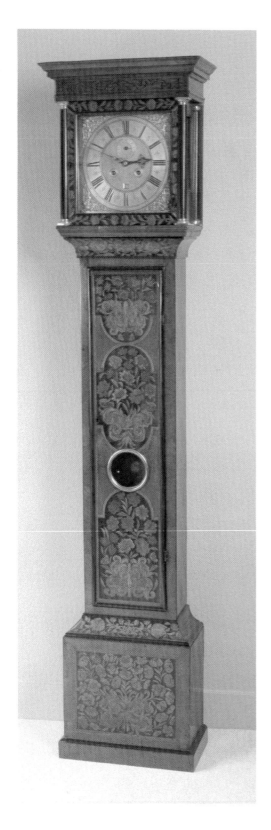

Figure 73. *Month Longcase Clock by Stephen Asselin, London, circa 1705.* This clock illustrates how different styles overlap. There is panelled floral marquetry but a concave, as opposed to a convex, mould beneath the hood. At this time Arabesque marquetry was also being made, usually in "all over" designs.

The relatively low position of the winding holes in the 12-inch dial usually indicates a month-duration clock and is brought about because of the necessity to provide an extra wheel in the train.

Figure 74. *Panelled Floral Marquetry Clock with 11-inch Dial, circa 1695-1700.* This has a shallow caddy top decorated in marquetry.

Figure 75. *George Etherington, London, circa 1700.* This clock has a relatively tall marquetry caddy, an "all over" panelled base, and a 12-inch dial.

## Square-Dial 55

Figure 76. *Wm. Sharpe, London, circa 1705-1710.* This clock has a fully developed two tier caddy top, typical of the Queen Ann period, which brings the clock up to seven feet nine inches high without the top block and finial or eight feet four inches with it. This reflects the larger houses being built in London at this time.

It will be seen that the mould beneath the hood is now concave. The month-duration six-pillar rack striking movement has a 12-inch square dial with wheatear engraving to the border and cherub and crown spandrels.

Figure 77. *Rare 1¼ Seconds Pendulum Clock by Benjamin Johnson, London, circa 1700.* The dial is 11 inches. Because the 1¼ seconds pendulum, which is made in two parts, is approximately five feet long, the lenticle has to be placed in the base which has a front opening door so that the clock may be regulated. The all-over marquetry, featuring tulips, is surrounded by a band of poker work. 6′ 9″ (206 cm) high.

## 56 Square-Dial

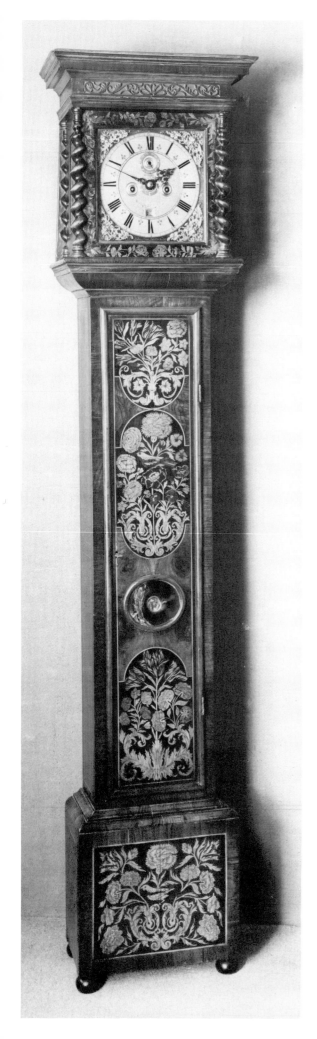

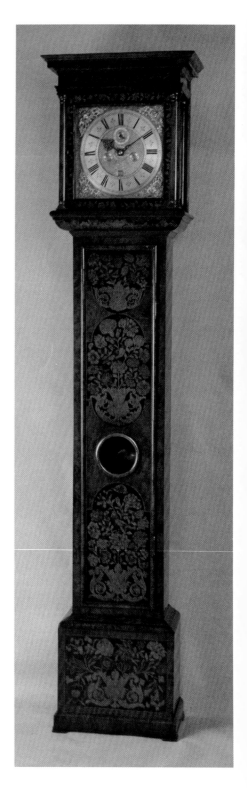

Figure 78. *John Hunt, Londini Fecit, circa 1695.* 6′ 6″ (198 cm) high.

Figure 79. *Richard Motley, London, circa 1695.*

Figure 80 A and B. *John Allaway, London, circa 1695-1700.* This clock has well detailed marquetry with a bird in each panel on the door and an 11-inch square dial with engraving to the matted center and between the spandrels. 6′ 9″ (206 cm) high.

Figure 81 A, B, and C. *"All Over" Finely Detailed bird and Flower Marquetry, circa 1695.* A month-duration longcase clock by John Andrews, London, with 6 latched pillars, external count wheel strike and an 11-inch dial with a Tudor rose in the center. A close-up of the top of the door shows how the marquetry has cracked across the top of it. This may or may not happen with early clocks and is due to the movement of the cross cleat (Figure 81 C) relative to the main body of the door. The cleat is placed to stop the door from bowing; however, as wood shrinks across the grain the main part of the door becomes narrower relative to the top cleat which itself shrinks away from the door.

**Figure 82.** *Charles Gratton, London, circa 1700.* A fine month-duration longcase clock by the eminent maker Charles Gretton who was master of the clockmakers' company in 1700. It has a 12-inch dial with twin cherub and crown spandrels and external count wheel strike.

Gretton appears to have favored very highly detailed floral marquetry such as this which is inhabited by numerous figures and birds. This style of floral marquetry most closely resembles the Arabesque which was to follow it.

**Figure 83.** *Thomas Speakman, London, circa 1700.* 7′ 2″ (219 cm) high.

## 60 Square-Dial

Figure 84. *Queen Anne Longcase Clock by John Wise, London, 1705.* This clock is particularly well proportioned with a long door, a double plinth, and a good caddy top. The concave mould beneath the hood is veneered in marquetry. 7' 6" (230 cm) high.

Figure 85. *Samuel Bowtell, London, circa 1705-1710.* This clock represents virtually the ultimate development of the square dial floral marquetry clock prior to the switch to other forms of marquetry, and more particularly a move back to walnut and the adoption of the breakarch dial. 8' 10" (244 cm) high.

## ARABESQUE MARQUETRY

Figure 86. *John Wise, London, circa 1700.* A slim, elegant and relatively early panelled arabesque marquetry clock with bold dark strapwork laid on a light ground, as is usually the case, however occasionally this is reversed. The 11-inch dial has engraving on the matted center.

Figure 87. *Peter Mallett, London, circa 1700-1705.* A particularly good and detailed example of "all over" arabesque marquetry with many figures and beasts inhabiting it.

## 62  Square-Dial

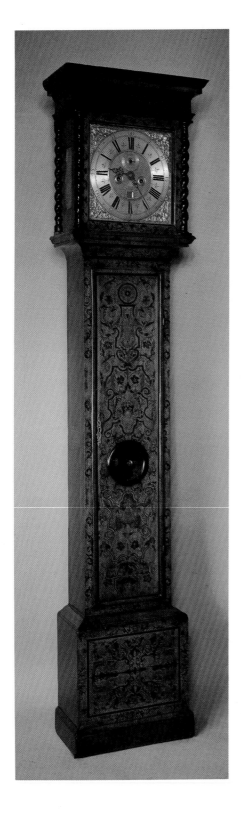
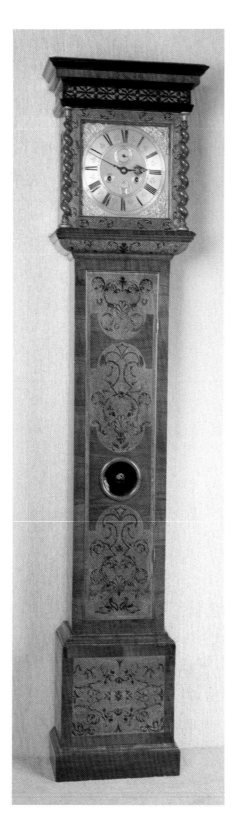
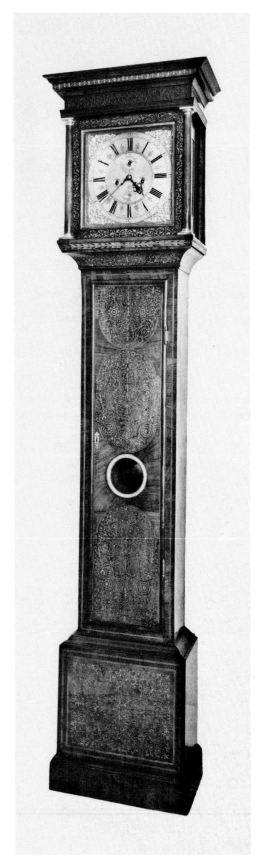

Figure 88. *William Sellers, London, circa 1705-1710.* A detailed arabesque marquetry such as this must have been time-consuming to produce than a floral marquetry case.

Figure 89. *Jos. Bates, Whits Alley, Holborn (London) 1700.* Bates was apprenticed to William Standish on 20th Jan 1678, became free of the Clockmakers Company on April 4, 1687, and died in 1706.

This clock illustrates the fine form of arabesque marquetry which was to eventually give way to the seaweed.

### SEAWEED MARQUETRY

Figure 90. *John Everell, London 1720.* The seaweed marquetry is far more delicate than arabesque. Despite the comparatively late date it was still sometimes panelled (or fielded) as in this case. 12-inch dial. 7' 2" (218 cm) high.

Figure 91 A, B, and C. *Thos Foster, London, circa 1705-1710.* This clock has a six-pillar month-duration movement with an external count wheel and a 12-inch dial with ringed winding holes, engraving between the spandrels, a cartouche around the date aperture and twin cherub and crown spandrels. Note the cracking on the base and the top of the door where the carcase of the clock has moved relative to the veneers, a sure sign that they are original.

64 *Square-Dial*

Figures 92 A and B. *The Drayton Tompion.* This clock, so called because it was in Drayton House, Northamptonshire for some 200 years, now resides in the Fitzwilliam Museum, Cambridge.

It is of year duration and has an 80-pound driving weight. The pendulum bob, visible through the decorative aperture in the lower part of the door, is backed by an engraved and silvered brass plate marked out in degrees to give the amplitude of swing of the pendulum. An unusual feature is two movable pointers which can be used to note the maximum position on either side. The very substantial movement has the train reversed with the barrel at the top and is provided with maintaining power.

The dial, apparently square, actually extends up to carry the dial for the year calendar which may be seen through an aperture when the hood is removed. There is a cutout in the center of the dial for the days of the week and the ruling diety.

The chapter ring, which shows mean time, is a 24-hour one graduated I—XII and 0—60 twice. The minute hand thus completes one revolution every two hours. A further ring immediately outside the main chapter ring gives apparent or solar time. This is made possible by rotating it backwards or forwards throughout the year, relative to the main ring. The control for this is a roller resting on the equation kidney which is mounted behind the year wheel in the arch, and is connected to the wheel which moves the solar chapter ring via a rod and rack. Photos Courtesy Fitzwilliam Museum, Cambridge.

# Chapter 3.
# Rare Clocks

Long Duration; Grande Sonnerie Striking; Night Clocks and Those Showing Seconds in the Arch

## LONG-DURATION CLOCKS

It was to be just a very short time after the introduction of the "Royal" or seconds-beating pendulum before clocks of month, three-month, six-month and even year duration were introduced. Month-duration clocks became relatively common and it was only after 1710 that they gradually went out of favor, although their production never ceased. Three and, more particularly, six-month clocks were made much less frequently and indeed it is likely that they are less common than year-duration clocks. This is probably because the natural cycles on which our lives run are the day, the week, the month and the year. Moreover there is obviously much more kudos to be obtained by making a clock of year duration than one which goes for three months.

Several of the foremost London makers such as Tompion and Quare made year clocks, two of the earliest recorded being those which were installed by Tompion in the Royal Observatory at Greenwich in 1675 (Figure 93). Other year clocks he produced were the equation clocks made for William III circa 1695 (Figure 473) and that known as the Drayton clock circa 1700 (Figure 92). Joseph Knibb evolved a series of three-month clocks (Figure 96) and an example of a year clock by Quare is in the Royal Collection. Seen in Figure 94 is an unusual year clock by Edmund Wright, London, with a single hand and a penny moon and another by Clowes is shown in Figure 157.

## LONG-DURATION CLOCKS

Figure 93. The beautifully executed dial of one of the year-duration clocks which Thomas Tompion made for the observatory at Greenwich. Note the extreme delicacy of the hands with the minute and seconds being counterbalanced.

These clocks are regarded by many as the first, and certainly the most important, precision clocks to be constructed in this country at that time. They were specially commissioned and paid for by Sir Jonas Moore for use by Flamsteed in the new observatory commissioned by the King in 1675.

The prime task of the clocks was to ascertain whether the Earth rotated at a constant speed, as no clocks had up until that time been made to go accurately or reliably enough to prove this. Tompion's clocks subsequently achieved this, and thus enabled work to go forward on the determination of longitude.

The specification of the clocks was interesting in that both were of year duration, both had maintaining power, possibly in part at least to prevent train reversal and damage to the escapement during winding, and both had 13-foot pendulums beating two seconds, and mounted above the movement. Unfortunately no details of the construction of the pendulums remain. Photo courtesy Trustees of The British Museum.

## 66 Rare Clocks

The reason for the relatively small number of year clocks produced is the extreme care which has to be taken in their manufacture. The wheel work has to be as fine and light as possible to keep friction to a minimum and the 'scape wheels are usually very small and delicate. The seconds hand is often omitted and the minute hand counterbalanced or even left off to keep the power requirement to a minimum.

Figure 94 A and B. *Year-Duration Clock by Edmund Wright, London* This interesting clock made circa 1705 has a year-duration movement with only a single blued steel hand to keep the power required to drive it to a minimum; However, time can still be read to the nearest minute because the outer edge of the chapter ring has 60 divisions numbered 10, 20, 30 and so forth between each hour. These are also delicate half-hour divisions and quartering on the inner aspect of the chapter ring.

An unusual feature for this period is the moon aperture below 12 o'clock with a separate opening for the lunar calendar which is advanced twice a day. The weight required to drive it is 65 pounds, which may seem excessive until you consider that this only represents 1¼ pounds a week as opposed to the ordinary eight-day longcase clock which needs 12-14 pounds. It is fitted with bolt and shutter maintaining power to avoid damage to the delicate escapement during winding.

The case is very substantial with sides 1½-inches thick to carry the heavy weight and reduce any movement to a minimum. It is decorated in "all over" arabesque marquetry inhabited by birds. The lower plinth on which it rests is a replacement.

Rare Clocks 67

Figure 95 A and B. *Year Clock with Equation of Time by James Reith* Although this clock, which was made circa 1720, has been rehoused in a case of considerably later manufacture, it is included here as it illustrates some of the finer points of the early year clocks.

It has a 60-pound driving weight and uses a two-part, 1¼ seconds, beating pendulum latched together in the center which, having to swing 20 percent fewer times than the conventional pendulum, reduces the total power required. Points to note are the very rigid movement with six latched pillars, a six-wheel train with all the wheels bevelled to reduce inertia, a small and delicate 'scape wheel with longnecked recoil pallets, and counterbalancing behind the dial for the minute hand.

One of the plates has been extended to carry the rise and fall mechanism for the pendulum which is activated by a hand in the arch of the dial.

The 14-inch dial is signed *James Reith London,* a maker about which little is written, but who was vice-director of the watch manufacturing business set up by Sully in 1719 for the Duke of Orleans.

The arch of the fourteen-inch dial has been reduced somewhat, presumably when it was fitted to its current case. It has urn and bird spandrels divided by engraved leaves and a matted center with a hand below 12 o'clock giving the equation of time against a curved silvered strip. There is an aperture through which a year calendar is displayed above 12 o'clock.

TRAIN COUNT

| Great Wheel | 96 | Barrel | 12¼ Grooves |
|---|---|---|---|
| 1st Intermediate Wheel | 80 | Pinion | 12 |
| 2nd Intermediate Wheel | 72" | | 8 |
| Center Wheel | 60" | | 8 |
| Third Wheel | 48" | | 6 |
| 'Scape Wheel (1-inch diameter) | 24" | | 8 |

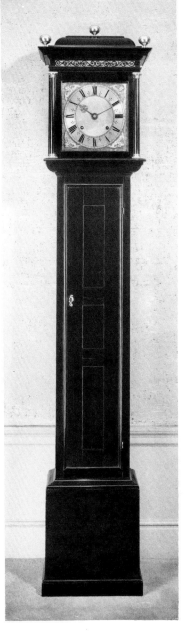

Figure 96 A and B. *Three-Month-Duration Clock by Joseph Knibb circa 1680.* A beautifully proportioned, panelled and ebony-veneered longcase clock by Joseph Knibb with 1¼-seconds pendulum. Roman

## 68  Rare Clocks

striking is present in which two bells are used, a small one to represent I and a larger one for V. Thus, at 4 o'clock there will be a single blow on the smaller bell followed by one on the big bell. It will be noted that in clocks employing this form of striking four is represented as IV not IIII as on most clocks. The use of Roman striking conserves a considerable amount of power, only 30 blows of the hammer being required in 12 hours instead of the usual 78.

The view of the movement shows the latching of the dome-shaped plates which were used on some of the early clocks; the delicate cock for the rear pivot of the 'scape wheel and the small cutout in the backplate to allow the pallets to be viewed and removed. The butterfly nut for pendulum regulation is seen above this with the suspension strip passing through a fork immediately below. The two bells for the Roman strike are seen above and on the backplate is the unusually divided count wheel.

It is known that Knibb made at least five of these three-month clocks, all with Roman striking. Photos Courtesy R.A.Lee Fine Arts

Figure 97. *Three Month Clock by W.Stokes* A striking arabesque marquetry clock of three-months duration with an unusual chapter ring in which the minutes are virtually the same size as the hours.

## GRANDE SONNERIE STRIKING

*Grande Sonnerie* striking means one bell or chime is struck on the first quarter, two at the second, three at the third, and four at the last quarter, each followed by the hour. It was introduced on spring clocks following the invention of rack striking circa 1675, although some had been made by Tompion with a separate spring to power the strike work a little prior to this. Grande Sonnerie striking on longcase clocks is less common, although Joseph Knibb made a small series; however all but one of these employed double six hour strike to economize on the power consumed. Tompion is known to have made one, No. 131, and a fine example by William Davis is shown in Figure 98. Very few Grande Sonnerie long case clocks seem to have been produced after circa 1710.

### GRAND SONNERIE STRIKING

Figure 98 A, and B. *Grande Sonnerie Longcase Clock by William Davis*. A well-figured flat-topped walnut longcase clock of good rich color with a quarter-veneered door, the border of which is crossbanded, as also is the base. It is one of the few *Grand Sonnerie* longcase clocks ever made. Symonds, in his book on Tompion, only records his having made one, No. 131. The Knibb brothers between them probably made three or four, although all but one employed double six-hour striking, and Daniel Quare may have made one or two. None are recorded in the Wetherfield Collection. This particularly fine example by William Davis, London would date around 1705. It has an eight-pillar movement with split plates and latching to both the dial feet and the pillars. To the top left of the dial is pendulum rise and fall, and to the right, regulation for silent; strike all (Grand Sonnerie); no hour on the quarter; and no quarters. 7′ 3″ (221 cm) high.

# 70 Rare Clocks

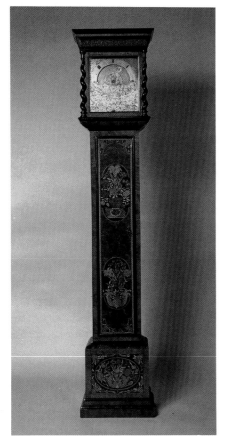

## NIGHT CLOCKS

The night clock, by which is meant in this instance a clock that could be used to tell the time at night by means of illuminated numerals, is believed to have been devised first by the Campari brothers in Rome circa 1660; however it was only to be a short while before they were being produced in England also. The simplest and earliest form consisted of a revolving chapter ring with perforated numerals and a point above through which light would shine from a lamp placed at the back; thus the relevant time could be read.

Some of these clocks had a conventional as well as a night dial but the majority of the clocks made in England in the 1660s-1670s were for night use only. The type most frequently seen, of the small number still in existence, have a curved semicircular slot at the top behind which a perforated disc revolves, the relevant hour, generally in Arabic numerals, appears on the left hand side of the cutout and over the next hour traverses the cutout, disappearing behind the right hand edge at the same time as the next numeral appears on the left. Above the cutout the four quarters are displayed by means of perforated Roman numerals while on the inner edge the minutes are indicated either by a series of fine perforations or scallops.

Illumination for the dials was usually provided by oil lamps mounted inside or occasionally behind the clocks; however this posed many problems as it created smoke and grease which was likely to foul the movement. To overcome this the mechanism often was enclosed. A free flow of air was needed for the lamp to burn satisfactorily, which meant that frets or holes at the back or underneath the case were necessary as well as some form of vent at the top to allow the smoke to escape.

Fire was another hazard for these clocks and possibly accounts for the relatively small number which still exist. Quite often when examining these clocks, charred woodwork is seen.

Several of the fine early makers produced night clocks, including East and Knibb, (four examples of the latter's work still being in existence). However, nearly all of these are table clocks and only two long case clocks of this type were known until now, both of which are illustrated here. Amazingly a third clock, by Tompion, has just come to light which is far more complex than the other two and was completely unrecorded until it was described and illustrated in Christie's catalogue of 5th July, 1989, prior to being sold.

The reason the night clock was made for such a short period in England (1660-1680), apart from the difficulties already mentioned, was that its purpose, i.e. to tell the time at night, was provided far more efficiently and safely following the introduction of pull quarter repeat work on bracket clocks in the late 1770s. These would tell you the time during the night to the nearest quarter hour.

## NIGHT CLOCKS

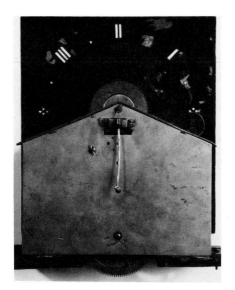

Figure 99 A, B, and C. *Longcase Night Clock by Edward East* One of only three longcase night clocks still known to exist. The movement circa 1670; the case of somewhat later origin. It is veneered in oysters of olive wood and decorated with panels of floral marquetry incorporating stained bone. The rising flat-topped hood has solid side doors which give access to the lamp shelf above the movement. This has eight fully latched pillars and shaped plates which are protected by dust covers to stop the ingress of dirt and smoke.

The four-wheel train has anchor escapement and is connected to the motion work operating the succession of Arabic hour numerals. The disc assembly carrying these rotates every two hours.

The dial, a little over square at 10¾ inches wide and 11½ inches tall, is beautifully decorated all over with bold floral engraving. Beneath the semicircular aperture at the top, it bears the signature *Edward East, Londini* and above it the quarter hours are fretted out in Roman numerals and the hour marks are indicated to the bottom left and right.

This clock was formerly owned by Mr. Hansard Watt and was in the J.S. Sykes collection. It was exhibited in the British Clockmaker's Heritage, Science Museum, 1952, No. 17 and "The Age of Charles II," Royal Academy, 1960-1961, No. 461. It was acquired by the British Museum in 1981. Photos Courtesy of the Trustees of the British Museum.

Rare Clocks 71

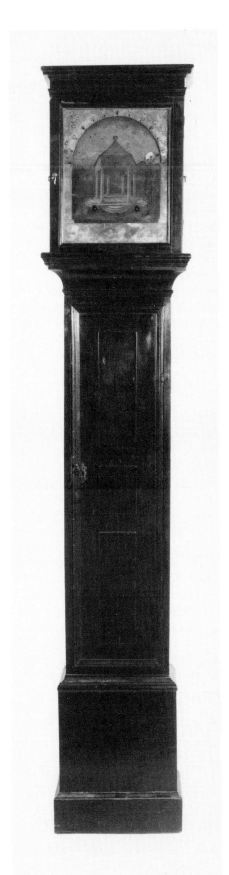

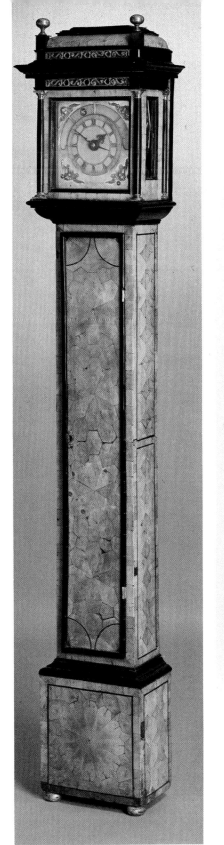

Figure 101 C.

Figure 101 D.

Figure 100. An ebony veneered night clock circa 1670 which is still believed to be in its original case. It has a two-train movement with latched plates and external count wheel strike. The way in which the time is displayed is similar to that shown on the East clock, but there is more Italian influence, the whole of the center being cut out to display a classical painted scene of a temple, beneath which are the two winding holes.

The panelled case of simple, classical lines has a rising hood beneath which somewhat surprisingly are concave moulds. It is signed *Joseph Knibb, Londini.Fecit*. Photo Courtesy Christie's, London.

## 72 Rare Clocks

Figure 101 A, B, C, D, E, and F. *Tompion Night Clock, circa 1678-1680.* We are extremely fortunate in being permitted by Christie's, London, to give the first full account in a book of this amazing clock by Thomas Tompion. In concept at least this is the most complex clock he ever made, incorporating the following principal features: It is only 6′1″ high to the top of the caddy, is of month duration with ting-tang quarter strike and Roman-rotation hour strike, employs a tic-tac escapement, has a brief-sounding alarm, incorporates a lunar or tidal hand, and above all is a superbly conceived and executed night clock.

It is amazing that this clock should have remained unrecorded until now. It has been in Canada for much of its life and is believed to have been the property of Sir James Douglas (1803-1877) a "Scotch West Indian" (the resourceful, energetic and intelligent son of a "free colored woman" and a Scottish merchant), fur trader, member of the Hudson's Bay Company, Governor of Vancouver Island 1851-1863, and Governor of British Columbia 1858-1864.

*The Movement.* This is of great complexity and in many respects has more in common with some of Tompion's finest bracket clocks rather than his longcases. Because of the limited space within the trunk, only two weights could be employed and thus one, that on the left, is used to power both the Roman-hour striking and the going trains, while the other dictates the ting-tang quarter strike. The size of the weights required, especially when one bears in mind the limited drop and that the clock is of month duration and has considerably more motion work to drive than a conventional clock, are quite modest. However, they leave no room for a conventional seconds beating pendulum and thus a tic-tac escapement with a short bob pendulum has been used.

*The description of the movement given here is that of Richard Garnier of Christie's who has studied it in great detail:*
It is unusually massive and basically rectangular in shape, the multi-divided frontplate being of skeletonized layered construction. There are latches to 14 of the 17 ringed pillars. The left weight gives dual drive to the outside countwheel with Roman notation hour strike on two bells via link rods and upstanding hammers. The going train of five wheels with tic-tac escapement embraces two teeth and short bob pendulum with holdfast hook. The right-hand weight powers the outside countwheel ting-tang quarter strike on two further bells via link rods and pendulous hammers. The backplate has a key squared right-angle drive setting for the cam-controlled serpentine-shaped detent releasing the pull-wound train of three wheels and a fly sounding seven blows on a fifth bell. This four-pillared subsidiary plated assembly is mounted above the double-wicked integral lamp, which is screwed to the top of the backplate. There are twin brass raked smoke flues originally connected to a once-present flu-cowl mounted in the top of the hood. A most ingenious feature is that when the alarm goes off it automatically snuffs out the lamp's two burners.

*The Case.* This is highly individual being veneered all over with boxwood oysters arranged in parquetry designs and having panelled sides. Nearly all the moulds are ebonized, thus accentuating the boxwood.

The carcass of the case is very substantial. It still has its lift-up hood complete with brass locking spoon and hood catch and there are giltwood bun feet to the front of the case. There are Knibb-style, gilt brass frets to the front of the base of the caddy and above the dial, and short, brass, ball finials resting on blocks.

At only six feet one inch to the top of the caddy, this is undoubtedly the smallest longcase clock Tompion ever made.

*The Dial.* This is some 8½ inches square and is signed along the bottom *Tho Tompion, Londini.* It has double-screwed, silver-winged, cherub spandrels and a skeletonized chapter ring some 5¾ inches in diameter. There are well-executed, blued-steel hour and minute hands and a third hand was designed to give the time of high water which was presumably used as there was no space left on the dial in which to fit a lunar calendar.

The cutout sector for the night dial, encircling the upper half of the chapter ring, is similar to those already described in Figures 99 and 100. There are bars dividing the hour up into the four quarters and behind these is a rotating, silvered ring bearing two opposing discs pierced with odd/even numerals respectively for the night hours. Each successive numeral traverses, during the hour, the cutout from left to right.

*Train Count*

Going train
(Barrel of 14 turns)
Great wheel 60
2nd wheel: 124 Pinion 14 carries alarm release cam.
3rd wheel: 100 Pinion 10
4th wheel: 90 Pinion 8
'Scape: 24 Pinion 6

Quarter train
Great wheel: 78
2nd wheel: 72 Pinion 9 carries 24 pins
3rd wheel: 60 Pinion 6 with hoop (one notch)
4th wheel: 60 Pinion 6 warning
Fly Pinion 6

Striking train
Great wheel: 78
2nd wheel: 90 Pinion 12 carries pins for hammers
3rd wheel: 60 Pinion 6 carries hoop (two notches)
4th wheel: 60 Pinion 6 warning (two pins)
Fly Pinion 6

*Motion Works.* Drive to night indications now missing. Drive to tidal hand on dial via a steel pinion, on the great wheel, of nine meshing with an idle pinion which drives the under-dial wheel of 60 teeth carrying the hand.

The movement, despite its complexity, has survived in remarkably complete and original condition. The apparent sum total of parts removed comprises:
—pull-pulley for 'alarm'
—blade spring for alarm drive
—one lamp nozzle
—night hour circlet drive-wheel/pinion
—blade to advance night-hour disc

74  *Rare Clocks*

## THE DISPLAY OF SECONDS IN THE ARCH

The 1720s and 1730s showed much innovation in the design and construction of clocks and many of the leading clockmakers vied with one another to produce the most important clocks, spurred on no doubt in many instances by their wealthy patrons who wished to be seen to have the finest available.

Besides the musical and astronomical clocks being produced, a considerable number were being made with different forms of dial layout. Probably the most successful maker in this respect was William Scafe (Figure 102) whose individuality, artistry and quality of work were second to none, his finest longcase clocks being those displaying seconds in the arch. Other makers who produced clocks with seconds shown in this way were John Topping (Figure 104), Francis Gregg and Ellicott (Ref: Figures 9/50 and 10/4, Tom Robinson, *The Longcase Clock*). Topping's and Gregg's work are so similar that it would seem likely that one made for the other; however, their workmanship, although of a high order, does not quite reach that of Scafe's nor does it have the artistic flare.

In most of the clocks the 'scape wheel is situated in the arch and on some of these, including that by Scafe, contrate wheels are used to connect this to the center wheel, although Topping employs wheels set at right angles to each other.

## SECONDS DISPLAYED IN THE ARCH

Figure 102 A, B, and C. *Walnut Longcase Clock with Seconds in the arch by William Scafe, London.* A beautifully conceived and executed clock full of originality by the eminent maker William Scafe. The Hon. B. Fairfax wrote to his nephew in 1727 of "One William Scafe a watchmaker born at Bushley near Denton (Co. York) served his time to his father, a blacksmith, but now the most celebrated workman, perhaps in London and Europe."

The walnut case is very well proportioned with applied columns to the hood door, herring bone inlay and crossbanding to the trunk door and base, and a double plinth. It would probably originally have had a caddy top.

The dial being highly individual in style makes it a little difficult to use for dating purposes. However, the omission of half-hour markings between the numerals, together with the use of half- instead of quarter-hour divisions, coupled with urn spandrels and the star engraving in the center of the large seconds ring in the arch, would suggest a date circa 1730-1735.

To the left of the arch is the regulation for strike/silent; on the right, graduated 5-25, that for fast/slow (pendulum rise and fall); the days of the month are indicated below 12 o'clock, and the days of the week, most unusually, numbered 1-7 in a large v-shaped cutout above 6 o'clock. Scafe's signature appears in another cutout rather like those used for displaying mock bobs on bracket clocks, but inverted. Another unusual feature is the way in which the spandrels are taken out to the edge of the dial, thus making the provision of a conventional mask impossible.

The seven-pillar movement is of the finest quality with beautifully shaped plates cast in one piece. The two most interesting features are the mechanism for the pendulum rise and fall and the seconds in the arch.

The center pinion has a contrate mounted on the rear end of its arbor which may be seen behind the pillar where the plate steps in. This drives a vertical arbor with a contrate mounted on top of it which engages the 'scape wheel pinion which is carried by a bridge.

The rise and fall mechanism makes use of a partially cut wheel mounted on the arbor to which the fast/slow hand is fixed. This wheel engages a rack operating on a vertical bar to which chops are connected which pass on either side of the suspension and thus vary its effective length.

# Rare Clocks 75

Figure 103 A, B, and C. *Oak Longcase clock by Major Scholfield.* An interesting mid-18th century oak longcase clock by Scholfield contained in a plain oak case with shallow arch and flat top with caddy above.

Unlike Scafe, he has featured seconds in the arch by means of a train of five wheels of equal count running up from the center wheel pinion to mesh with that of the 'scape wheel.

The plates are cast in one piece with a tall, shaped, central extension going up to carry the escapement. Because of this, the bell has had to be mounted to one side. Two of the six pillars are latched and count wheel strike is employed.

The highly individual 12-inch dial has a seconds ring 4¼ inches in diameter in the arch with dolphins on either side and there are cherub spandrels to the four corners. The chapter ring has decorative half-hour marks with quartering on its inner aspect. Immediately inside this the dial plate is grooved and within this is a raised and silvered brass chapter ring engraved 1-30.

Loomes in *Lancashire Clockmakers* records two Major Scholfields, the father who was born in 1707, and worked in Manchester and Rochdale where he died in 1783. It would have been this man who made this clock. The son was born in 1749 and died in 1813. 7′ 8″ (234 cm) high.

## 76  Rare Clocks

Figure 104 A, B, C, D, and E. *Longcase clock by John Topping with Equation of Time, Year Calendar and Seconds Shown in The Arch.* A fascinating tortoiseshell lacquer longcase clock by John Topping, London, who frequently signed himself "Memory Master," circa 1730.

The dial layout is a particularly fine one. To keep that rare feature, a seconds ring in the arch, as large as possible the rectangular part of the dial has been extended to some 14 inches in height, as against a width of 12 inches. This also facilitates the provision of larger than usual strike/silent and pendulum rise and fall (fast/slow) subsidiary rings. The spandrels are of the Turbaned head variety.

A large cutout in the bottom half of the dial center reveals the year calendar which also shows equation of time.
Engraved on the disc from within out are:

| | |
|---|---|
| First Line. | Sun Slower—Sun Faster. |
| Second Line. | The equation of time shown in minutes. |
| Third Line. | The months of the year with their number of days in Roman numerals. |
| Fourth Line. | The day of the month. |

Nicely balancing this in the top half of the dial center is Topping's recessed signature. The calendar may be set by means of a key inserted through the chapter ring a little below 1 o'clock. The train layout has had to be modified to clear the year disc and this is indicated by the high position of the winding squares.

The excellent quality of the dial is revealed in many ways such as the engraving between various rings, the finely chased and matted center to the seconds ring and the delicate hands made in such a way as to mask the year calendar to a minimum while they are over it. The plates are unusual in having three extensions:

a. That on the left which carries the mechanism for pendulum rise and fall, the cranked lever supporting the pendulum being moved by an eccentric cam, part of which may be seen behind the square for fast/slow and between the two plates.

b. That on the right to which the bell and the strike/silent mechanism is fitted.

c. A central support which carries the vertical drive to the 'scape wheel, the 'scape and pallet arbors and the fork which passes on either side of the suspension spring.

The original bracket fixing the top of the movement to the backboard is still present. The case is of excellent proportions, the caddy top, complete with original finials nicely balancing the taller-than-usual dial. 9′ (274 cm) high.

# Rare Clocks

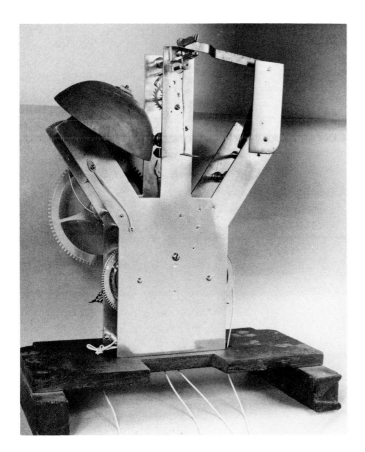

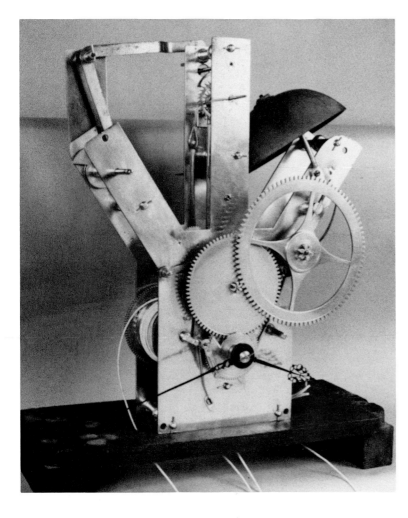

78  *Rare Clocks*

Figure 105 A and B. *Astrolabe Clock by Thomas Tompion circa 1677.* It is clocks such as this which illustrate so well the meteoric progress of Tompion's career, bearing in mind that he was making clocks of this complexity only six years after he came to London. It was owned by the late Mr. Ernest Prestige and was presented to the Fitzwilliam Museum in 1948.

*The Movement* The movement is of month duration and has a very early form of anchor escapement with the 'scape wheel revolving in an anti-clockwise direction, possibly to allow for the very unusual train. It is fitted with maintaining power but no shutters are provided.

As it would be impossible to wind the clock through the center of the dial, extensions have been fitted to the bottom of the plate on which are mounted separate winding arbors which are geared to the main arbors.

*The Dial* The dial has a 24-hour (2X I-XII) chapter ring indicating mean time with a straight steel hand for the minute and a graduated hand carrying the sun at its' tip for the hours. The chapter ring, two inner rings and the spandrels are made of silver.

Attached to the central ring is a moon hand which indicates lunar time on the chapter ring. Within the central ring is an aperture to show moon phases and engraved on it are the age of the moon and the time of High Water at London Bridge.

The rotating fret within the chapter ring has engraved on its' rim a year calendar, the zodiacal calendar and the ecliptic for the position of the sun. Behind the fret may be seen the curved lower border of the perceived horizon and engraved above this is a scale of azimuth lines.

The space immediately below the horizon colored light blue represents twilight and below this again is a dark blue crescent the edge of which represents the crepuscular line. When the sun drops below this there is total darkness.

Marked on the fret are a number of stars and thus as it rotates it can be seen when these cross the horizon and become visible. Their position can then be read off during the night against the azimuth and altitude scales engraved on the plate.

*The Case.* The case is of superb quality, being decorated with olivewood oyster veneers set in panels and with delicate floral marquetry to the four corners of the door and the top of the base and a bird set among flowers in the middle of the door. The fine capitals to the hood pillars would have been built up from several castings and then soldered together and fire gilt. The hood is still of the lift-up variety and bears a cresting of somewhat later origin. Photos courtesy of The Trustees Fitzwilliam Museum, Cambridge.

# Chapter 4.
# Astronomical and Equation Clocks

The movements of the heavenly bodies relative to each other have been studied for thousands of years, and quite early on astronomers were able to predict the position of various stars and planets, but often only after long and complex mathematical calculations. To simplify this procedure, which was of great value in astrology, models were constructed on which some of the movements could be shown, and as early as the 8th century A.D. the Tantric Monk I-Hsing constructed armillary astronomical models controlled by a water escapement. In Islam, gear trains were devised to show the relative rates of the sun, earth and phases of the moon.

The most famous early Astronomical clock to be made in England was that of Richard of Wallingford (1296-1336), an Abbot of St. Albans. The clock was unfinished at his death, but was completed by Laurentius de Stokes and William Walsham. It was an elaborate astronomical model with an astrolabe dial and mechanisms for showing the motions of the sun, moon and planets besides demonstrating the eclipses.

Another fine early astronomical clock was that of Giovanni de Dondi (1318-1389) a physician in Padua. He is recorded as having built it entirely himself over a period of sixteen years. It was the first self moving version of the complicated models based on Ptolemaic planetary astronomy. It was called an Astrarium, representing the universe with the motions of the five planets and the moon. Dondi used internally cut and elliptical wheels in his design. Several models of this clock have now been built. Another fine astronomical clock which is still around for all to see after 500 years is at Hampton Court.

In the 16th and 17th centuries a considerable number of clocks with astronomical indications were produced at Augsburg and Nuremberg in Germany. Some of the first to be designed in England were the complex spring or table clocks made by Watson and a longcase clock with an astrolabe made by Tompion (Figure 105): However, it is interesting that for no obvious reason many of the fine astronomical clocks produced in England in the 18th century were made, not in the capital, but the provinces. Examples of provincial makers which spring to mind, are Cockey of Warminster, Lister of Halifax, Barker of Wigan and Budgen of Croydon.

The diversity in which they displayed the astronomical information on the dials of their clocks, so clearly seen in the following pages, is amazing and makes one wonder at their skill and ingenuity and their knowledge of mathematics and astronomy.

## THE EQUATION OF TIME

Due to the changing speed of the earth in its' elliptical course around the sun, caused by variations in gravity and the inclination of its' axis, the length of the solar day (the time taken between two successive passages of the sun over the meridian, timed at mid-day) varies constantly throughout the year. It is, at a maximum, fast 16½ minutes on or near November 4th, and 14¼ minutes slow on February 12th, and four times a year the sun's (Solar) and our (Mean) Time agree; these are on April 16th, June 15th, September 2nd and December 26th. The difference between Solar and Mean time is known as "The Equation of Time."

Although astronomers had been aware for at least a thousand years that the earth did not keep a constant speed as it passed around the sun, it had not been known by the general public. It was to enable the speed of the Earth's rotation as it passed around the sun to be measured, that Tompion's year clocks were commissioned for Flamsteed's new observatory at Greenwich. Although there were some problems with these clocks in the early years, they undoubtedly enabled Flamsteed to achieve the results he required and made it possible for him to concentrate on such matters as finding the longitude at sea.

It was only with the introduction of the seconds beating longcase that people began to realize their clocks behaved erratically when compared with the sundial.

Figure 106 A. *Edward Cockey's prototype Astronomical clock as it stands today in the Great Hall at Longleat housed in a replacement oak case.* Photo courtesy of the Marques of Bath.

To overcome this equation tables were produced (Figure 453) showing the difference between Mean and Solar time so that this could be taken into account when they were correcting their clocks.

An alternative solution provided by some clockmakers, of whom Fromanteel was undoubtedly one of the first, was to provide the clock with a movable minute ring which could be set daily. Another approach was to provide the clock with a large year calendar with the time equation marked on it for the days of the year such as that seen on the clock by Topping (Figure 104).

There is much debate and confusion as to who first devised the equation cam, which was to enable clockmakers to show both Mean and Solar time without having to refer to tables. Tompion in 1695 produced a year clock for William III showing Mean and Solar time, and Quare in the same year made one for King Charles II of Spain for which the equation work was provided by Williamson. However, it is thought by some that the idea originated with Huygens who is known to have been acquainted with both men.

One of the most outstanding equation clocks was the double-dialed (back to back) clock made by Williamson circa 1720-25; one dial showing, amongst other things, solar and the other mean time. An excellent description of this clock is contained in Alan Lloyds book *Some Outstanding Clocks Over 700 Years*. iii

Williamson undoubtedly devoted much of his time to equation clocks, producing the equation work for all of Quare's clocks, always mounted as a separate unit within the movement. However, when he made his own equation clocks the work for this was incorporated in the design as in Tompion's clocks.

The recorded claim by Williamson that he made all the equation clocks up until the time of his pronouncement needs to be treated with some caution. For instance, it obviously wasn't true in Tompion's case. However, it may be that he was actually referring to his solar time clocks (Figure 126) which were undoubtedly invented and made by him.

In these an eccentric cam or disc rotates once a year and as it does it raises or lowers the pendulum, the suspension strip of which passes through a slotted brass block. In this way its effective length varies throughout the year, and therefore the timekeeping, enabling it to keep solar time. It can thus be checked directly against a sundial.

## SIDEREAL TIME

One other form of time measurement needs mentioning here and that is sidereal time, which is used by astronomers. It is measured by the successive passage of stars, infinitely far away, across the meridian and is 3 minutes 56 seconds shorter than the mean day.

Because it has a fixed relationship with mean time, it is far easier to measure and calculate than solar time. Indeed all that is required to make a clock keep sidereal time is to shorten the pendulum to the required length.

As it is useful on occasion to know both mean and sidereal time, clocks have been devised to fulfil this purpose, one of the first being made by Tompion using a single dial (Figure 127). Quare made a clock circa 1710 with separate chapter and seconds rings for both mean and sidereal time, and in 1836 Walsh of Newbury made a clock with two separate dials. This one is now in the Science Museum, London.

## REFERENCES

(i) House D., "The Tompion Clocks at Greenwich and the Dead Beat Escapement," *Antiquarian Horology*, Dec. 1970 & March 1971.
(ii) Symonds R.W., *Thomas Tompion. His Life and Work,* Batsford Ltd., London, New York, 1951.
(iii) Lloyd, H.A., *Some Outstanding Clocks Over Seven Hundred Years,* Antique Collectors Club, Reprint 1981, Pages 76-85 and 248-255.

## GENERAL REFERENCES

Asprey & Company. *The Clockwork of The Heavens*. Private Publication. 1973.
Bassermann, Jordan, Bertele. *The Book of Old Clocks and Watches.* George Allen and Unwin Ltd., 1964.
Lloyd. H.A. see above Mayer, Maurice. *The Clockwork Universe.* Smithsonian Institution, Washington, D.C. and Neal Watson Academic Publications, New York 1980.

## THE RESTORATION OF E. COCKEY'S ASTRONOMICAL CLOCK

We were very grateful for the support and help given by the National Maritime Museum when we restored this clock and were fortunate in that when we first visited them their clock had been dismantled for restoration. We were thus able to examine it in detail and take moulds of the dial furnishings which were missing on our clock so that we could make accurate copies. It was also interesting to be able to put the two movements alongside each other for comparative purposes.

The case in which the Cockey we were restoring was housed was so completely out of keeping that there was nothing to do but scrap it. Although John Martin's research led us to believe that the clock we were restoring may have been Queen Anne's, we felt that this would have been far too great an assumption to make. Moreover, all we would have had to go on was one drawing and thus we decided to copy the clock at Greenwich which we knew to be original and obviously as approved of by Mr. Cockey.

Several visits to Greenwich over the years proved to be essential as we went into the restoration in greater depth and more craftsmen became involved in it. Initially, overall photographs were taken together with general dimensions and from these rough plans could be drawn up. A further visit was then necessary to check all the detail dimensions and draw up the final plans. When these had been produced it was realized that a large number of detailed photographs would be necessary if the craftsmen involved, for instance those doing the wood carving and decorating the case, could do accurate copies of the original and "get a feel" for the work involved.

The final stage, after the case was completed, was examining the finish on the clock in detail to try and achieve the same results, and here we were surprised to find that on looking at the paintwork with a magnifying glass that different surfaces had been employed in different areas, presumably to heighten the effect of the decoration. Full-size drawings were also made of various parts of the decoration to avoid the distortion which occurs in photographs.

Although the overall concept of the decoration was copied it was felt that no two original cases would have had exactly the same design and thus the best part of a week was spent looking at books on classical decoration to decide which would be most suitable. 11′ 10″ (361 cm) high.

Figure 106 B. The movements of the Longleat clock on the left and the clock described in detail here during restoration of both movements.

Figure 107. The National Maritime Museum Clock. This is the most original of all four as its case still carries all its original ornamentation and decoration. Photo courtesy of the Trustees of the National Maritime Museum.

## EDWARD COCKEY, COUNTRY CLOCKMAKER EXTRAORDINARY

Edward Cockey was born at Warminster, a small town in Wiltshire, in 1669. His father ran a very successful business there as a brazier, making and supplying brass and copperware throughout the district. Research to date has failed to find any details of his apprenticeship as a watch and clockmaker, although there seems no doubt that he must have had a very good training. Surprisingly, however, he was more frequently referred to locally as a brazier, but this may be because he worked for his father and then inherited his business and pursued it as well as clockmaking.

Much of Edward Cockey's life revolved around the great house of Longleat, the seat of the Weymouth family, which stands about two miles from Warminster, and there seems little doubt that the greatest part of this trade was conducted with the Longleat estate. His father had been supplying all the kitchen ware, brass, copper and pewter ware for the gardens and horse harness etc., from around the middle of the 17th century. Edward took over the running of the business around 1700 and became owner after his father died in 1711.

Edward Cockey was married in October 1695 in Totnes in Devon where their first child was born. After that their children were born in Warminster and baptized at the Minster church there, including Edward junior. He obviously ran the family business diligently and one wonders how much he might have wished to devote more time to his own trade as a clockmaker.

The earliest of his recorded clocks can be dated to the last five years of the 17th century, being an eight-day longcase clock with ten-inch square dial, but it certainly appears that his thoughts must have been turning to his monumental astronomical clocks by about 1695 when he was 26, and this presents researchers with their biggest puzzle. We have no idea where Cockey learnt his trade. It certainly could not have been in London as the Apprentice Register of the Clockmakers Company contains no-one of that name. It is felt that it must have been in the West Country and the town of Totnes seemed a likely place, as he spent some time there, but nothing has come to light as yet. To learn the trade of clockmaking is one thing, but to acquire the knowledge necessary to design a highly sophisticated and complex astronomical clock is another, and there were very few master clockmakers around the provinces at that time capable of imparting such knowledge.

In the Great Hall at Longleat stands Edward Cockey's first astronomical clock (Figure 106 A and B). It is undoubtedly his prototype. The clock is of one month duration and includes a striking train. It also has many major alterations which were made at the time that the movement was constructed, thus indicating a high degree of experimental work. There is evidence to show that the clock was originally in a fine lacquered case, but today it is housed in a plain oak case dating from around 1770. How the original case was lost is not known, but recent research into the archives at Longleat reveals that the clock was probably bought by Lord Weymouth in 1707 after much negotiation with Cockey.

Undoubtedly, however, this clock was the prototype for his three other early astronomical clocks. The first of these was probably that presented to Queen Anne. Contemporary accounts show that this was first set up in St. James's Palace in London in the Queen's apartments around 1705, but who saw Cockey's prototype in Warminster and commissioned him to make this clock is yet another mystery. It must have been considered quite exceptional to have been accepted by the Queen for her private apartments and may have prompted Lord Weymouth to try very hard to buy the prototype on which it was based from the man he had never thought of as more than his brazier and supplier of hardware. It is probably no coincidence that Edward Cockey did not take over the responsibility for looking after the clocks and watches at Longleat until 1707—merely a previous lack of realization of his ability in their field perhaps.

As was stated previously, apart from the Longleat clock and that given to Queen Anne there are two other examples of Cockey's great early astronomical clocks. One is well known and stands in the National Maritime Museum at Greenwich (Figure 107) and the other is, at the time of writing, undergoing restoration in London (Figure 108). These two clocks are the only ones of the four to retain their original cases, and only the National Maritime Museum clock has its original case decoration—the lacquer having been stripped from the other one some time in the 1940s.

It was thought for some time that the National Maritime Museum clock was the Queen Anne clock, but this now seems to be very unlikely as the style of case is totally different from that shown in a contemporary illustration in the Royal Inventory, whereas the layout of the movements of all three clocks, other than the prototype, are very similar. All three are timepieces only and of three-months duration. The plates are all obviously cast from the same patterns and the dial furniture is the same, in fact, Cockey, following the modifications to his prototype, seems to have standardized his design with the only exception being the style of engraving of the chapter rings which differ considerably among all four clocks.

An intriguing mystery exists over the whereabouts of Queen Anne's clock. In 1809 a fire occurred in St. James's Palace and a large amount of furniture was removed from the affected apartments and stored in a part of the establishment known as "The Ride" and, included, was Edward Cockey's clock. This furniture must have stayed there for a very long time, for in 1835 the Cockey was included in an inventory of Royal Clocks and it still stood in "The Ride." However, between 1835 and 1850 the clock disappeared as a hand written note on the inventory states: "Memorandum Nov. 1 1850, I never could discover what became of this clock." What happened then to Queen Anne's clock? It must surely be one of the four clocks extant.

By a process of elimination we know that it cannot be the Longleat clock—which has been in that house since 1707. It seems most unlikely that it is the National Maritime Museum clock as that is the only one in its original case complete with decoration, and bears little resemblance to the Royal Inventory illustration. The clock that is presently undergoing restoration in London does have its original case, although stripped of its decoration. Also, this case bears a striking resemblance to that in the Royal illustration. However, the whereabouts of this clock have been well documented and it has, over many years become known as the Carrington clock because it was, for a very long time, in the possession of the Carrington family. Indeed it was in their possession at the time that the Royal Inventory was made in 1835 when Queen Anne's clock stood in "The Ride," and there is a note on that inventory to the effect that Lord Carrington has a similar clock.

We are left, therefore, with the clock described in detail in this book. This clock was purchased in 1979 and efforts to trace its previous history have all been unavailing.

The case in which it was housed, the pendulum, weight, hands and some of the dial fittings were all incorrect and had to be discarded and a major work of restoration was undertaken which required careful examination of the other extant Cockey astronomical clocks in order to ensure the accuracy of the work. The case was made as a copy of that of the National Maritime Museum clock in the interests of authenticity. Whether this is the clock that stood in St. James's Palace we shall probably never know, but it does seem the most likely contender to have Royal Provenance.

Edward Cockey's four great early astronomical clocks were certainly made before 1720, but during the period up to about 1740 he does appear to have been very busy with his clockmaking as there are many examples of fine longcase clocks extant, some with complications such as lunar and tidal dials in the arch or quarter chiming, and almost without exception they are housed in the lacquered cases which were popular at that period (Figure 110). He has also left us bracket clocks and a number of watches which appear to be of the finest workmanship.

During the period from 1707 until 1743, Edward Cockey looked after the clocks and watches of Lord Weymouth when, at the age of 74, he presumably decided to retire, but during this time he had never ceased to supply the estate with all its hardware requirements. Indeed by 1710 he had extended his supplies business to that of wines and spirits and he appears to have become a major purveyor of liquor in Warminster by about 1720. His son Edward Junior entered the business about this time and seems to have taken an ever-increasing share of the load which probably enabled his father to devote the time he wanted to his clockmaking. Edward Junior was not apprenticed as a clockmaker and appears only to have been concerned with running the family business. At his death in 1786 he is referred to as a wine merchant which, perhaps, is some indication of the steady change of direction that the business took over three quarters of a century.

To return, however, to Edward Senior, there exist two further fine astronomical clocks bearing his name. The first is in the British Museum (Figure 111), and the

Figure 108. *The "Carrington Clock"* This clock currently is undergoing restoration as all the original lacquer work was stripped off some years ago. The case style of this clock is very similar to the case of Queen Anne's clock. Photo courtesy of Sotheby's, London

84  *Astronomical and Equation*

Figure 109 A, B, and C. *Astronomical Longcase Clock by Edward Cockey, Warminster.* This clock dates from the first decade of the eighteenth century. It has a movement of three-months duration with recoil escapement and one and a quarter seconds pendulum. The driving weight housed in the Corinthian column to the front of the case is of ninety pounds and the clock is wound through a contrate wheel driving on to a wheel mounted on the front of the barrel. Access to the winding square is through the left hand side of the hood.

The dial plate is 20-inch square and carries a 24-hour chapter ring. Within this is set a dish which rotates once in 24 hours. This dish is painted to represent the day and night skies and carries a gilt representation of the sun. The dish revolves behind a painted fixed shield which fully obscures the celestial scene between the hours of 8:45 P.M. and 3:45 A.M. Appearing from each end of this shield are blued steel shutters which rise and fall according to the season and ensure that the sun rises and sets at the correct time each day.

The sunrise/sunset shutters are controlled through arms which ride on large cams which are each attached to one of two nine-inch diameter wheels with 365 teeth rotating once a year. Fitted to the bottom shield is a ten-inch diameter brass ring decorated with wheatear engraving which acts as a guide to the inner edges of the sunrise/sunset shutters. Within this fixed ring is a blued-steel, wavy-edged ring which rotates with the celestial dish and attached to this is a cast gilt figure of Father Time. His hand indicates the hour, while the minutes are shown by the long gilt center-sweep hand.

Within the blue wavy ring are set three concentric silvered dial rings. The outer is a year calendar, the next represents the ecliptic divided into the 360 degrees of the Zodiac, and within that is the 29 ½ days of one lunar month.

The calendar and Zodiac scales are fixed together and are read from the pierced gilt pointer fitted to the blue ring at 90° from Father Time. These two rings were moved eleven days relative to each other during the eighteenth century to correct for the change, in 1752, to the Gregorian calendar.

The age of the moon, read from the inner of the three rings, is indicated by the right foot of father time. Set within the lunar calendar in a blued plate with gilt stars is a moon ball and this rotates about a radial axis to show its phases.

In the center of the dial, set on a silvered plate engraved to show clouds and rain, is a representation of the earth. It is interesting to note that, viewed from the earth, the sun and moon always maintain their correct relative positions in the Zodiac.

The design of the four large spandrel ornaments is unique to Cockey and is thought to represent the seated figure of Queen Anne.

The clock is housed in a reproduction case which has been carefully copied from that of a similar clock by Cockey in the National Maritime Museum.

## 86 Astronomical and Equation

Figure 110. *Lacquered longcase clock by E. J. Cockey*. A lunar and tidal dial is in the arch. An example of his conventional work, circa 1730. Photo courtesy of the National Trust.

second in the collection of Lord Harris. There is considerable mystery surrounding these clocks as they are both much later than the other four and cannot date before about 1740. The British Museum clock is far more sophisticated than the early clocks, but the workmanship is very similar. The Harris clock has a very different feel about it and gives the impression that it has been largely made by different hands. We do not know when Cockey died but we have reason to believe that he lived possibly into his eighties. Could those two clocks have been the work of his retirement years? Perhaps he completed his masterpiece, the British Museum clock, and only commenced work on the Harris clock.

Interestingly, we know that even though not a clockmaker himself, Edward junior employed a journeyman clockmaker and perhaps it was this man who completed the construction of the Harris clock after Edward senior's death. Once again we are unlikely ever to know.

Edward Cockey, clockmaker, remains today an enigma. How was it that this member of the family of braziers in a small market town was able to think out and make clocks of immense technical brilliance, clocks of great complexity and made with such superb skill and artistry as to better the work of many of the best London makers in an age when communications were difficult and the dissemination of knowledge and information slow? He was, without doubt, a Country Clockmaker Extraordinaire.

Figure 111. *Dial and hood of the later Astronomical clock in the British Museum.* Illustration courtesy of the Trustees of the British Museum.

Figure 112 A, B, and C. *Astronomical Clock by William Webster.* A well-proportioned and relatively small walnut longcase clock with rotating globe moon and lunar calendar in the arch by William Webster, Exchange Alley, London. He was an eminent maker who was Thomas Tompion's apprentice and journeyman. An advertisement, reproduced in *Thomas Tompion, His Life & Work* by R.W. Symonds, reads as follows:

> On the 20th Instant Mr. Thomas Tompion, noted for the making of all sorts of the best Clocks and watches, departed this Life. This is to certify all Persons whatever Quality or distinction that WILLIAM WEBSTER at the Dyal and Three Crowns in Exchange Alley, London, served his Apprenticeship and lived as a Journeyman a considerable time with the said Mr. Tompion, and by his Industry and Care is fully acquainted with the Secrets of the said Arts. (This announcement appeared in *The Englishman* Nov. 21-24, 1713; *Mercator or Commerce Retrieved, being Considerations on the State of the British Trade, &c.* November 21-24, 1713; *The London Gazette.* Nov. 24-28, 1713; and *The Post Boy,* Nov. 21-24, 1713.

Examples of Webster's work are in many famous collections. Interestingly there is at least one bracket clock by Webster with a virtually identical dial.

The eight-day seven-pillar movement is of fine quality, having rack striking and large Y-shaped extensions to the plates to carry the mechanism for strike/silent and pendulum rise and fall (fast/slow) to the top left, and strike/silent to the top right. In the arch is a recessed rotating globe moon, one half black and the other half silvered, set against a star-studded background. Around the periphery of this is a disc marked out with the age and phases of the moon, against which a hand rotates. A square at 12 o'clock permits the moon to be set correctly by use of a key. 7' 2" (218 cm) high.

## 88 Astronomical and Equation

Figure 113 A, and B. *A Complex Astronomical Clock by Christopher Pinchbeck.* A magnificent and complex astronomical clock in a walnut case by the eminent maker Christopher Pinchbeck, London. Displayed in the arch of the fine quality dial are: the phases of the moon, the time of high tide in 29 European ports, and the lunar date. The upper left dial shows the position of the sun and the southing of 23 stars, and that on the upper right gives the moon's position in the Zodiac by means of a rotating disc. The bottom left dial shows the months, and that on the bottom right the day and its ruling deity. The oversquare dial (13-by-12 inches) has Indian-head spandrels obviously specially cast for the dial, and dolphin spandrels in the arch. The sides of the dial have a nicely engraved border. There is a strike/silent lever above the XII. The blue steel hour and minute hands are well cut and faceted. The complicated movement of month duration has three spiral drives for the various astronomical indications.

The slim walnut case has a breakarch hood with frets and side windows, tapered pillars and capitals to the hood door. The mouldings throughout are of the highest quality. The trunk door, sides and base have a delicate herringbone inlay of walnut together with cross banding to the trunk and base.

Christopher Pinchbeck was well known for his astronomical clocks and for his alloy of zinc and copper: Pinchbeck Gold. An astronomical clock by him is at Buckingham Palace. A similar complex clock by him is shown in Tom Robinson's book *The Longcase Clock* p.233, but the signature is simpler and in a curved sector in the dial center, whereas this clock has a rather magnificent engraved signature plaque at the bottom of the dial. Christopher Pinchbeck was born in 1670 and died in 1732. This clock may be dated circa 1725-1730. 7' 9" (236 cm) high. (Ref.: Pages 316-319 for an account of Pinchbeck's life and work.)

Figure 114 A, B, and C. *Astronomical Clock by Thomas Budgen, Croydon.* A well proportioned walnut longcase clock, the caddy top with fretwork below and further fretwork above the breakarch. It has applied pillars to the hood door and a square topped trunk door with cross grain mouldings around it, circa 1730.

The astronomical dial is, so far as the author is aware, unique. In the arch set beneath a domed glass is a globe moon and below this and standing out from the dial, a lunar calendar. There is wheatear engraving to the border, dolphin spandrels in the arch and mask spandrels in the four corners.

The clock has a most unusual but very practical 24-hour chapter ring numbered 1-12 twice, 12 noon being at the top, and 12 midnight at the bottom. The minutes are shown in large Arabic numerals on its inner aspect and the minute hand, shown here too long and extending to the hour ring, going round once an hour. The hour hand carrying the sun at its' tip revolves once a day and most unusually a third hand carries the moon and shows its correct position relative to the sun, revolving roughly every 24 hours 50½ minutes. Thus, at new moon, the moon hands position is just behind that of the sun and at full moon it is exactly opposite it. There is a seconds ring below 12 o'clock and the lower dial inside the chapter ring gives the day of the month at its periphery while in the center is indicated the moon's position in the Zodiac. Within cut-outs on either side of the seconds ring are shown the days of the week with their ruling deity and the months of the year with their zodiacal sign.

Mounted against the chapter ring on either side, centered on 6 A.M. and 6 P.M., are two curved strips marked with the signs of the Zodiac which are used to calculate the times of sunrise and sunset. Each strip carries six signs with four divisions, that on the left has Cancer to Sagittarius (June-November) and that on the right Capricorn to Gemini (December-May). The scales are carefully calibrated to allow for the day-to-day changes in the times of sunrise and sunset. To find these out firstly, the Zodiac is read from a pointer on the right slot carrying the year calendar. A line is then taken from the dial center to the corresponding reading on the left strip and this automatically gives the time of sunrise. A similar reading through the right strip gives the time of sunset.

Figure 115 A, B, and C. *Walnut Longcase Clock by Gandy of Cockermouth Dated 1757.* This astronomical clock by Gandy is contained in a walnut case with a quarter-veneered and cross-banded door and banded base. At first sight it looks a little top heavy but this is so that the astronomical dial can be displayed to the maximum advantage. The lower part is 14-inches square and the arch is far larger than normal.

The chapter ring is beautifully engraved as is the whole dial. It is relatively wide to accommodate all the information. At its periphery is a year calendar for the center-sweep date hand, and within this are depicted the signs of the Zodiac each most unusually preceded by the word "enter." The Roman hour numerals are enclosed by a heavily arcaded double line and between them are very decorative half-hour marks. There is quartering on the inside of the chapter ring.

The dial center bears initials below 12 o'clock, presumably those of the first owner. At the bottom is Gandy's signature and surrounding this and the ringed winding holes is floral engraving.

The chapter ring in the arch is basically for 24 hours but with a sector missing at the bottom. The Roman hour numerals are marked IIII—XII and XII—IIII. The periphery of the ring is marked out in degrees to give the sun's amplitude in the heavens, and the inner border bears the signs of the Zodiac. At the top is engraved *The Cause of the Vielssitude* (Variations?) *of the Seasons.*

In the center of the arch is a separate dial bearing the signs of the Zodiac at its periphery and inscribed at the top "The Winter Signs of the Zodiac." Within this are calendars for the age of the moon and

## THE ASTRONOMICAL CLOCKS OF CUMBRIA

During the 18th century there were many fine clockmakers in Cumbria and at least three families, all on the coast and within 50 miles of each other, produced astronomical clocks very similar in concept. There were the Gandys who worked at Cockermouth and Maryport, the Thompsons of Whitehaven and Cockermouth, and Edward and John Harriman of Crosby and Workington.

John Penfold in his excellent account of Cumbrian clockmakers gives a detailed history including the genealogies of the three families but sometimes omits to give the maiden names of the wives which is unfortunate as one strongly suspects that there must have been at least some intermarriage between these families.

To put matters in context, it must be remembered that at the time the towns in question were important ports, although much of their business was lost gradually due to the greater attraction of larger ports and difficulties which silting brought about.

## THE GANDY FAMILY

James Gandy is first reported in Cockermouth in 1726. His two sons, Samuel and James II, joined the business in due course and presumably took over some time before their father died in 1779. Samuel had eight children, but only one, James III, seems to have survived and carried on as a clockmaker. Samuel died in 1803 and appears to have been the last of the Gandys to have worked in Cockermouth. His son James III is recorded in Maryport where he married in 1785. Sadly he died there in 1793 at the age of only 34.

Several clocks are recorded as having been made by them besides that shown here. Most are eight-day clocks and according to Penfold at least two were musical. He sums up the family well by concluding with the words: "One always feels that there may be something unusual about a Gandy Clock."

## EDWARD AND JOHN HARRIMAN

Edward the father was born around 1700, possibly in Crosby. He married in 1729 and moved to Workington in the 1730s. His first wife died in 1740 and he remarried shortly afterwards. His eldest son, John, born around 1742, followed him into the business, taking it over after his father's death in 1776. He continued what appears to have been a thriving concern until his own death in 1812.

A sign of the hard times in which John lived is given in the account that in 1786 he had six watches stolen. The men caught for this crime were sentenced to death at the Carlisle assizes in September of that year.[1]

time of high water, presumably at Cockermouth. In the center is an engraved moon disc.

Shutters rise and fall throughout the year between the inner and outer dials and thus show the sun which circles within the outer chapter ring as always rising and setting at the correct time. Unfortunately when the dial was photographed the sun had set for the day, but it may be seen at around 7 o'clock in the overall photograph. 7' 3" (221 cm) high.

Figure 116 A, B, and C. *Astronomical Clock by Harriman of Workington Circa 1780.* The dial is some 14 inches wide and 20½ inches high, having an unusually large arch to accommodate the astronomical work. There are large cherub spandrels, wheatear engraving to the border, a beautiful floral engraved dial center with blued-steel, ringed winding holes and a highly decorative name plaque. The Dutch-style chapter ring with wavy minute marks has well-executed half-hour divisions and is marked on its inner aspect 1-31 for the date.

Within and below the arch is a large ring some ten-inches in diameter which intersects the chapter ring. Its periphery is marked in degrees to show the sun's position in the heavens and engraved *Amplitude Occa* with effigies of the sun preceding these captions. Between the outer and an inner ring, a rotating disc carries the sun around. The disc has a blue ground and is decorated with clouds. Between the two rings are shutters which are decorated with scenes of

---
[1] Renfold, JB. *The Clockmakers of Cumberland.* Brant Wright Associates Ltd., 1977.

people, buildings and horses. These rise and fall throughout the year so that the sun always rises and sets at the right time and pointers attached to the shutters give the actual time when this occurs.

The left side of the inner ring is marked in degrees to show the sun's declination north or south of the equator and the right side indicates the sign of the Zodiac. Within the inner ring the moon rotates with a silvered scale at its periphery and below the moon two globes are depicted with particularly fine floral engraving incorporating dolphins.

The walnut case has a swan neck pediment with large brass paterae and a finial between. The hood pillars are reeded and have wooden capitals and there is a cross-grain mould above the door. Both the trunk and base have reeded quarter columns and are decorated with quartered walnut veneers surrounded by crossbanding with a raised panel on the base. The clock rests on bracket feet. 7′ 9″ / 7′ 5½″ (237/227 cm) high with/without finial.

## 92 Astronomical and Equation

## THE THOMPSON FAMILY

Because Thompson is a very common name in Cumbria, and for various other reasons mentioned in his book, John Penfold has found it difficult to establish an exact genealogical tree for the family. It is believed that William Thompson established the business in the 1730s and probably died in 1777. Joseph, possibly his younger brother, moved to Cockermouth in the 1740s and died there in 1768. Clocks signed by him at the same period sometimes bear the name Cockermouth and sometimes Whitehaven, so he presumably worked in both places.

Other members of the family who worked in the businesses were Joseph and his son Plaskett who took over from his father in 1841.

Figure 117 A, B, and C. *Walnut Longcase clock by Thompson, Whitehaven, circa 1760.* Although this clock is basically similar to that by Gandy, the dial layout is considerably simpler and because the dial is smaller the proportions of the case, which is most attractively inlaid, are more pleasing.

The main dial has a well-engraved center with serpents curling around the winding holes and a bird between them. There are mask spandrels and center-sweep seconds, minute, hour and date hands.

Within the arch, the outer ring, besides showing the hours against which the pointers on the shutters can be read to give the times of sunrise and sunset, is also marked *40° East -45°* on the left, and *40° West -45°* on the right. It is inscribed around the top, from left to right, with the sun's sign and *Amplitude Ortive, Meridies and Amplitude Occa.*

In the center on the left is marked at the bottom *North* and then *20° -0° -20° declination,* and on the right with a symbol of the sun followed by *Place VS* and then the signs of the Zodiac marked in degrees. It bears the maker's name plaque in the center.

## 94 Astronomical and Equation

Figure 118 A, B, C, and D. *Astronomical and Tidal Dial Longcase Clock by Philip Lloyd, Bristol.* This is a particularly interesting astronomical and tidal-dial longcase clock very similar to that made by Ferguson in 1770 to show the tides at London Bridge. Engravings of castles are on either side of the arch and in the center is an extensive cutout, behind which is a harbor scene with Men of War, Merchantmen, rowing boats, cliffs and the quayside. In the foreground, two boats, an admiral's barge and a ferry, automatically rise and fall with the tides.

The 24-hour dial is divided into I-XII twice and has a date aperture at the bottom. There are center-sweep seconds and minute hands and the hours are shown by a sun pointer attached to the periphery of the outside ring of the central disc. This ring has I-XII marked on it twice for the time of high water, and within is shown the age of the moon. Both of these are read by a moon hand, here partially hidden by the seconds hand which is attached to the edge of the inner rotating disc. This hand gives the relationship of the moon to the sun and earth.

The rotating lunar disc has a hole in its periphery through which the moon's phase may be seen. Inscribed around its edge is the moon's age, the time of her southing, and time of high water. Within this the state of the tides: low water, half ebb, high-water, half flood and low water, marked twice, are shown. These are indicated by a hand attached to the central earth disc.

### PHILLIP LLOYD OF LAWHADEN AND BRISTOL

Philip Lloyd started life in the small village of Lawhaden in Pembrokeshire (South Wales). He later moved to Bristol which was at that time a commercial and cultural center for the people of South Wales as it was then easier for them to travel by boat to Bristol than to go over the mountains, since the valleys run down to the sea. (It is quite remarkable that he produced the astronomical clock illustrated in Figure 119 A, B, and C at that time.) It would seem likely that even when Lloyd was in Lawhaden he obtained materials in Bristol, and also possibly specialized help, for instance with the engraving of the beautifully executed dial.

Figure 119 A, B, C, and D. *Astronomical and Equation Clocks.* This clock, made by Philip Lloyd around 1770 is built to a design by James Ferguson, F.R.S., (1770-1776), a brilliant, self-taught astronomer who invented many aids to the teaching of astronomy and travelled widely to lecture on the subject.

In the center of the dial is a fixed disc representing the Earth, and in this design the moving assembly rotates about the North Celestial pole. The time of day is indicated on the 24-hour dial by the fleur-de-lis attached to the sun. This fleur-de-lis is fixed to the annular ring, graduated in 29½ divisions which represent the moon's age. Within this ring is a circular plate on which the months and days of the year are engraved, and within this are the divisions of the ecliptic. This plate revolves relative to the lunar ring once in a Sidereal year (366 days) and thus the day of the year is indicated under the sun pointer.

Within the astrological circle on the center plate are shown the major first, second and third magnitude stars visible in the latitude of London, together with ecliptic, equinoctial and tropics. The age of the moon is indicated on the lunar ring under the wire of the moon itself, and the latter revolves fully in one lunar month.   Marked on the glass of the door is an ellipse which represents the horizon relative to the latitude of the observer. All the stars that are visible within this ellipse are visible in the heavens at that time. As a star appears within the area from the left it is rising, and as it moves out of this area to the right it is setting. Likewise the sun and the moon rise and set accordingly as their relative wires cut the horizon.

Many of the star names on the dial have changed their spelling over 250 years as below:

For Speia read Spica

Syrius Sirius

Mahcar Mankar or Mankab

Fomalhant Fomalhaut

Lyrais Vega in the constellation of Lyra Markal Markab

Alcair Altair

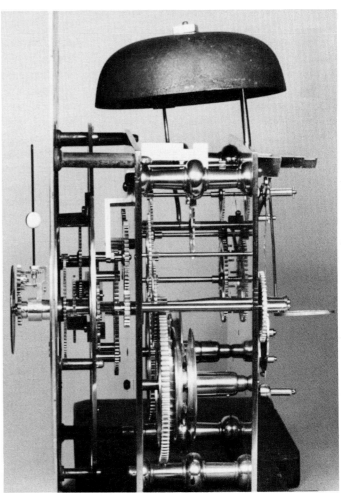

## 96 Astronomical and Equation

Figure 120 A and B. *Astronomical Longcase Clock by Thomas Lister (Junior), Halifax, circa 1790.* This clock has an inlaid mahogany case of individual style, an eight-day movement with an unusual pin wheel escapement and outside pinned countwheel. The extended motion work is connected by a series of massive high count wheels to the astronomical dial below. This is colorfully painted in the center with the signs of the Zodiac which are bordered by an annular calendar ring which revolves against another ring giving the lunar date. All this moves against a 24-hour (I XII twice) dial.

The upper 12-inch engraved and silvered brass dial bears Lister's signature in an attractive cartouche set against a classical scene in the upper half of the arch. 6' 6" (198 cm) high.

Photo courtesy Sotheby's, London.

## Astronomical and Equation

Figure 121 A, B, C, and D. *Astronomical Clock attributed to Finney, Liverpool.* This clock, attributed to Finney of Liverpool, has one of the best astronomical dials we have seen. It appears uncluttered and yet all the information is clearly displayed and the engraving, while fine, is easily read. Particularly attractive features are the two discs in the arch, one for the moon rotating clockwise half a turn every 29½ days, and one for the heavens, as we look south, rotating anti-clockwise half a turn every 18½ years.

# 98 Astronomical and Equation

## THE CASE

This has survived remarkably well over the last 200 years and can only be described as of superb quality, the beautifully chosen "book matched" veneers being laid on oak. It would seem likely that it was made by Gillows, although this cannot be substantiated.

The swan-neck pediment with wood paterae has black glass decorated with gilt birds and foliage and three carved wooden finials. There are two fluted columns with gilt wood capitals to each side of the hood and some moulds beneath it are lightly decorated with carving.

The fluted columns on either side of the trunk are beautifully executed and there is applied fretwork below these and to the top of the base which is panelled and has canted corners. There is delicate carving on the case, tiny bunches of grapes for instance, being found on the top of the trunk.

## THE MOVEMENT

As originally conceived this was a two train clock with the chiming train added shortly afterwards. This was obviously specially made for the clock and has been executed to the same standards. Note for instance the way in which its winding square is on exactly the same level as the other two and lies dead center of the minute ring.

Very little is known of the history of this clock. It sustained minor damage from bomb blasts during World War II when the hood glasses, including the black glass, were broken. (A sign of the austerity of immediate postwar years, when glass was at a premium, is that this black glass was replaced, quite skillfully, with pieces cut from old gramophone records.)

## THE DIAL

The principal dial displays seconds, hours and minutes and also has a long gilt calendar hand. This indicates the day of the year on a calendar ring engraved outside the minute circle. On this calendar are engraved, and filled with red wax, the principal fixed festivals of the church calendar. Set outside this fixed calendar ring is a further adjustable ring engraved with the movable festivals. Within the inner edge of the chapter ring is a further adjustable scale engraved with the zodiacal calendar.

Engraved around the inner edge of the dial in the breakarch is a half azimuth circle showing South in the vertical position. Within this is a continuously rotating lunar dial, the indication repeated on each 180° and engraved 1-29½ days on its inner edge, and with the 12 hours repeated every 90° on its outer edge. This latter outer dial, when read against the azimuth indication, gives the time of the moon's southing (curiously, it is worth noting that the engraver has made a mistake in engraving one of the four "8" as a "7"!). In addition, the use of a blued-steel setting hand pivoted and locked at the arch center enables the observer to set the tidal indication for any particular location. If, for instance, at full moon, high tide occurs when the moon is in the SW, then the hand is set to SW on the azimuth dial and from then onwards, the time of high tide is read from the outer dial of the continually rotating lunar ring.

In the center of the arch a complete disc is engraved as a double celestial sphere. Its outer edge represents the ecliptic and is divided twice over into the twelve 30° divisions of the Zodiac. The disc is separately driven in the reverse direction to the lunar ring in order that the moon's position in the Zodiac remains correct and the moon's nodes will have the required retrograde motion along the ecliptic. This disc will complete one half rotation (representing one complete rotation of the celestial sphere) in 229 mean synodic months or approximately 18½ years. This seems to be a compromise between the ancient Greek concept of the Metonic cycle of 235 mean Synodic months and the later proposition of Saros giving the Synodic revolution of the moon's nodes as 223 mean Synodic months.

The Golden Number, which will automatically be forecast by this clock, is the number of the year in the lunar or Metonic Cycle. It was used in particular by the Ancient Greeks to determine the days for religious festivals. The name is derived from the fact that the number is believed to have been displayed in gold in public places.

In any event, at the end of this period, the sun and the moon will again occupy similar positions relative to the moon's nodes as they had at the beginning and the pattern of eclipses will repeat the same order.

Knowing the length of a Draconic month (time required for the moon to travel from one node to the next) to be 27.212 mean Solar days, and by setting the sun hand from the Zodiac calendar within the main chapter ring, it should be possible to determine the next eclipse, although this is made difficult because no provision appears to have been made for a line-of-nodes hand. However, this facility is presumably intended as the eclipse limits are stated above the arch.

In the Gregorian Calendar, Easter Day is the first Sunday after the 14th day of the calendar moon, immediately following the Vernal Equinox about 21st March; if the 14th day is a Sunday, the Easter Day is the following Sunday. This rule ensures that Easter falls soon after the Passover, a fixed date in the Jewish lunar year. In this clock, the Vernal Equinox is clearly marked on the celestial dial in the arch by a black dot at the First Point of Aries and this can be read in conjunction with the lunar calendar. In addition to the indications already described, there are engraved on the discs various other lines, the significance of which is not so far fully understood.

Figure 122 A, B, C, D, E. *Astronomical clock by William Robb, Montrose.* We spent over a year trying to piece together the fascinating history of this fine longcase clock by William Robb of Montrose (in Scotland) which has such a superbly designed and possibly unique Astronomical dial, and even at the end of that time its' history was by no means complete. It bears a plaque inscribed "Formerly the Property of the Nice Family" and was presented to the Historical Society of Pennsylvania, U.S.A., by Mrs. O.H.P. Conver and Mrs. Anne N. James sometime between November 9 and December 14, 1868.

No such family as Nice are recorded as having lived around Montrose and there would appear to be two explanations for this; either that the Nice family acquired it from someone else who brought it over from Scotland or that their original name was Neish and this became corrupted to Nice, a common enough occurrence at that time.

The Neish family lived in and around Montrose in Georgian times and are believed to have emigrated to Pennsylvania in the 1770s.

William Robb came from a wealthy family owning much property. He appears to have emigrated to America in 1777 but is recorded as working in the Burgh of Montrose as a watchmaker from 1756-1808 and this is backed up by the fact that he is listed in the Burgh Court claims in 1772, 1774, 1785, 1792, 1799 and 1804, either claiming for unpaid work or rent from tenants. At first it was thought that there might be some confusion between father and son but this proved to be false and it is only an article titled "Business in Philadelphia during the British Occupation" which gave rise to the probable explanation.

On September 26th 1777 Lord Cornwallis approached the Quaker City (of Philadelphia) and took command but prior to this the leading Whig Merchants vacated their places of business and withdrew to Lancaster or York. Within a month British transports brought to the city a host of merchants and artisans who pre-empted the deserted premises and a William Robb, presumably the man we are interested in, moved into those of William Redwood. However the following spring British troops withdrew from the city and it is likely that after this things would have become uncomfortable for William Robb and he subsequently returned to Scotland, although he is still recorded as working in Philadelphia in 1779.

It seems probable that William Robb made the clock in Montrose and either delivered it there or, because it was such a complex piece, took it out to America personally and set it up himself. It is also possible that he made it in Philadelphia, although the fact that it has Montrose on the dial probably indicates otherwise. The clock is illustrated with its' presentation plaque in *The Old Clock Book* by N. Hudson Moore which was published in 1911.

By tradition the clock passed down through the Nices who were a prominent and well-documented Pennsylvania family until 1868 and was then given to the Society where it remained until it was disposed of in 1966. It was referred to in a NAWCC bulletin in 1956. The clock remained in the States until it was repatriated in 1987.

## 100  Astronomical and Equation

Figure 122 A.

Figure 122 B.

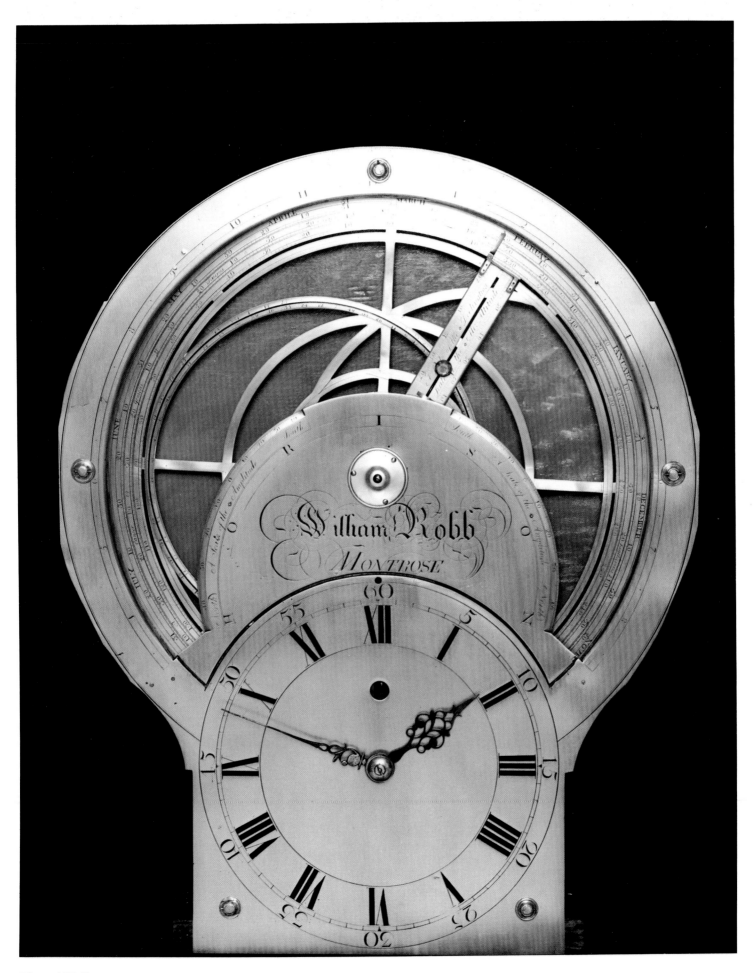

Figure 122 C.

## 102 Astronomical and Equation

### THE CASE

This is typically Scottish in many ways being made of solid mahogany and having a tapering trunk. However, it is unique in concept with a very large hood basically circular in form, to accommodate the large astronomical dial. It is surmounted by giltwood finials on either side and a celestial globe in the center. In the illustration in *The Old Clock Book* it had a central finial in the form of an eagle but this had been removed prior to our receiving the clock because it was not original. On the top of the trunk door are two engraved linked plaques representing the earth's hemispheres and it seemed likely to us that the original finial on top would have been a celestial globe to compliment these and so this is what we provided. Certainly something relatively substantial had been fitted there in the first instance.

The clock has a panelled base and this rests on unusually tall, outwardly splaying bracket feet. 7′ 4″ (224 cm) high including globe.

### THE MOVEMENT AND DIAL

The movement has an inverted train with the 'scape wheel at the bottom of the plates and the pallets below this. A train of wheels runs up from the movement to the center of the dial terminating in a very large wheel which drives the astronomical work.

The flat engraved and silvered brass dial consists of two parts: (a) the lower portion with arched top and Roman numerals, on which hours and minutes are indicated and above which is William Robb's signature, beautifully executed; and (b) the astronomical dial some 17 inches in diameter.

A large arched ring at the top of the dial is engraved with the hours of daylight, hours from 3:15 A.M. to 8:45 P.M. Within this is a rotating calendar ring indicating, from the outside: months of the year, days of the month, zodiacal day of the year, and the 360° of the Zodiac.

A central solid arch represents the earth and its perimeter, the perceived horizon, and between and behind this and the outer ring is depicted the sky by day, painted on a wooden board.

The movable calendar ring carries an applied ring to represent the ecliptic. The whole ring rotates once in a calendar year, carrying this ecliptic ring with it.

In front of the ecliptic ring is a cursor with a pointer at its outer end. The cursor carries a representation of the sun and makes one full rotation in 24 hours. The sun has a pin on its reverse side which engages a slot in the ecliptic ring thus carrying it across the painted heavens during daylight hours at the correct altitude throughout the year.

The cursor also carries two scales, one showing the sun's declination north and south of the Celestial equator (marked with an E) and the other showing its altitude in degrees above the horizon and these are read from the small pointers on each side of the sun.

Two further scales are engraved to the left and right of the earth's horizon. On the left is seen the variation to north or south of due east for the rising sun and on the right the same variation with respect of due west for the sun's setting. These are read from the slot in the ecliptic ring.

Figure 122 D.

## Astronomical and Equation

All these variations can be checked by setting the calendar ring at the four cardinal points bearing in mind that on this clock, ALL READINGS MUST BE TAKEN AT MID-DAY, this being the point at which the date is read.

First, the calendar ring is set at the Summer Solstice (i.e. 90° on the zodiacal calendar and about 21st June). The cursor will then indicate that declination is zero and the sun is rising at its maximum amplitude north of east and setting similarly south of west.

Next, the calendar is rotated through 180° to the Winter Solstice when once again the declination is zero and the amplitude readings reversed to south of east and north of west respectively.

The calendar is now rotated 90° to the Vernal Equinox (i.e. the 1st point of Aries or 21st March) when it will be seen that the sun is at its maximum declination North of 23½° and also its maximum altitude of 57°. Also from the scales on the horizon it is rising due east and setting due west.

Finally, the calendar is taken round 180° to the Autumnal Equinox when we have the maximum southern declination and the minimum altitude of approximately 10°, and again the sun rises due east and sets due west.

Engraved around the ecliptic are two scales. The outer one indicates the time of sunrise and is read against the left bevel of the cursor, the inner one gives the time of sunset and is read presumably against the right bevel of the cursor. Once again, it is necessary to read it at 12 noon exactly.

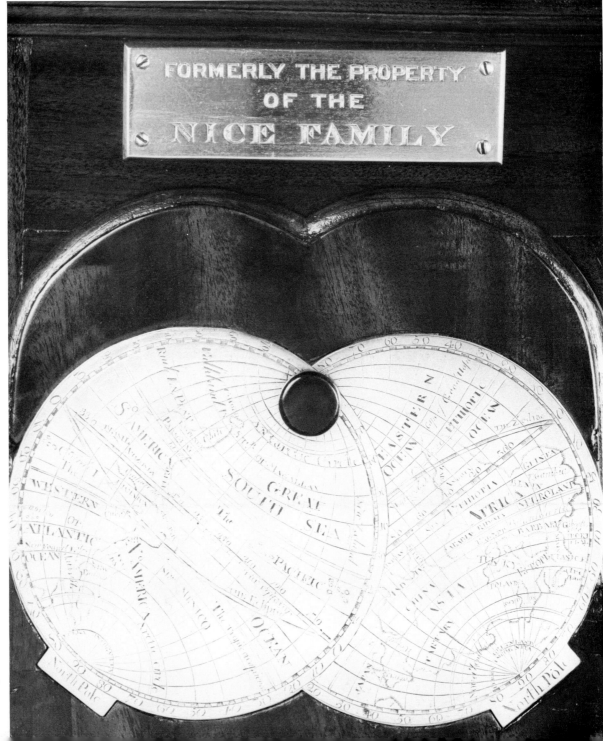

Figure 122 E.

# EQUATION CLOCKS

Figure 123 A and B. *Equation Clock by Daniel Delander.* This fine longcase clock by Delander was restored by us several years ago and this process is described in some detail here as it gives an insight into the care which must be taken when restoring fine antique clocks.

Beside posing the usual routine problems with the casework and movement which were relatively straight forward to solve, it also presented us with two further queries. The first of these concerned the restoration of the equation work, much of which was missing. Here we were, as is often the case, misled by various components which had in the past been replaced. Once these were disregarded, we could recognize Delander's original concept and the solution became clear.

The second query concerned the two pulleys at the top of the case which were obviously original but were now fulfilling no useful purpose. At first it was thought they may have originally been used in a barometer, but subsequent investigations revealed that they were almost certainly part of a hygrometer. The pulleys were required so that a long line could be used which would expand or contract appreciably with changes in humidity and thus move the hygrometer hand which would almost certainly have been incorporated in a barometer.

At this stage, this problem was put on one side and the rest of the work on the clock completed, when a further unusual fact appeared. The clock required a pendulum 7- to 8-inches shorter than normal if it was to keep time. Our first reaction was that the clock had been altered and that possibly it had the wrong number of teeth in the 'scape wheel or that one of the wheels in the top end of the train had been changed. However, a detailed examination revealed that this was not the case and that all the components were original.

It now became apparent that Mr. Delander had worked out the whole design most carefully and had used a short pendulum so as to leave adequate space below it for the hygrometer/barometer and thus we had confirmation of our original diagnosis.

## THE CASE

The case of this clock is of the finest quality, carefully chosen and beautifully figured burr walnut veneers being laid on an oak carcase. The proportions too are excellent. All the moulds are cross grain. There is a caddy top surmounted by three finials and with a fret below. The hood door has applied brass capped pillars and its original glazing. There is glass to either side of the hood and frets immediately above.

The sides of the trunk and base are panelled, there is a double plinth and a rectangular trunk door. This and the rudimentary breakarch would suggest a date of manufacture of circa 1715. 9′ 4″ (284 cm) high to top of finial. 8′ 8″ (264 cm) high excluding finial.

## THE MOVEMENT AND DIAL

The month-duration movement of this clock is of excellent quality having five latched pillars and bolt and shutter maintaining power. The 13-inch square brass dial has a very shallow arch on it in which are displayed three rings which are, from without in: (a) year calendar, (b) equation of time, and (c) sun slow/sun fast. Beneath this is the inscription *The Equation of Natural Days*. To either side of this are most unusual spandrels depicting heraldic birds. The four urn, corner spandrels are also of fine quality and individual design. The chapter ring has quartering on its inner aspect, half-hour diamonds, and half-quarter marks on the outer edge.

The clock is signed by Daniel Delander and is indeed typical of his work, a similar but possibly slightly later clock being illustrated on page 122 of Tom Robinson's book on the longcase clock.

Delander was an exceptionally fine and ingenious maker being famed for his equation work and also his duplex escapement. Examples of his work are in numerous collections and museums, a year-duration equation clock, for instance, being in the Wetherfield Collection. He started work in 1692, gained his Freedom of the Clockmakers Company in 1699, and died in 1733.

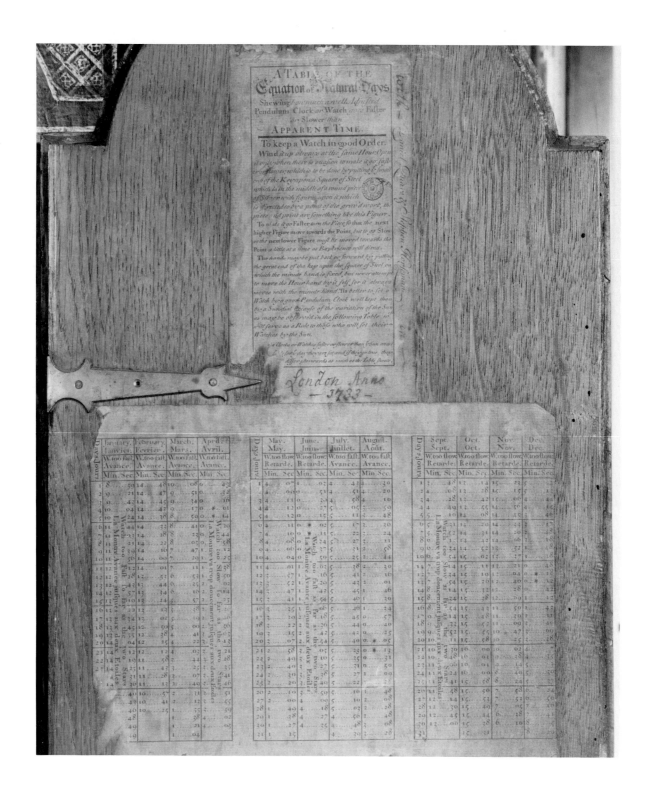

Figure 124. A table of the equation of natural days pasted inside the door of a lacquer longcase clock, (Ref. Figure 71) signed by Quare and Horseman and dated 1733.

Figure 125. *Tompion's Equation clock* with a year calendar aperture set within the equation dial. It bears his signature both on the main dial and immediately above the year aperture. He presented it to the Pump Room, Bath in 1709.

Figure 126 A, B, C, and D. *Solar Time Clock by Joseph Williamson, Circa 1725.* One of only two solar time clocks by Williamson which is thought to still have its original cam for raising and lowering the pendulum.

In the arch of the 12-inch dial is the year calendar marked on its inner aspect with the months and outside this, the days. It has a center-sweep hand carrying around an effigy of the sun. Immediately below 12 o'clock is Williamson's standard inscription *Horae Indicantur Apparentes Involutis Aequationibus, Joseph Williamson Londini.* There has been much debate about an exact translation of this but in essence it probably means "The apparent (solar) hours are shown by means of complex equation work."

In the lower half of the dial is a "Penny Moon" with lunar calendar above. If this clock is compared with that illustrated in *The Longcase Clock* by Tom Robinson pages 180-184 it would seem that it was probably made a little later and is an advance on this in that by using a center-sweep seconds hand he has been able to eliminate the seconds ring and move his inscription up to this position, thus leaving space for the moon at the bottom of the dial.

In the rear view of the movement may be seen the spiral drive to the wheel which is connected to the solar time disc. A stud resting on the edge of this moves the pivoted L shaped lever on which it is mounted in or out and in this way raises or lowers the pendulum which is attached to the other end.

The movement is contained in a beautifully figured beveled mulberry wood case with applied pillars to the hood door.

110 *Astronomical and Equation*

Figure 127 A, B, C, and D. *Thomas Tompion Siderial and Mean Time Clock, No 483.* A rare and important siderial and mean time month-duration regulator signed *Thomas. Tompion London* on an oval silvered plaque beneath which are the joint signatures *Thomas Tompion* and *Edward Banger, London.* The concealing of Banger's name would seem to indicate a date of manufacture for the clock of 1708-1709 as it was in 1708 that Banger left Tompion's employ.

## THE DIAL

The 12-inch square dial has a finely matted center with a seconds ring below 12 o'clock and a winding hole above 6 o'clock with a shutter connected to the maintaining power which is actuated by a lever situated just below 2 o'clock.

There are two chapter rings, a conventional inner one with half hour markings and quartering on the inside and an outer one which has minutes shown on its' inner aspect and rotates twice a year.

As presently set up, the pendulum beats mean solar seconds and thus shows mean time on the inner ring and siderial time on the outer rotating ring. To read siderial time is, as described by Neilson[1] a relatively complex affair involving the subtraction from the meantime shown of the difference, counted anticlockwise, between XII on the rotating and fixed rings. Had the ring rotated anticlockwise, siderial time could have been read off directly.

In view of the foregoing it would seem more likely, as Todd suggests[2] that the clock originally beat siderial seconds, particularly as the outer ring rotates twice in 366¼ not 365¼ days. If this supposition is correct then the fixed chapter and second rings will show siderial time and the hour and minute hands, if extended, would show mean time on the outer rotating ring.

The function of the gilt minute hand is also a matter for some debate. Neilson suggests that it could have been used as a manually set equation hand. However another suggestion[3] is that it may have been intended to indicate siderial time for a given longitude East or West of the clock's location. The adjustment allowed via the curved slot in the gilt hands boss would encompass any two longitudes as far apart as Greenwich and Plymouth, and if the hand was extended it would register local meantime for the given comparative longitude.

## THE MOVEMENT

The single-train movement has six latched-plate pillars, four feet for the separate dial center, and a further four for the outer dial plate which is supported by extensions from the front plate of the movement Figure 127 D. In this picture may also be seen the drive to the outer chapter ring terminating in a spiral at its periphery. Figure 127 C shows the four rollers on which the chapter ring runs, the shutter work and the 288-tooth hour wheel which meshes with the 24-leaf pinion of the 144-tooth minute wheel. This runs on an arbor which extends through the plates and then carries a 288-tooth first wheel of the differential gearing which in turn engages a second wheel of 144 teeth.

At the end of this wheel arbor a worm engages a 15-leaf pinion at the bottom of the vertical arbor which runs up to drive the outer chapter ring.

The brass rod pendulum which is suspended from the backplate has a calibrated rating nut and is provided with a beat scale. A securing bracket is used to fasten the movement to the backboard.

## THE ESCAPEMENT

This is, in many ways, the most interesting feature of the clock in that it has been fitted with what has always been known as a Graham dead beat escapement. However, there can be no doubt that the clock was made several years before Graham is attributed with having invented the dead beat. Various explanations have been put forward for this, the principal one being that Graham changed the escapement, probably after Tompion's death. However the report of the Research Laboratory for Archaeology and the History of Art of Oxford University which is quoted verbatim below seems to contradict this and makes it extremely likely that the escapement fitted is original to the movement. If this is so, then either Tompion invented this form of dead beat or Graham did so whilst still in Tompion's employ and before he became a partner. Of the two the former possibly seems the more likely as Tompion appears to have been well aware of the deficiencies of the recoil escapement as early as 1675 when, working in conjunction with Townley, he fitted a form of dead beat to the year clocks at Greenwich[4].

### Report of the Research Laboratory for Archaeology and the History of art of Oxford University

*Tompion Longcase Clock (?1709).* The composition of thirteen brass parts of the gear train of the above clock were analyzed by x-ray fluorescence analysis at Oxford; the following elements were estimated: iron, nickel, zinc, arsenic, lead, silver, tin and antimony, as well as copper. The object of this analysis was to determine whether the 'scape wheel had a similar composition to the other parts.

Further these results were compared to another series of 15 analyses of a gear train from a Graham longcase clock dated to 1722.

*Results.* The impurity composition of the 'scape wheel was very similar to all the other parts of the Tompion examined; moreover the general deviation of all the parts from the means for the Tompion was also very small. When this result is taken into consideration with the fact that no tampering or redrilling of the ' wheel arbor bearing holes is evident, there seems every likelihood that the 'scape wheel is contemporary with the rest of the mechanism.

It should also be pointed out that with the analysis of the Graham, although the spread of composition was greater, the actual values were in three instances very similar to the Tompion average composition. This would show that some of the materials were used in Tompion's time.

*Conclusion.* Although no positive decision can be given as to the date of the Tompion clock 'scape wheel and hence whether Tompion did construct the first dead beat escapements, the evidence does point in that direction. If the opposite was true and a new dead beat 'scape wheel had been fitted by Graham, it is a considerable coincidence that the latter used brass of almost identical composition and managed to use the same pivot holes for the 'scape wheel when the change from anchor escapement was made.

### Footnotes

[1] Neilson M., "Important Siderial Regulator by Thomas Tompion and Edward Banger No. 483 c1709." *Antiquarian Horology,* Vol. X, No. 2, Spring, 1977, pp. 214-216.

[2] Todd W. Communication in *Antiquarian Horology,* Vol. X, No. 3, Summer, 1977, p. 366.

[3] Christie's *Catalogue of Important Clocks and Fine Watches,* 16th Dec., 1982, pp. 62-65.

[4] Howse D., *The Tompion Clock at Greenwich and the Dead Beat Escapement," Antiquarian Horology,* Dec., 1970, & March, 1971, pp. 164-172.

# Chapter 5.
# Walnut Breakarch Longcase Clocks
### (including various other woods)

## THE MOVEMENT AND DIAL

The use of a simple square dial had from its inception imposed limits on the amount of information which could be clearly displayed; although this was eased a little by the increase in its size, first from ten to 11 inches and by 1700 to 12 inches. A seconds hand was usually provided below 12 o'clock and a date aperture above 6 o'clock. When subsidiary dials were fitted for instance for pendulum rise and fall (fast/slow), strike/silent or calendar work (days of week, days of month, months of year) the commonest place to fit them was in the two top corners (Figure 45 B) but on occasions all four corners were used.

Thomas Tompion was one of the first to realize the restrictions of the square dial and as early as 1695 had added a semicircle carrying a year calendar onto the top of the dial of the year-equation clock he made for William III (Figure 473).

By around 1715-1720 the breakarch dial, usually 12-inch square with an arch on top, giving an overall height of around 16 ½ inches, had come into general use. Some of the early breakarch dials were made in two parts which were held together by straps at the back and it is also likely that some square dial clocks were upgraded and modified at the same time. However, by 1720-1725 nearly all the breakarch dials were being cast in one piece. The main exceptions to this are clocks where there is only a circular strip of brass in the arch, as for instance with clocks displaying the months, moon phases, or having automata, which is fabricated separately and added to the square dial.

In the early stages, all that usually was displayed in the arch was the maker's name and this gradually gave way on most London clocks to the provision of strike/silent regulation. The name was then displaced either to the bottom of the chapter ring, a plaque, or an arched strip applied to the center of the dial or sometimes carried across the top of the arch.

Increasingly from 1720-1725 onwards a rotating lunar disc, sometimes called a rolling moon, was provided in the arch; although this is far more commonly seen on Country than London-made clocks. Other features displayed in the arch were automata such as a ship or "old father time," which rocked back and forth in time with the pendulum. With musical clocks a center-sweep, tune selection indicator was often fitted in the arch (Figure 226) and on some of the finer examples a colorful scene was depicted with, for instance, musicians apparently playing their instruments, their arms, heads, etc. being linked to the motion of the pendulum, or occasionally being contrived to coincide with the striking, chiming or music being played by the clock.

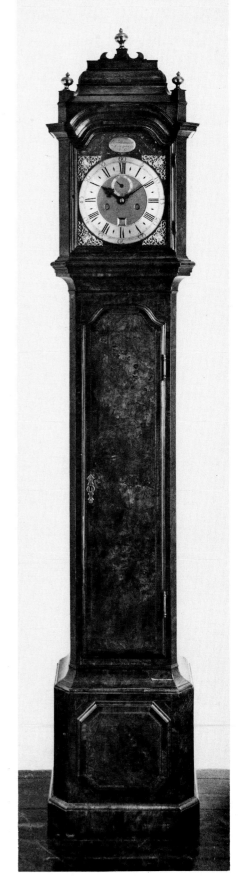

Figure 128. *Daniel Delander, London No. 17.* Delander was a fine and highly individual maker, having made year and equation clocks, but he is probably best known for the use of the duplex escapement on longcase clocks, the shape of the dials he employed on many of his clocks, and the style of case he favored. This clock is a typical example of his work.

The dial has what might be termed a flat-topped breakarch, being almost transitional between the square and the standard breakarch dials. It will be seen that this shape is not just repeated on the mould above the dial but on the trunk door also. Pillars have been omitted from the hood and the front corners of the case are canted from top to bottom, including all the moulds and plinth. The base panel has also been shaped to match in with these.
Photo courtesy R.A. Lee Fine Arts.

## THE EQUATION OF TIME

By the end of the 17th century, in part at least because of the greatly improved accuracy of clocks, increasing attention was being paid to the difference between *solar* or *sun's* time and *mean* time, in which the variations in time due to the earths uneven progress around the sun are averaged out. The difference between solar and mean solar time is known as "The Equation of Time" and this was yet another feature which it was possible to display on a clock dial. It is discussed in considerable detail in the chapter on astronomical and equation clocks.

## THE CASES

The walnut case with a breakarch dial gradually evolved from those containing a square dial. One of the earliest the author has records of was ordered in 1709 and delivered in 1714.

The cases evolved from a flat-top style to one with the caddy placed on top of a breakarch. The trunk door at first remained rectangular but gradually assumed the breakarch form used on the hood, and occasionally more decorative shapes were used. In time the caddy top was changed to a shallow pagoda which was to evolve into the full blown pagoda seen on London mahogany longcase clocks circa 1760.

In the majority of cases walnut longcase clocks were originally surmounted by caddy tops. Examples are also found with breakarch tops. In those where the top mouldes are thin, it is likely that a caddy top originally rose above them; but where the top mouldes are substantial, it is probable that the clock was originally as seen.

Although by 1750 mahogany had largely supplanted walnut as the wood of choice, walnut clocks continued to be made for many more years, usually in styles similar to those of the mahogany cases currently in fashion.

Whereas the pillars on either side of the hood were almost always applied to the hood door until circa 1725-1730, after this period they became freestanding, and few London clocks made after 1735 are seen with applied pillars.

## CADDY TOP

The longcase clocks of the Queen Anne/Early George I period which still retain their original caddy tops are among the most perfectly proportioned ever produced, and have a dignity seldom achieved by other designs. Unfortunately the number which still exist is relatively small, and moreover they are almost always tall clocks, often exceeding eight feet six inches or even nine feet in height to suit the beautiful houses being built in London at that time, and thus they cannot be accommodated in most modern homes.

Figure 129. *Eight-Day Walnut Longcase Clock by Windmills.* A walnut, caddy-topped, longcase clock by Windmills which still features a lenticle in the trunk door, something which was going out of fashion at this period.

Figure 130 A, B, C, D, and E. *Month-Duration, Quarter-Chiming Longcase clock by Windmills.* The clock seen here, just over nine feet in height, has a month-duration movement which chimes the quarters on six bells. Figure 130 D shows the drive to the month calendar in the arch which has a 12-sided star chased out in the center, a decorative feature favored by Windmills and a few other fine makers at this period. Figure 130 E shows the movement with the calendar wheel removed, thus exposing much of the strikework including the hour and quarter racks. Points to note on the case are the beautifully chosen veneers, the deep plinth quite often seen around 1705-1720, the square top to the door which was to gradually give way to the breakarch, and the excellent proportions of the caddy, right up to the top finial.

A particularly interesting feature is the original setting up instructions of Mr. Windmills pasted inside the trunk door and now, sadly, somewhat damaged.

The history of the clock is known from its beginning, a very rare occurrence. It was ordered in London by Robert Wrightson of Cusworth in 1709 and delivered in 1714. Robert Wrightson was legal adviser to the Court of the King's Bench. The clock was originally housed in the Elizabethan mansion at Cusworth which Robert Wrightson had purchased from Sir Christopher Wray in 1669. In 1740 Robert Wrightson's son, William, decided to build a new mansion which is the present Cusworth Hall. The Windmills clock was moved from the old house in 1745 and stood in the Square Hall Lounge at Cusworth Hall from that time until the death of Robert Cecil Battie-Wrightson's sister, Mrs. M. L. Pearse, under the will trust of their mother, The Lady Isabella Georgiana Katherine Battie-Wrightson. Her ladyship came originally from Burghley house, Stamford and was the eldest daughter of William Alleyne Cecil, the third Marquis of Exeter.

After the death of her brother, Mrs. Pearce removed the clock to her house at St. Germain Lodge, Hastings and later returned it to Cusworth. Mrs. Pearse was the last surviving member of the Battie-Wrightson family. This clock has thus remained in the same family since its date of making and purchase in 1709. There was a sale of much of the contents of Cusworth Hall in 1952.

Seen alongside the quarter chiming clock is a small oak longcase with walnut crossbanding and caddy top. It is signed J. Windmills, London and is of month duration.

116  *Walnut Breakarch*

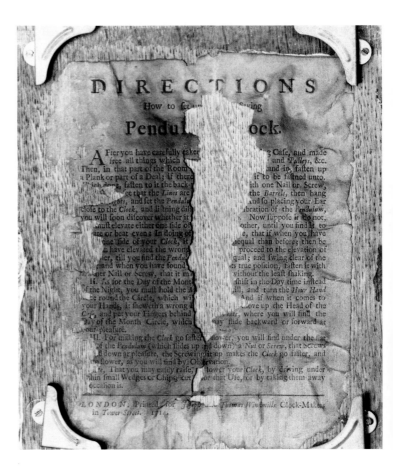
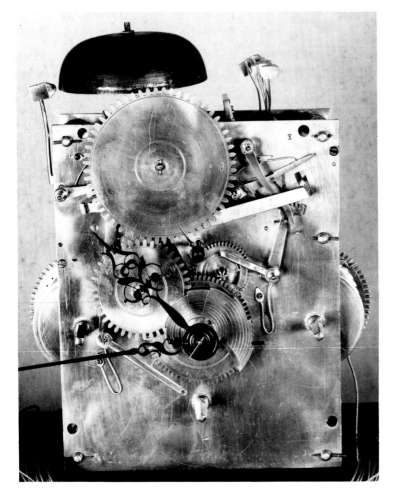
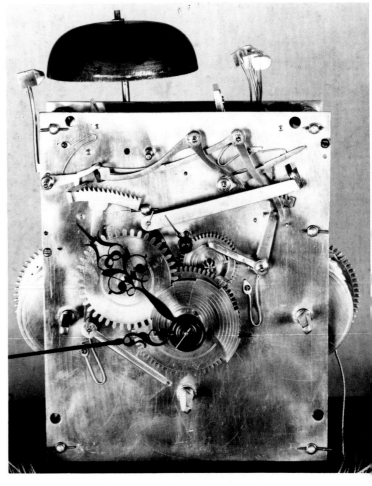

## Walnut Breakarch 117

Figure 131 A and B. *John Ellicott FRS. Circa 1725.* A flat-topped, walnut longcase clock, probably originally with a caddy, by the eminent maker John Ellicott, London. Not the relatively early use of strike/silent regulation in the arch, still with the square-topped door.

The photograph of the movement gives some idea of its quality and one of the large rollers for the date ring is visible on the left. It chimes and repeats the quarters on eight bells and the return spring for the strike/silent regulation may be seen on the backplate.

*118 Walnut Breakarch*

Figure 132. *Charles Tucker London circa 1725-1730.* A walnut longcase clock by Charles Tucker without the upper part of the caddy top. It has a well-figured case with crossbanding with stringing to the base and the border of the trunk door which has an arched top. There is fretwork to the front of the top of the hood and the pillars are applied to the door.

An attractive feature is the small calendar dial in the arch which mimics the main one even down to the hands. 7′ 11″ (241 cm) high.

Figure 133. *Thomas Warner, London, circa 1725-1730.* A beautifully figured walnut longcase clock with double plinth and a crossbanded door and base. All the smaller moulds are taken out of the solid in cross grain, a lengthy procedure, and the larger moulds are cross veneered. There is wheatear engraving to the edge of the dial and around the name plaque. 7′ 3″ (221 cm) high.

Figure 134. *Walnut Longcase Clock by John Ellicott 1730.* A walnut longcase clock very similar to that seen in Figure 129 but with an arched top to the trunk door and signed on a plaque above the two date apertures. An interesting feature of these is that the one for the month turns to a blank on the 29th February and on advancing the date ring to the 1st March, it turns to show MAR in the aperture.

Figure 135. The relatively early appearance of the moon disc in the arch of a well-proportioned walnut longcase clock with an attractive cartouche around the date aperture.

Figure 136. *John Sapley, Edinburgh 1775.* A superbly figured, flat-topped, walnut longcase clock produced in Edinburgh long after they had gone out of fashion in London. Note that it still has a square-topped door, somewhat shorter than on a London clock and with no cross grain moulds around it, although it is crossbanded as also is the base. The hood pillars are light with no brass capitals and the shell inlay has probably been added at a later date.

The dial has an engraved center and seconds ring with the name plaque in the arch.

## 120 Walnut Breakarch

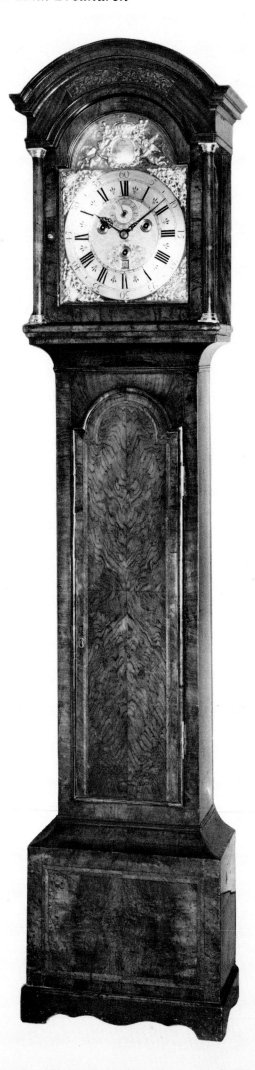

### THE BREAKARCH HOOD

Figure 137 A and B. *Quarter Chiming Clock by William Sellers, London, circa 1725.* A fairly small walnut longcase clock with a fret to the top of the arched hood, quartered veneers to the trunk door which are surrounded by herringbone inlay and crossbanding and a panelled base. The plinth would have originally been straight along the bottom edge and may have had a second tier.

The quarter-chiming movement has a very fine 12-inch breakarch dial with cherubs in the arch on either side, two upholding the name plaque which has a crown on top and two bearing scepters. Interwoven around these is floral engraving and there is wheatear engraving to the border of the dial. Matching the decoration in the arch are crown and twin cherub spandrels in the four corners. The winding holes are ringed and there are birds and foliage engraved on either side of the date aperture.

Figure 138. *David Hubert, London, 1725.* A substantial walnut longcase clock with freestanding front hood pillars, a feature that started to come into use in London around this time. It originally probably had rear quarter columns also. The door is decorated in beautifully figured veneers and it is crossbanded. The base has at some time been reduced, evidence of this being the lack of crossbanding along its lower border. There is strike/silent regulation in the arch.

David Hubert, an eminent maker, became Free of the Clockmakers' Company in 1714 and rose to become Master in 1743. 7′ 9″ (236 cm) high.

Figure 139. *Richard Colley, London, Circa 1725-1730.* A walnut longcase clock in very-much-untouched and original condition with freestanding pillars, raised panel to the base, a double plinth and cross grain moulds. The finials on top are replacements. 7′ 10″ (239 cm) high.

Figure 140. *Richard Burputt London circa 1730.* A walnut longcase clock with relatively light moulds to the breakarch trunk door and a flat crossbanded base. It has the name plaque in the arch, half hour divisions and quartering on the inside of the chapter ring. 7′ 7″ (231 cm) high.

## 122 Walnut Breakarch

Figure 141. *Musical Clock by Joseph Barber, London, circa 1745.* A superb walnut musical longcase clock by Joseph Barber, London. The eight-day three-train movement plays four tunes on ten bells with 15 hammers, the tunes being "March," "Minuet," "Ally Croaker" and "Miller of Mansfield." The 13-inch gilt brass dial has a raised chapter ring with Roman numerals and large second ring at XII and date calendar at VI, with Indian head spandrels. There are subsidiary dials for strike/silent and tune selection and a lever at III for chime/silent.

In the arch is a rolling moon with lunar calendar and a fine signature running in a sector across the entire arch. The case, of exceptionally well-figured walnut, has crossbanding to the door, base and hood with fine ebony and boxwood stringing. The breakarch hood has freestanding fluted and brass strung columns, something which had just started to come into fashion and was to become standard on the best London mahogany longcase clocks. Joseph Barber was apprenticed to William Webster in 1721.

Figure 142. *Robert Le Bond, London, Circa 1755.* A mid-18th century walnut longcase clock with the name plaque in the arch, a feature which was disappearing in London at that time. It has brass-strung front and rear columns to the hood and crossbanding to the door and base. 7' 8" (234 cm) high.

Figure 143. *William Lovelace, London, circa 1755.* A beautifully figured walnut clock of excellent color and patination, but a little unusual with no crossbanding to the trunk door. There is strike/silent regulation in the arch. 7′ 6″ (229 cm) high.

Figure 144. *John Morgan, Mommouth, circa 1750-1760.* A finely figured mid-18th century walnut eight-day longcase clock of very good color and full of character by John Morgan of Monmouth. The base, which stands on a double plinth, is panelled and in the center of this is a well-executed painting of a spray of flowers surrounded by tulipwood crossbanding. The trunk door has a conch shell set into its center and is attractively shaped at the top.

To either side of the trunk door are relatively delicate rectangular pillars which are fluted and have carved wooden capitals. The hood has brass capped columns to either side and a breakarch top beneath which is fretwork.

The rack striking movement has an attractive 12-inch brass dial showing moon phases and high water at Bristol Key in the arch. There is a raised chapter ring, seashell spandrels, second hand and date aperture with engraving around it. 7′ 10″ (239 cm) high.

*124 Walnut Breakarch*

Figure 145. *John Behoe, London, circa 1745.* This clock is typical of the beautiful quality walnut cases being produced at this period, still with cross grain moulds, crossbanding, panelled base with double plinth and freestanding brass capped and strung hood pillars. Moon phases, a little more popular at this period, are shown in the arch with fine floral engraving above, and the sun and earth depicted below. The finials are replacements. 7′ 9″ (236 cm) high.

Figure 146 A and B. John Crolee, London, circa 1750. A fine, mid-18th century walnut longcase clock having a very decorative dial with some Dutch influence, for instance in the provision of two triangular cutouts in which are shown the days of the week with their ruling deity and the months of the year and zodiacal signs. The fitting of an alarm to the dial center is also a Dutch feature. Note the fine flambeau finials which were starting to go out of fashion at this time, and the strike/silent regulation next to 9 o'clock. 7′ 11″ (241 cm) high.

# Walnut Breakarch

Figure 147. *Walnut longcase clock circa 1760.* A walnut longcase clock with a broken-arch top originally surmounted by three finials and with strike/silent regulation in the arch. 7' 10" (239 cm) high.

Figure 148 A and B. *Musical Clock by William Porthouse of Penrith, circa 1780.* A two-tune musical clock, the pin barrel, hammers and bells for which may be seen in the view of the movement, together with the external fly and pin barrel return spring.

The 13-inch brass dial has a well-engraved center with the large cherub spandrels which staged a revival in the North at that time, and a painted moon disc in the arch.

Whereas in London only the finest burr walnut was generally used, particularly up until 1750-1760, in the country far less highly figured veneers or even solid walnut, probably often available locally, were used.

This clock is typical of the well-proportioned clocks which were being produced in Cumbria at this period. The shaped door is crossbanded as is the raised panelled base. Note the decoratively turned hood columns and the quarter columns to the trunk. 7' 10" (239 cm) high.

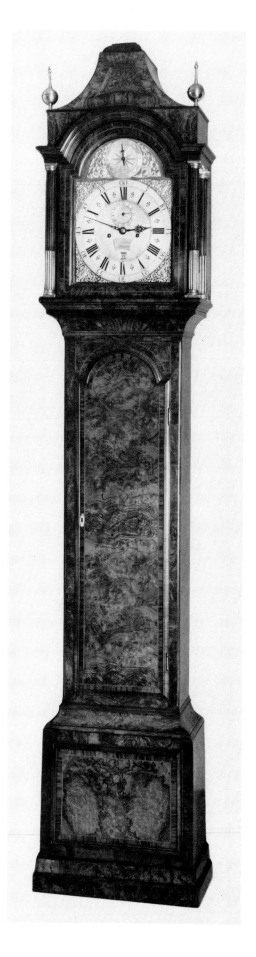

Figure 149. *Walnut Longcase Clock by George Wentworth, London, circa 1745.* Note the recessed panel to the base. 7' 5" (226 cm) high.

Figure 150. *John Hopkins, Deptford, 1780.* A well-figured walnut longcase clock much in the style of the mahogany cases being produced at that time. It has an all-over engraved and silvered brass dial with strike/silent regulation in the arch.

Figure 151. *Peter Dupont, circa 1730.* A fine walnut longcase clock with a raised panel to the base, brass-strung freestanding pillars to the hood and strike/silent regulation in the arch. This clock is interesting because of the early and rudimentary form of the pagoda.

## Walnut Breakarch

Figure 152 A, B, and C. *Astronomical clock by Thomas Clark, Warrington, circa 1750.* The brass dial, with raised chapter ring and unusual urn spandrels has wheatear engraving to the border. The 13-inch square section has the early attached arch with dolphin spandrels and a Lunar Equation Dial. This consists of three concentric rings; the outer one showing the days of the week four times. This ring is rotated manually to adjust the calendar day with the lunar date. The second ring shows the signs and degrees of the Zodiac and is numbered 0-360. The inner ring is the lunar date and numbered 0-29 ½. The phases of the moon are shown in the upper section of the ring, the rest of the arch being attractively engraved.

The center-sweep hands indicate the age of the moon and calendar day with the Zodiac. Interestingly there are indications on the moon disc with concentric rings and symbols for the days of the week. The chapter ring has quartering and half-hour divisions and half-quarter marks. The winding holes are ringed and have a wide engraved border. The date aperture also has a decorative chased surround.

The breakarch case is of good, slim proportions and is made of solid walnut.

## DIFFERENT WOODS

Figure 153. *Yew Wood.* An anonymous, country-made longcase clock constructed entirely out of solid yew wood circa 1770. If it were a veneered case, it would usually be more highly figured with more burrs in it. Yew is a slow growing tree and supplies of good timber were not plentiful in the 18th century. For this reason, it was often used in small pieces such as tea caddies, writing slopes and occasionally bracket-clock cases. It is also quite often used to decorate pieces, for instance, as inlay or crossbanding. Only two London-made yew wood longcase clocks have been seen by us in the last 20 years.

When first cut, the timber has quite strong reddish streaks but it mellows to a deep yellowish brown color.

Figure 154. *Burr Elm.* Elm has a similar disadvantage to walnut in that it is relatively soft and readily attacked by the worm. However, much early furniture was made from it and it takes on a lovely patina with age.

Pollarded elm in particular is sometimes used to veneer clock cases. It is highly figured and looks similar both in color and figuring to some walnuts and even pollard oak. It has a warmth and life which is particularly attractive.

The elm clock shown here, with relatively dark walnut crossbanding, has an eight-day, five-pillar movement with a center-sweep seconds hand and pump-action quarter chime on two bells. It is signed on a plaque in the arch *Joseph Oxley, Norwich,* circa 1750.

Figure 155. *Maplewood.* Maplewood was used very little in the 18th century, coming into fashion at a later date and being used in particular for decorative pieces especially after 1840-1850. It is occasionally seen on bracket clocks but very seldom on longcases. The example shown here, despite the apparent earliness of the dial, would probably date circa 1770-1780. It has a penny moon and lunar calendar in the arch and is signed by Thomas Waller of Preston.

*Mulberry wood.* Mulberry is undoubtedly one of the most attractive of all woods, being a little like burr elm in appearance, but of a different color. It was used quite frequently on furniture mostly around 1700-1730. Figure 45 is a particularly good example.

130 *Lacquer*

## THE SQUARE DIAL

Above: Figure 157 B, C. See case Figure 157 A, page 132. The movement is of massive construction with latches for the pillars, shaped plates, bolt and shutter maintaining power, and a six-wheel train, the last two wheels of which run in a subsidiary plate on which the equation cam is mounted externally (Figure 157 C) with the rack which activates it immediately above it. The under-dial rack is controlled by a cam situated behind the backplate and geared to an equation disc calibrated in five-minute intervals.

It is thought that the equation work was added to this clock in the early 18th century. 6′ 7″ (200 cm) high. Photo courtesy Sotheby's. London.

Opposite page:
Figure 156. A dark blue lacquer longcase clock with much more delicate decoration than that seen in Figure 158. Birds and insects fly amid trees, buildings, foliage and flowers. 6′ 11″ (211 cm) high.

# Chapter 6.
# Lacquer or Chinoiserie Decoration

The term *lacquer work* is applied to furniture which is decorated with gilded gesso, usually raised, laid on a Japanned background. *Gesso* is plaster of paris mixed with glue and then applied to furniture and various other items and shaped or carved or incised to create different raised designs which are then either gilded or painted. *Japanning* is the decoration of an article with a varnish, usually colored, which giveshe resin from various species of the gum tree and when hard is highly resistant to wear and many solvents.

The first lacquer work was probably brought to Europe via Venice and Genoa whose merchants had a monopoly on trade with the East until the end of the 15th century. Vasco de Gama sailed from Portugal around the Cape of Good Hope, reached India, and founded a Portuguese colony. From then on Portuguese ships, which largely replaced the land convoys, were to dominate trade with Japan and India until British and Dutch ships started to compete with them nearly a century later.

Oriental lacquer work may be divided into two types: (a) flat and (b) raised. In both forms various colored backgrounds were used. Although lacquer work is thought to have been used in Japan and China as early as the third century B.C., its importation to England only started in the early 17th century following the formation of the East India Company whose employees brought back examples. These were usually small items such as boxes, furniture being rarely shipped at this stage. However, wall panels were imported and cut up to form screens, when inevitably the original designs were mutilated. Few pieces still exist which were imported before 1660.

Among the early items were small boxes containing tea. For reasons of economy these were not usually decorated with true lacquer work but covered and decorated in paper which was then protected by a coat of varnish. A reflection of this occurs in the pictures often applied to English tavern and occasionally longcase clocks (Figures 180 and 181) in the 18th century in which a painted print is usually pasted onto the trunk door, and then protected with varnish, the rest of the case being decorated with conventional English lacquer work.

The wall pictures (Makimonos) from the third period of Japanese wood engravings (1680-1710) were often mounted on canvas, protected with varnish, imported into England and used to make folding screens. These and other imported items were to provide the designs which many English craftsmen copied during the following century.

The import of lacquer work from the Orient, always a somewhat hazardous procedure, varied greatly with the fighting and piracy waged at that time among the fleets of England, Holland, Spain and Portugal. It may well have been this which prompted, in 1688, the publication by John Stalker and George Parker of *A Treatise in Japanning and Varnishing*. This was a complete rediscovery of those arts, containing the best way of making all sorts of varnish for Japan, wood, prints and pictures, the method of gilding, burnishing and lacquering with the art of gilding, separating and refining metals, and so forth.

What followed was the publication of many other books on the subject in England during the next 80 to 90 years. The title of the book by C. K. in 1697 *Art's Masterpiece, or a Comparison for the Ingenious of Either Sex* gives an idea of the English attitude to lacquer work at that time, it being regarded as a suitable hobby for young ladies and gentlefolk. Indeed, in the publications of the time advertisements appeared offering to give instruction in the art. However, this gradually gave way to a more serious and commercial approach to the subject as the demand for lacquer furniture rapidly increased.

*132 Lacquer*

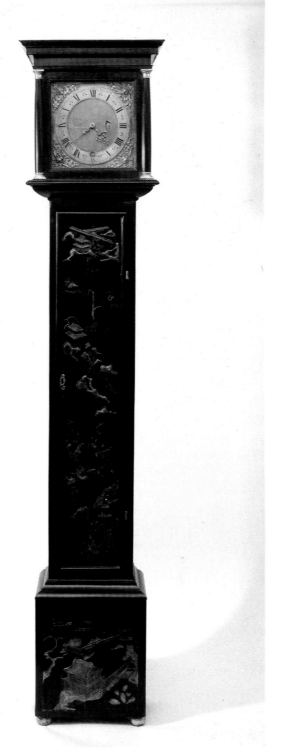

Figure 157 A. *Year-duration, black-lacquer, longcase clock by James Clowes, circa 1685* A particularly early example of a lacquer case. The period, the bold style of decoration applied to the front, and its' quality all suggest that this is of Dutch origin. The proportions of the case are excellent with a long slim trunk and matching base which rests on gilt bun feet. The capitals to the hood pillars are silver.

It is rare to find a year-duration clock of such small size and with so relatively light a case. Because most year clocks have to support a very heavy driving weight, they are usually far more substantial.

The ten-inch dial is signed along the bottom plate *James Clowes Londini*. It has silver hands, cherub and leaf spandrels, a cutout inside 3 o'clock for the equation and opposite this the inscription "Apparent time."

Figure 158. A good quality, dark green lacquer, eight-day longcase clock by Henry Fish, London, who is recorded as working from 1730-1775. However, this clock would appear to have been made around 1720.

The case, which has applied brass capped pillars and glazing to the sides of the hood, is attractively decorated with gilt gesso work depicting flowers, foliage, people, a building beside a lake, and many different kinds of birds. Interesting features are the raised work which is not just confined to the trunk door but extends to the hood and the sides of the case, and the bold concept of this decoration.

The good five-pillar movement has rack strike and a 12-inch square brass dial with a raised chapter ring which has half-hour marks, quartering on its inner edge, a seconds ring, date aperture with engraving around, ringed winding holes, and twin cherub and crown spandrels. 7' 0" (213 cm) high.

Figure 159. A good tortoiseshell lacquer longcase clock by John Hawkins of Soughton (Southampton) who is recorded as working on the west side of the High Street at the corner of Butcher Row in 1723. Loomes suggests that it may well be the same John Hawkins who was born in 1681, apprenticed to Cornelius Herbert from 1695 to 1702 but not freed, and who was subsequently known to work in London. This particular clock would date circa 1715-20[1]

The five-pillar movement of this clock is of typical London quality with rack strike and a fine 12-inch square brass dial. The raised chapter ring has quartering on its inner edge and fleur-de-lis half-hour divisions. There is well-executed wheatear engraving around the border of the dial, and the center is matted and decorated by four engraved birds, two carrying leaves. The center of the seconds ring has sunburst engraving and there is an attractive cartouche, again with birds around the date aperture.

The case has a flat top and a rectangular trunk door with a brass ring let into it surrounding the lenticle which is original. The hood door is glazed with old green tinted glass and has applied pillars.

Birds are again featured on the lacquer decoration flanking the trunk door which has country scenes depicted on it with sprays of flowers, a building and people seated around a table. 7′ 1″ (216 cm) high.

Figure 160. *Deep red lacquer longcase clock by J.Barron of Hermitage circa 1720.* The pine case has applied pillars to the hood door, a double plinth and a long trunk door which is attractively decorated with birds, buildings, people and trees.

## 134 Lacquer

Figure 161. A small black lacquer longcase clock by the eminent maker Daniel Delander with a shaped top to the dial, typical of the maker and bearing his signature with floral engraving on either side. The trunk door is decorated with typical oriental scenes with a ship at the bottom and two sailing ships on the base. 6' 8" (203 cm) high.

Undoubtedly early English lacquer work was usually of poor quality when compared with that of the Orient. This was due in part to the different materials that were used. For instance, with Japanese lacquering the surface is built up by using repeated layers, each of which is left to harden and then is rubbed down before another one is applied, thus achieving a beautifully smooth and hard surface. With much of the English work, the ground was prepared by painting and applying subsequent coats of varnish. Similarly, instead of using gold leaf, brass dust—known as "gold dust"—was often substituted, this being applied on top of gold size. Moreover, the raised gesso was frequently not incised and the designs were poor.

Whereas in England lacquering was little used until 1710-1720, in Holland they had been developing this process since the 1650s and by 1680 were producing some fine work.

Few lacquer clocks were made prior to 1700 and of those that were the cases are probably of Dutch origin, the lacquer work usually being of fine quality, often fairly bold in concept and markedly different from the later work produced in England. Interestingly, although very few early lacquer clocks still exist the condition of the decoration on these, at least in the author's experience, is markedly better than those made from approximately 1720-1770.

Although lacquer cases started to be used towards the end of the square-dial period, the vast majority have breakarches and their production continued concurrent with the walnut, and later the mahogany, cases until the 1770s.

The carcases were usually of oak on the London clocks but sometimes of pine; this order being reversed with the country clocks.

Several books containing excellent designs for lacquer work appeared around the mid 18th century, which gives some idea of how popular lacquer work was at this time. Examples of these are: (a) W.J. Halfpenny's *New Designs for Chinese Temples, Triumphal Arches, Garden Seats, Palings, etc.*. [Even the people depicted are good Chinese figures] and (b) Edwards and Darly's 1756 *A New Book of Chinese Designs*.

Chippendale, in his *Director*, illustrates lacquer work and it is likely that it is this which gave rise to the "Chinese Chippendale" designs in his book. Manwaring and others followed his example.

It has been suggested that much of the lacquer furniture and clock cases were sent out to the Far East for decoration. However, this only seems to have happened infrequently, judging by the very small amount of Oriental lacquer which is seen on English pieces and by those records which still exist from that period.

Sending something out to the Far East was a hazardous procedure and it could well have taken up to five years before the pieces returned, which probably made it economically inviable in the majority of cases.

Most lacquer clock cases have, because of damage and flaking of the decoration over the years, been restored to a greater or lesser degree. A particular problem is the fact that gesso dries out over the centuries, and becomes detached from the case.

Lacquer work is occasionally seen from the Victorian and Edwardian periods, but it is usually markedly different from the earlier examples.

### References

Cescinsky, H., *English Furniture of the 18th Century*, pages 174-226, George Sadler & Co., London, 1909.

Cescinsky, H. & Webster, M. R. *English Domestic Clocks* Waverley Book Co. Ltd., London.

Stalker and Parker *A treatise in Japanning and Varnishing*, first published in 1688, reprinted 1971.

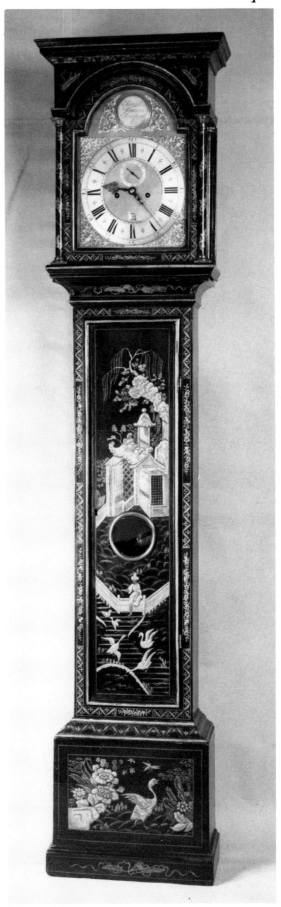

Figure 162. *Dark green lacquer clock by Cornelius Herbert, London Bridge, circa 1720.* Note the relatively early features such as the retention of the lenticle in the trunk door with its' square top, the wheatear engraving to the border of the dial, the ringed winding holes, the cartouche around the date aperture and the beautifully executed spandrels in the arch containing Herbert's signature.

Figure 163. *Green lacquer longcase clock by Samuel Weston of Stratford, circa 1720.* Stratford is just east of London. Bold raised gilt designs feature birds.

# 136  Lacquer

The caddy top case is surmounted by three pear-shaped finials, and there is glazing to either side of the hood and applied brass capped pillars to the hood door. There is a substantial mould around the breakarch trunk door which houses a Vauxhall plate mirror, a sign of affluence at that time.

The silvering of large sheets of glass started in Italy in the late 16th century and most early mirrors are of Italian origin. It was not until 1670 that production started in England with the establishment of a plate glass factory at Lambeth by Dawson, Bowles and Co under the Duke of Buckingham's patronage in Vauxhall square, hence the name Vauxhall plate. At this period the glass was blown, thus limiting its sheet size to a maximum of three feet six inches. The long shallow bevels applied to the edge were produced by moulding the glass while it was still semi-molten, grinding and polishing not coming into general use until 50 years later.

Originally, silvering was achieved by floating a thin layer of mercury onto the back of the glass, applying a thin leaf of tin over this, and then applying pressure to expel the excess mercury. However this process was eventually abandoned because of the poisonous effect of mercury.

At this time the manufacture of plate glass had not come into use and the sheets of glass were made by whirling molten glass around a cylinder until it covered its inside surface. The glass was then removed from the cylinder, cut longitudinally, heated and pressed flat.

In the late 17th and early 18th centuries glass was extremely expensive, initially because of a heavy tax repealed in 1698, but also because of the limited production, and thus it was usually used only in small pieces.

**THE CADDY TOP**

Figure 164 A and B. Green Lacquer Longcase clock with Vauxhall Plate mirror circa 1720. A fine quality, green lacquer longcase clock by Ben Goddard, London who was apprenticed to his father in 1692, became Free of the Clockmakers Company in 1701, and continued working until 1726. A longcase by him is in the Mosque of Achmet in Constantinople.

Figure 165. *Well-proportioned tortoiseshell lacquer longcase clock by John Thompson, St. Catherine's Dock, London, circa 1725.* The five pillar movement has a 12-inch dial with the name plaque in the arch. The chapter ring has quarterings on its' inner aspect and decorative half-hour marks.

A rare feature is the clockmaker's original trade label inside the trunk door, thus proving conclusively that the movement and case have always been together.

Figure 166. *Green lacquer longcase clock of month-duration by Charles Goode, working from 1686 until he died in 1730.* Many of his clocks are of month duration. This clock is interesting in that it has some transitional features, for instance the dial was made in two parts, as were some of the early breakarch clocks, and it still retains the square top to the trunk door and the lenticle which were to disappear within the next five-to-ten years. In the arch is a year calendar. 7' 1" (212 cm) high.

## 138  Lacquer

Figure 167. *Tortoiseshell lacquer longcase clock by Walter Morgan of Hereford circa 1730.* Note the relatively early appearance of the moon in the arch, and that the trunk door now has a breakarch top. Sadly, Mr. Morgan went bankrupt in 1743, a fate that befell quite a few clockmakers. 7′ 0″ (213 cm) high.

Figure 168. A black lacquer longcase clock in relatively untouched original condition, except probably for a missing caddy. The dial is an attractive one with female mask spandrels, engraved decoration around the date aperture, engraving around the arch, good half-hour marks, and a name plaque in the arch signed by George Avenall of Farnham with beasts rampant on either side and a crown above.

The Avenalls were a large family of clockmakers who worked in and around Farnham in the 18th century and George was probably the first, circa 1725. 7′ 0″ (210 cm) high.

Figure 169. A tortoiseshell lacquer longcase clock with applied pillars to the trunk door by the eminent maker Markwick-Markham famed for his musical and complex clocks for the export trade. Circa 1730-1735.

Figure 170. A mid-18th century, light green lacquer longcase clock by the previously unrecorded maker John Fry of Kilmersdon. Like many clocks made in or near Bristol, it is tall and elegant. The design above the breakarch of the hood is most unusual and apparently inspired by classical architecture. Whether or not it originally had some superstructure above this is difficult to know. The hood pillars are still applied to the hood door, something which had gone out of fashion in London by that time.

The 14-inch dial is typical of those being made in Bristol at that time, incorporating moon phases and giving high water at Bristol Key.

## THE BREAKARCH TOP

This style of case first appeared in the 1720s. Note that the curved moulds around the top of the hood are more substantial and fully developed than those seen just above the dial on the lacquer clocks illustrated earlier.

Figure 171. A particularly interesting red lacquer longcase clock by George Neale of London who was first apprenticed to Peter Wise and then turned over to Stephen Horseman in 1726. This particular clock has all the original equation-of-time tables on the door (Figure 124) and the back of the case. It is fascinating that they are Quare and Horseman's tables and written on these in longhand is George Neale's name and details of his apprenticeship. It is dated 1733. It is difficult to think of a more positive way of proving that the case and movement have always been together and of dating it.

The dial still retains the half-hour marks between the Roman numerals, although these are now small, and the quartering on the inside of the chapter ring, all features which were at this time starting to disappear. There is a name plaque in the arch. 7' 1" (212 cm) high.

Figure 172. A dark red lacquer longcase clock with well executed designs to the trunk and base. It is signed by the well known provincial maker John Buffet of Colchester, most unusually on either side of the strike/silent ring. He was active 1721-1735.

*Lacquer* 141

Figure 173. A small country-made lacquer longcase clock by Joshua Brace of Chepstow, circa 1760. It is interesting to notice the late use of the half-hour marks, quartering on the inside of the chapter ring, and the matted dial center, all features which had gone out of fashion in London thirty years earlier. In the arch a bird is engraved together with the inscription *Tempus Fugit*. 6′ 11″ (212 cm) high.

Figure 174. A country-made black lacquer longcase clock by Thomas Dicker of Silchester, a village south of Oxford, circa 1750-55. It, like the clock seen in Figure 18, still has the early style of chapter ring. It also has a sunburst in the arch and, most interestingly, a series of circles engraved on the inner aspect of the chapter ring, a feature which was popular with Quaker clockmakers, particularly those around Oxfordshire, at that time. 7′ 0″ (213 cm) high.

Figure 175. A tortoiseshell lacquer longcase clock signed across the top of the arch by Humphrey Sellon of London, circa 1765. Note that the chapter ring has no half-hour marks or quartering on its inner edge. The scalloped border to the recessed seconds ring is also a comparatively late feature. The finials would appear to be out of period for the case. 7′ 9″ (236 cm) high.

## THE PAGODA TOP

The pagoda top first made its appearance in the late 1720s being in effect an inverted caddy top. The earlier examples tended to be fairly short but they gradually increased in height over the next fifty years and are seen in their final stage of development in the London mahogany longcase clocks being made around the 1780s.

Figure 176. A green lacquer longcase clock by John Burgess of Gosport, with a man-of-war being depicted near the bottom of the door, an appropriate feature as Gosport is a naval port. Note the attractive treatment of the signature on the recessed plaque in the arch with fine floral engraving below and a band of wheatear engraving above.

Opposite page:
Figure 177. A tortoiseshell lacquer longcase clock by Henry Baker of Malling, circa 1770, showing moon phases and lunar calendar on a star-studded disc in the arch. It was shortly after this that painted moon discs started to appear on brass dials. The spandrels in the four corners show girls representing the four seasons, a feature usually seen only on country clocks. It is interesting that even at this late period, when all the additional marks had disappeared from the chapter ring, he persisted in the use on nearly all his clocks of an engraved cartouche around the date aperture with one or two birds on either side.

Figure 178. A relatively fully developed pagoda-topped lacquer longcase clock, circa 1770, possibly originally on a double plinth, with fretwork incorporated in the front of the pagoda. Although this was a standard feature on mahogany pagoda topped clocks, the fret was usually omitted on lacquer longcase clocks.

The 12-inch dial has, as would be expected at this period in London, strike/silent regulation in the arch. Immediately above the date aperture is a plaque bearing Thos. Hall's signature, a common way to sign the clock at that period. 7' 10" (239 cm) high.

Figure 179. A well proportioned green lacquer longcase clock by Abel Panchaud, Oxford Street (London) with a typical five-pillar movement and dial. Note that even at this comparatively late date the pillars are still applied to the hood door, as opposed to being freestanding, a common feature on lacquer cases.

Wooden finials, usually three in number as in this case, were mounted on the pagoda top of lacquer cases, but the center one was often removed at a later date to reduce the height of the clock. 8′ 2″ (249 cm) high.

Figure 180. A dark green lacquer longcase clock which is unusual in that there is floral decoration, typically English, around the trunk and hood doors and on all the moulds. It has two scenes of courting couples painted on it. The upper one is of a gardener handing a posy to a young girl, and the other is of a fisherman and a girl. There is a similar scene on the base with a couple seated near a lake with a ruin in the background. These pictures are usually overpainted prints which have been pasted onto the case, a technique also used on tavern clocks.

*Lacquer* 145

Figure 181. A most impressive lacquer clock with well-executed religious scenes on the trunk door. On the base is a copy of a painting by Raphael in the Queen's Collection titled *St. Paul Preaching* or *Athene*. It is signed on a plaque in the arch by Greg Sewell (Senior), London, circa 1760. 7' 11" (241 cm) high.

Figure 182 A. A dark green lacquer musical longcase clock. The pagoda-topped case has a fret in the front to let the sound out and finials on either side. The hood door has applied pillars and the sides are glazed. The delicately executed lacquer work depicts buildings, people, a garden scene, a bridge and on the base a man in a boat crossing over to a tower. As is usual the painting on the center of the door is typically English. It depicts a fisherman and his lady to whom he is proudly showing the fish he has just caught.

Figure 182 B, and C. The substantial three-train movement chimes the quarters on six bells and plays either a polonaise or a minuet on eight bells and 16 hammers at each hour. The 12-inch brass breakarch dial has tune selection in the arch, a seconds ring, date aperture and a strike/silent lever at 9 o'clock. It is signed in the sector above 6 o'clock by Charles Coulon of London. 7′ 7″ (231 cm) high.

# Chapter 7.
# "London" Mahogany Longcase Clocks

The fashion for walnut furniture including clocks, which held sway from around 1675-1735 gradually started to decline after this period. This was in part because of the loss through disease of many of the walnut trees, particularly in France, and also the rising level of importation of other timber from overseas.

England's trade with the West Indies and America was rapidly increasing at that time and thus it was only natural that ships would wish to fill their holds for the return journey, partly for economic reasons but also because it increased the stability of the ship.

The timber which they bought back, in particular mahogany from Cuba and Honduras, was much to the liking of cabinetmakers and their customers. It was easy to work, available in long and wide boards, and was attractively figured. Thus it could be used in the solid, for which many walnuts are not really suitable, and also in the form of a veneer where the maximum decorative effect was required.

One of the principal disadvantages of walnut is that it is highly prone to attack by worms, as also is the pine carcase on which it is often laid, and for those reasons the bases of many walnut clocks have rotted and been either shortened or replaced. Mahogany, on the other hand, is highly resistant to worms, so this fact and its inherent stability, particularly if it is carefully chosen and correctly treated in the first place, which has resulted in the remarkably high survival rate of mahogany longcase clocks.

It is difficult to know exactly when mahogany first started to be used in England and certainly this is a matter of much debate, but it is unlikely that its introduction occured much after 1720. The earliest mahogany longcase clock the author has seen was signed and dated 1728. The earlier pieces tend to be made from the solid, and probably the most decorative clocks with the finest veneers were produced around 1760-1785 (Figures 185 and 189).

A cause for some confusion is the use of red walnut, mainly in the period 1720-1735 when English and French walnut was in short supply and mahogany had not established its pre-eminent position. This timber, although as its' name implies a fairly deep red color when first cut, tends to fade to a similar color to that of faded mahogany. It is very dense and does not usually have much figuring, which differentiates it from most other walnuts, and if one looks at the unpolished and protected surfaces, such as the inside of the trunk door or beneath a table, then its original color may be seen.

## THE PAGODA TOP

Figure 183. Thos. Budgen, Croydon, Circa 1755-1760. A typical "London" mahogany longcase clock of this period with a well, but not highly figured, case. The pagoda top has a wooden fret to the front, and fluted and brass strung columns to the hood which has glazed sides. There is a raised rectangular panelled base with a double plinth and a shaped, lower border.

The dials had by now lost all their extra ornamentation such as ringed winding holes and half-hour marks and this clean design suites the mahogany cases excellently.

Croydon is 12 miles south of London, and it was usual for clockmakers who were in such close proximity to the city to produce clocks virtually identical with those made there. 7' 9" (236 cm) high.

148 *"London" Mahogany*

The following six illustrations show the fully developed London mahogany clock with brass capped, reeded, and brass strung quarter columns to either side of the trunk. Note on the clock in Figure 188 also the narrow chapter, seconds, and strike/silent rings. 8' 0" (244 cm) high.

Figure 184. *Conyers Dunlop, London, 1760.* A fine quality longcase clock signed on a plaque above the date aperture by Conyers Dunlop, an eminent maker who became Master of the Clockmakers Company in 1758. The finials, as so often occurs, are replacements and a little too small (see Figure 188).

Figure 185. *John Monkhouse. London, circa 1770.* A classic London case with beautifully figured veneers of a fine rich color.

## "London" Mahogany

Figure 186. William Mason, London, circa 1780. It had been the fashion in the early 18th century to sometimes employ ebony moulds on clock cases. This fashion was revived, particularly on some of the clocks made after 1775. Here we see its' use being put to dramatic effect by emphasizing the beautiful quality of the mahogany veneers.

The height of this style of longcase clock tended to increase as the century progressed, this one being eight feet nine inches to the top of the finial. It is, of course, for this reason that so often the top block and finial have been removed today.

Figure 187. *John Cowell, London, circa 1790.* This clock probably represents the London clock and cabinetmakers' efforts to produce the finest possible clocks, stimulated in some measure by increasing competition following the start of the decline in popularity of the longcase clock in London.

The movement is of the finest quality with dead beat escapement, a massive brass cased bob to the pendulum with a graduated rating nut, pinned pulleys and carefully matched weights, that for the going train being only eight and one half as opposed to the usual 12 pounds. The brass dial has that attractive feature: raised chapter and strike/silent dials which are enamelled.

The case is extensively inlaid, the panel in the base being omitted to show the inlay to maximum effect. John Cowell was working from 1759-1799. 8' 2" (245 cm) high.

## 150 *"London" Mahogany*

Figure 188. *John Long, London, 1770.*

Figure 189. *John Baker, London, circa 1770.* Note the superbly chosen veneers to the trunk door. 7′ 10″ (239 cm) high.

Figure 190. *Crump, Canterbury, circa 1780.* The all-over engraved and silvered brass dial probably started to be introduced on London clocks in the 1760s and within ten years had become quite common. Here we see a particularly good example with the strike/silent regulation to the right of the arch nicely balanced by the month calendar on the left. There is delicate floral engraving to the four corners and within the arch.

The case is a little shorter than usual with a shallower pagoda and a squatter base. The shape of the top of the trunk door was quite common at this period. 7′ 5″ (231 cm) high.

Figure 191. Richard Reed, Chelmsford, Circa 1760. A very similar clock to that seen in Figure 183 but with a few differences. For instance, there is a cast brass fret to the pagoda, Corinthian capitals to the hood columns (not generally used at this period), and scalloped corners to the base panel. A slight "country" influence is achieved by the birds engraved on the matted dial center near the winding holes and the decoration around the date aperture. Chelmsford is 30 miles east of London. 7' 9" (236 cm) high.

Figure 192. John Hovil Rotherhithe, (London), circa 1760. A London mahogany longcase clock with the standard strike/silent regulation in the arch above which is a curved strip carrying the makers' name, always an attractive feature. The seconds ring is recessed which is an advantage in that there is less chance of the seconds and hour hands fouling each other. The fret in the pagoda and the finials are replacements. 7' 10" (239 cm) high.

## 152 "London" Mahogany

To appreciate the design of longcase clocks made in London in the Georgian period, it is important to realize how rigid fashions were at that time and how precisely the designs of any particular item, be it a house, a clock, or a table, tended to conform; moreover, at any one time there were probably only a handful of cabinetmakers supplying cases to the various clockmakers in the city, and the majority of these are likely to have used similar pattern books.

The pagoda-topped longcase clock is thought of by many as the classic London-made mahogany longcase clock and probably originated from the caddy top via walnut clocks such as that seen in Figure 151 which has a form of pagoda. By 1740 or possibly a little later, it was fully developed. Standard features included a pagoda top with either a brass or wooden fret in front, reeded and brass strung columns to either side of the hood, a raised panelled base and a double plinth. The top of the trunk door was usually in the form of a breakarch to copy the shape of the dial, had an attractive mould around it, and was decorated with well-figured veneer.

The clock movements were nearly always of five-pillar construction with a recoil escapement, brass cased lead weights, and a pendulum with substantial brass cased lead bob. The brass dials were almost invariably 12-inch square with a breakarch on top giving an overall height of around 16½ inches. In the earlier clocks there was usually, but by no means always, a name plaque in the arch and the main dial had a raised engraved and silvered brass chapter ring, a date aperture above 6 o'clock, and either a recessed seconds dial or a seconds ring below 12 o'clock. On the slightly later clocks the name plaque in the arch was replaced by strike/silent regulation and the signature was displaced to a variety of positions such as the chapter ring, a curved strip inside the bottom of this, a plaque above 6 o'clock (Figures 189 and 193), or a strip around the arch (Figure 188).

Reeded and brass strung quarter columns probably first appeared on either side of the trunks of these clocks in the 1760s and added considerably to their appeal, making them look slimmer and more elegant. Another minor variation was the scalloping out of the corners of the raised base panel.

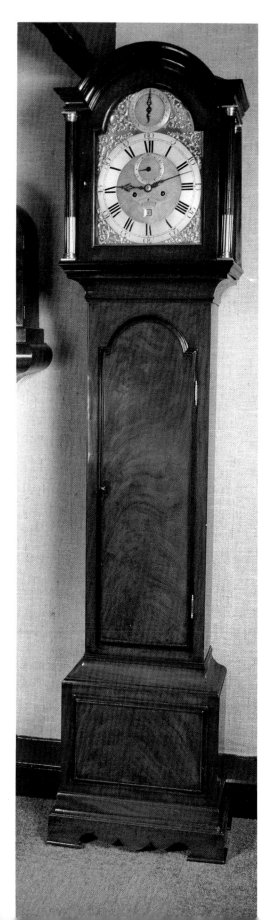

### THE BREAKARCH TOP

Because pagoda tops are so tall, clocks made with them are sometimes later reduced in height, either by making the top removable or by completely removing it. However there are also London mahogany clocks made with simple breakarch tops and it is by no means easy to differentiate between these two groups. A further group of clocks have fretwork, etc. and three finials on their tops.

Figure 193. *Theodore Morrison, London, circa 1760.* A well-proportioned clock with simple breakarch top, raised panel to the base, and a double plinth. 7' 2" (218 cm) high.

As the century progressed cases tended to become smaller, and the designs simpler. The pagoda, for instance, often was omitted, although in many of the clocks seen nowadays it has been removed to reduce the overall height to allow it to fit into modern accommodation.

A further development was the added decoration above the breakarch as seen in Figure 195 and the beautiful cases, such as those made by Rich for Mudge and Dutton, and other makers. These have a shaped block on top with a single central ball finial. To increase the elegance of the design the wooden hood pillars were omitted and the corners on either side canted and brass strung (Figure 204).

All the changes in case design which occurred during the second half of the 18th century were, in the main, following the changes brought about first by Chippendale and later Hepplewhite, Sheraton and other leading furniture designers.

## DIALS

By 1765 there was a tendency to again use simple square dials like those used prior to 1710; however, these had all-over engraved and silvered brass. Probably the most famous proponent of this style was Vulliamy (Figure 207). Towards the end of the century the square dial was replaced by the circular dial, either flat or occasionally dished.

Exactly when painted dials first started to be produced is not known with any certainty. The earliest the author has seen was on a longcase clock by Eardley Norton dated 1776 but they were being advertised in Birmingham at least four years before this. On London clocks they were usually either plain white all over or occasionally had delicate flowers painted in the four corners and the breakarch.

An attractive and much more durable alternative to painted dials was hard enamelling; however this was both difficult and costly to produce and was only used infrequently and by a handful of the better makers. Occasionally, enamelled dials were used in conjunction with a conventional brass dial, when a most pleasing effect was achieved (Figure 187).

Figure 194. *E. Smith, Richmond, circa 1775.* A mahogany longcase clock of fine quality with quarter columns flanking the trunk and strike/silent regulation in the arch. 7′ 4″ (224 cm) high.

Figure 195. *Samuel Lines, Luton, circa 1775.* An unusually small longcase clock of London style, standing only 6′ 9″ excluding the top finial, but base, trunk, and hood are all in proportion. The only thing which betrays a slight country influence is the smooth engraved dial center and the large date aperture. Luton is approximately 40 miles north of London.

154 *"London" Mahogany*

Figure 196. *Bowen, Glasshouse Street (London), circa 1765.* The practice of using chapter rings and subsidiary dials such as the strike/silent with a solid center (i.e. not cut out) probably started in London around 1760 and also was used in the provinces. However the fashion obviously never caught on, as the number seen is quite small. 7′ 1″ (216 cm) high.

Figure 197. *Hector Simpson, London, circa 1790.* In this clock is seen the all over engraved and silvered brass dial without any decoration which came into fashion at about the same time as the painted dial and was preferred to it by the London makers. The flat silvered brass dial is also found on country clocks. Note that the sides of the trunk have canted and brass strung corners instead of quarter columns. 7′ 1″ (216 cm) high.

"London" Mahogany 155

Opposite page:
Figure 198. *John Howell, London, 1780.* An interesting clock with some unusual features. It has pump-action quarter chime on eight bells for the first three quarters, a finely engraved solid (not cut out in the center) chapter ring with a center-sweep date hand, a strike/silent ring, and solid floral spandrels with a continuous bead running around their edges.

The top of the hood is unusual for a mahogany clock of this period and would originally have had a central finial.

Figure 199 A and B. *John Hubbard, London, circa 1785.* An attractive engraved and silvered brass dial in which the plainness is relieved by an engraved moon disc and calendar in the arch, with the sun and earth represented on either side, recessed seconds ring with scalloped border and a semi circular cutout for the date. There is also decorative engraving to the corners and center of the dial.

The movement is of month duration and a rare feature is Hubbard's trade label showing the equation of time which is pasted onto the inside of the backboard.

Figure 200. *William Carter, London, circa 1760.* A relatively small London mahogany longcase clock surmounted by three blocks and finials with scalloped woodwork between. 7' 1" (2 16 cm) high excluding finial.

## 156 "London" Mahogany

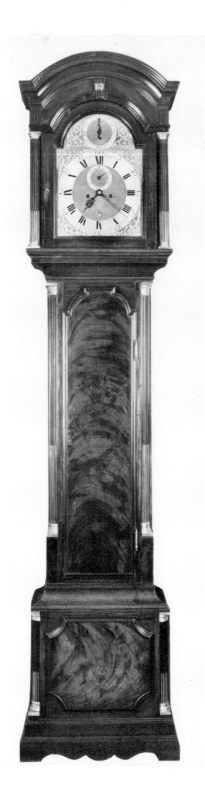

Figure 201. *William Trail, Paddington (London), circa 1800.* A good mahogany longcase clock with a plain engraved and silvered brass dial having seconds and date hands. Already a little of the feel of the regency period is starting to appear. 7′ 7″ (231 cm) high.

Figure 202. *Longcase Clock by Newman Peachey with Quarter Columns to the Base.* Occasionally brass capped quarter columns are provided not just on either side of the trunk but on the base also, producing what is probably the most impressive of "London" cases and in its' time the most expensive also which is probably why this feature is generally seen on musical clocks which cost far more than an ordinary striking clock. Like most London made cases it has a mahogany carcase and an oak backboard. 7′ 11″ (241 cm) high.

Figure 203. *Henry Jenkins, London, 1775.* A fine longcase clock similar to that seen in Figure 20 with beautifully figured flame mahogany veneers to the trunk door and a moon shown in the arch set against a star-studded background.

Henry Jenkins, an interesting maker, published a book in 1760 entitled *A description of several geographical and Astronomical Clocks.* 7′ 7″ (231 cm) high.

## "London" Mahogany 157

Figure 204. *The Omission of Columns to the Hood.* These cases, in which the columns are omitted from the hood and replaced by reeded and brass strung canted corners, are probably the most aesthetically pleasing of all London mahogany clocks. In the majority of cases the sides of the trunk are treated in a similar fashion and it is usual to place a shaped block on top of the clock to carry a single central ball finial. This clock is signed by William Smith, London and dates circa 1795. 7' 5" (226 cm) high including finial.

Figure 205. *Gravel and Tolkein, London, circa 1810.* This clock has a similar case to that seen in Figure 204 but with an "all over" engraved and silvered dial with a center-sweep date hand and recessed seconds ring. The combination of a silvered dial and this case style compliment each other well.

Gravel and Tolkein, an eminent partnership, succeeded Eardley Norton, famed for his fine musical clocks. 7' 2" (218 cm) high excluding finial.

### THE RETURN OF THE SQUARE DIAL

The vast majority of London makers had abandoned the square dial in favor of the breakarch by 1715, although a few, such as George Graham stayed faithful to the square dial until the mid-18th century. Also it was still used by many country clockmakers, including virtually all those who made works of 30-hour duration.

Around 1770 the square dial had started to make a reappearance, with makers like Vulliamy, for instance, adopting it. In the majority of cases it was used as a flat engraved and silvered brass dial, but on occasions a chapter ring and spandrels were also used.

Figure 206. *Late 18th Century Month-Duration Timepiece by John Billings, London.* A particularly small (six feet three inches) and well-proportioned, single-train longcase clock with a ten-inch square engraved and silvered brass dial. It has a panelled base and rests on a double plinth. The front corners of the hood are chamfered and reeded and the broken architectural top has a central finial.

## 158 "London" Mahogany

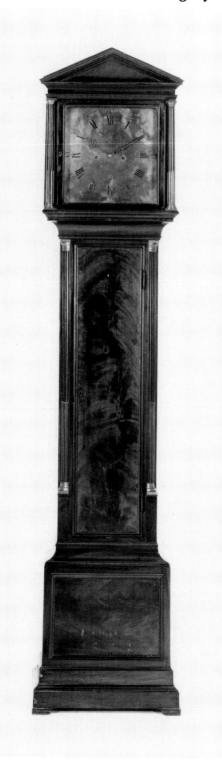

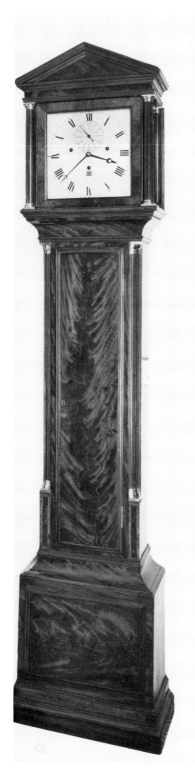

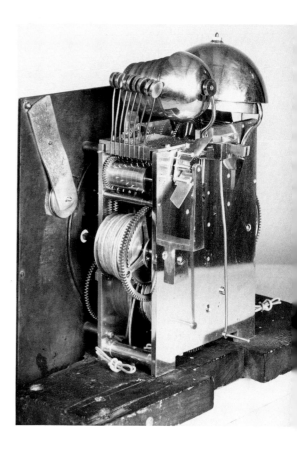

Figure 207. *Justin Vulliamy, London.* A good George III mahogany longcase clock in a style very much favored by Vulliamy. The 12½-inch flat brass dial, much in need of resilvering, has a seconds hand, calendar aperture and strike silent regulation at 3 o'clock. The hood has an architectural top and fluted and brass strung pillars with quarter columns to the trunk to match, a raised base panel and a double plinth. Thus it will be seen that the case, apart from the hood, is the same as the standard breakarch London mahogany longcase clock being made at that time. 7′ 2″ (218 cm) high.
Photo courtesy Sotheby's, London.

Figure 208 A and B. *Unsigned Quarter-Chiming Longcase Clock, circa 1780.* Both the movement and case of this clock are of the finest quality. The movement, which chimes the quarters on eight bells, has many attractive features. Note the large adjustable rollers for the date ring, the long crutch to the dead beat escapement, the unusually thick plates, and the two bells which are struck almost simultaneously for the hours giving a most pleasing effect. It will be seen that the dial is made up of two layers, an inner one on which all the feet, brackets, etc. are mounted and an outer one which, because nothing is fixed to it, has a virtually perfect surface.

It will be seen that the case of this clock is virtually identical with that in Figure 207. This and the construction of the movement suggest that it was probably made by Vulliamy. 7′ 2″ (218 cm) high.

## "London" Mahogany

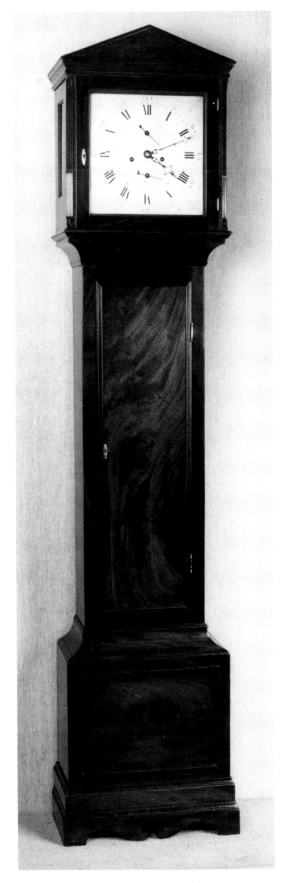

Figure 209 *Robert Favey, London, circa 1790-1800.* A simple, pleasing mahogany longcase clock with canted and brass strung corners to either side of the hood and a 12-inch square flat brass dial. 6′ 11″ (211 cm) high.

Figure 210. *Barwise, London, circa 1820.* A very attractive, striking mahogany longcase regulator, if such it may be termed. The case of well-figured veneers has an architectural top to the hood with ebony stringing and quarter columns with gilt brass capitals. The sides are glazed to reveal the movement. The solid, rectangular trunk door has delicate mouldings and reeded quarter columns to either side with gilt brass capitals. The base is decorated with a Greek key design.

The substantial, six-pillar movement has heavy plates with Graham dead beat escapement, maintaining power and the lovely feature of a double bell with two hammers for the hour strike, both striking simultaneously to give a most melodious sound, a feature also seen in the clock in Figure 208 A and B.. The wooden rod pendulum is suspended from the backplate and has a very heavy brass bob with a large graduated rating nut. A nice touch is the pulley offset for the brass-cased striking weight.

The 14-inch square silvered brass dial is signed *Barwise, London* and has engraved Roman numerals and a minute ring with Arabic five-minute divisions. The hour and minute hands of blued steel are nicely fretted and chamfered. The large seconds ring with Arabic numerals is below the XII and a strike/silent lever is above it.

John Barwise, an eminent maker, starting work in 1790 and became Chairman of the British Watch Company. He died in 1842 having made many important pieces; a regulator of his stands in St. James's Palace. 7′ 2″ (218 cm) high.

# 160 "London" Mahogany

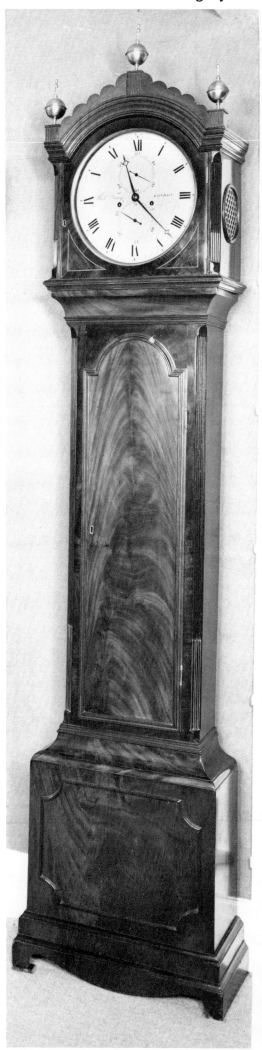

## THE CIRCULAR DIAL

The circular dial began to replace the square dial around 1800 and was in general used for the entire following century; although the use of the square dial never completely ceased. Initially most of the London-made clocks had engraved and silvered dials but by 1810-1820 many had switched to the painted dial.

The case styles continued to be very similar to those of the previous 50 years (see Figures 210 through 217) but with the appropriate modifications to the hood. Overall they were usually appreciably smaller clocks than their Georgian counterparts.

Figure 211. *Richard Farley, London, circa 1810.* A typical, high-quality longcase clock of the early Regency period, maintaining similar features to the earliest mahogany clocks such as the panelled base, although this is not raised, the double plinth, and the canted and brass strung corners to the hood and trunk. Note the circular sound frets. The silvered dial some 12 inches in diameter, is well laid out with large seconds and date rings and delicate blued steel hands very different in concept to those used on the George III clocks.

Opposite page:
Figure 212. *Robert Roskell, Liverpool, 1811.* The case of this clock is virtually identical to that seen in Figure 217 and was almost certainly made by the same cabinetmaker. However, it has a convex painted (as opposed to a silvered) dial. It will be noted that all the five Regency clocks shown so far have full-width opening hood doors but by 1820 it was to become increasingly common to have just an opening bezel.

Robert Roskell of both London and Liverpool was one of the most talented of the 19th-century clock and watchmakers. Much of his fame lay in the precision of his watch and chronometer manufacture. However, he also produced some longcase clocks, usually favoring the beautiful case style seen here. This particular clock is signed *Liverpool No. 4* but we have had a virtually identical clock signed *London* and this is obviously where these cases were made.

## "London" Mahogany

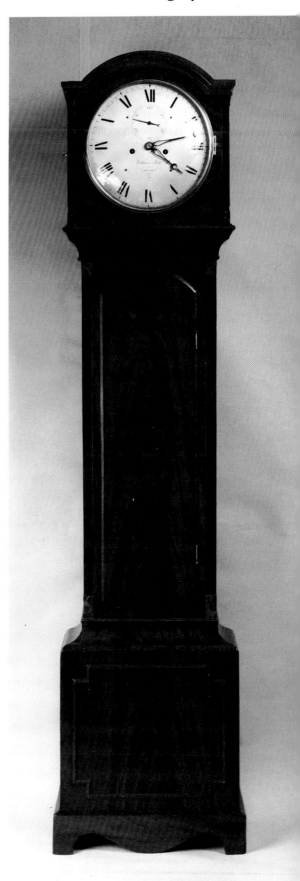

Figure 213. *Gladding & Co, Brighton, circa 1820-1825.* This clock has a fine quality case, still with a full-width, opening-hood door and a 12-inch circular silvered dial. It has several mechanical improvements over most of the longcase clocks previously produced. These include the provision of a dead beat escapement, maintaining power, and a wooden rod pendulum with heavy brass cased bob. It is because of these features that clocks such as this are sometimes termed "striking longcase regulators." 6' 7" (201 cm) high.

Figure 214. *William Pitt, London, 1825-1830.* Several features on this clock indicate that it is later than those illustrated so far. For instance, there is no hood door (just the front bezel opens), the sides of the hood are canted but not reeded or brass strung, there is only a simple arch to the trunk door, no mould is around its edge, and the treatment to the top of the hood has been simplified.

162 *"London" Mahogany*

Figure 215. *Bromley, Horsham, circa 1810.* This clock is similar to that seen in Figure 211, also with a 12-inch silvered dial, although there is no date ring. There is only a single plinth and the sides of the hood and trunk are canted but not brass strung, the kind of minor simplifications one would expect of a clock made around forty miles from London. A Greek key design is used on the base. 6′ 9″ (206 cm) high excluding finial.

Figure 216. *Le Plastrier, Deal, circa 1815.* A very similar clock to that seen in Figures 211 and 215. The use of Arabic numerals on a clock of this period and style is unusual, although they are seen on some country painted dial longcase clocks and they were in common use in France in the previous century.

"London" Mahogany 163

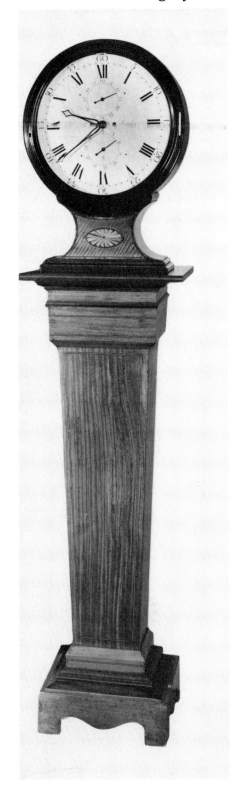

Figure 217. *Salmon, Pimlico (London), circa 1810-1815.* This clock represents what many would regard as the classic Regency London-made longcase clock; it is probably even more elegant than somewhat similar examples with a breakarch dial. It has a 12-inch circular silvered dial still with the earlier style of hands, canted corners to the hood, but brass strung quarter columns to the trunk. There is no separate raised base panel, the effect being created by a mould laid in Greek key design on top of the base which has been veneered all over. A nice touch is the way in which the top mould is reversed at its upper edge to run in towards the central block. 7′ 1″ (216 cm) high.

Figure 218. *Three-month timepiece by James Ely, London, circa 1820.* A solid mahogany timepiece of three-months duration with a slightly convex, ten-inch silvered dial. It has provision for fast/slow which is achieved by raising or lowering the pendulum by means of a graduated silvered nut above it. 6′ 8″ (203 cm) high.

Figure 219. *Haswell, London, circa 1800.* An alternative to the longcase clock was (mainly in late Georgian and Regency times) a pillar on which a bracket clock was placed. Probably the commonest style in which this was used was the balloon. The French had used this technique quite extensively since around 1700.

164  *Quarter Chiming and Musical Clocks*

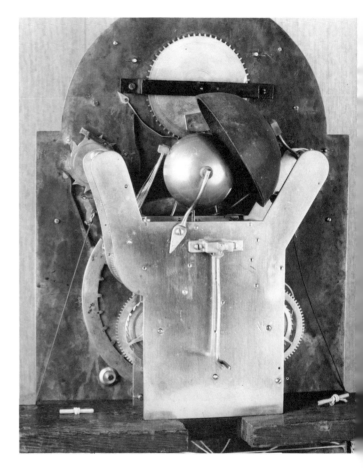

## QUARTER CHIMING CLOCKS

Figure 220 A, B, and C. *Quarter-Chiming and Tidal-Dial Longcase Clock with Year Calendar by Charles Clay, London, Circa 1725.* A fine quality longcase clock by Charles Clay who is justly famous for his musical and in particular organ clocks.

The case is a good example of the fully developed breakarch style. There is a breakarch top with substantial cross grain moulds and a sound fret below, free standing pillars to the hood, and frets to either side. The trunk door, which repeats the breakarch of the hood and dial, is surrounded by crossbanding and has cross grain moulds. The base panel is quartered and stands on a double plinth.

The spectacular and very carefully laid-out dial has a rectangular part 14 inches high and 13 inches wide, the extra height permitting much larger subsidiary dials than would normally be possible. To the top left is regulation for strike, silent quarters and silent all, much in the style used by some of the finer makers in the 17th century, and to the top right are the months of the year.

A superb feature of the clock is its silvered ring in the arch, 7¼ inches in diameter, which shows the age of the moon and the time of high water, presumably at London Bridge. In the center is a star-studded disc which revolves and displays the various phases of the moon.

The rear view of the seven-pillar movement shows the eight bells for the chime with the hour bell partially covering them nestling between the hinged extensions on either side which carry the calendar work and the strike/silent regulation. 8′ ½″ (241 cm) high.

# Chapter 8.
# Quarter Chiming and Musical Clocks

The first recorded chiming clock known to have been made in England was the weight-driven clock by Nicholas Valin dated 1598, which is now in the British Museum. Valin, like most of the early English makers, was a Huguenot who had emigrated from Flanders.

The number of chiming longcase clocks produced in the 17th century was very small and it was not until the 1720s that they started to be made in significant numbers. The same holds true for musical longcase clocks, although a spring-driven musical clock had been made by Fromanteel as early as the 1660s.

The heyday of chiming and musical longcase clock production in London was between 1750 and 1780 when a relatively large number were made, many for export, by such eminent makers as Eardley Norton and Ellicott. However, the manufacture of these clocks continued in the country until the 1850s.

Considerable confusion exists between *striking, chiming* and *musical* clocks and thus it may help at this stage if we try to define them.

When a clock sounds only the hours it is known as *hour striking*. The word *hour* is often omitted from the term. When it sounds the quarters on one or two bells, it is called *quarter striking*. When it sounds the quarters on four or more bells, most commonly six or eight, it is known as *quarter chiming*. The word *quarter* is often omitted.

A clock which plays a tune at the hour, usually after it has struck, is called a *musical* clock. Usually it will play a single verse every hour or three verses every three hours. However, confusion may still occur because in the 17th and 18th centuries the term *chime* was applied to a musical clock, and for this reason a chime/silent lever will frequently appear on a musical clock, an apparent contradiction in terms.

Quarter-chiming and musical clocks generally have three trains, one for the going, one for the striking and one for the chimes. Sometimes, however, there are only two and in these instances the last quarter chime (i.e. that on the hour) is generally omitted. The mechanism used to achieve this is known as "pump-action quarter chime." In this the pin barrel activating the quarter chime is moved (pumped) over prior to the hour so that the chiming pins do not engage their hammer tails. Instead, pins are brought into operation which operate the hour bell.

A similar technique is occasionally used on three train clocks to provide both music and quarter chime (Figure 233), but again because of the pump action the fourth quarter has to be omitted.

Occasionally four driving weights are used on longcase clocks, one each for the time, hour strike, full quarter chime and musical trains of the clock.

In the majority of early clocks, prior to circa 1710, only a small number of bells were used to sound the quarters and thus they are really quarter striking clocks. After this period quarter chiming became much more popular. This was sometimes on four or six bells, more commonly on eight, and occasionally ten or 12 were used for the better clocks.

Many quarter chiming and musical clocks have been upgraded and changed over the years so that it is now common to find a 17th-century longcase clock made with quarter striking on two bells converted, only ten to 20 years after it was made, to quarter chiming on six to eight bells.

Musical clocks were similarly upgraded by the music barrel being repinned to play popular music of the day rather than music which had gone out of fashion 100 years before. It is not uncommon for a quarter-chiming clock to have been converted to a musical one or vice versa.

Sometimes a third train has been added to a normal striking clock, making it either musical or chiming. This usually is obvious because of the irregular position of the winding squares in the dial due to the alteration of the movement.

In Victorian times the center of the dial was sometimes cut out and replaced with a new dial center complete with a chiming or musical movement, thus overcoming the problem caused by existing winding holes being in the wrong place for a new movement.

The installation in the mid-19th century of the chiming clock popularly known as Big Ben in the clock tower of the Palace of Westminster (where the two Houses of Parliament meet) produced a great surge in the demand for chiming clocks, particularly those with Westminster chimes.

Until this time all chiming clocks had bells, although a few organ clocks had been produced in the 18th century with wooden or metal pipes. However, from the 1850s onward sets of coiled gongs which produce a deeper and, some would say, a more melodious note became increasingly common.

A popular combination in some of the better clocks was a choice between a chime on eight bells or Westminster chimes on four gongs. A single gong was also frequently used for the hour, even when bells were used for the chimes (possibly to provide a better contrast to the bells).

Towards the end of the 19th century, straight rod gongs were sometimes used in place of the coiled variety. It was at around this time also that the ultimate in chiming clocks was produced, which made use of tubular bells graduated from two to five feet long. Undoubtedly these simulated the beautiful chimes of church bells more closely than anything which had been used previously (Figure 223).

Figure 221 A, B, and C. *Quarter-Chiming Clock by William Lee, Leicester.* A well-proportioned, walnut flat-topped clock, probably originally with a caddy. It has cross grain mouldings throughout the case, frets above the arch, applied pillars to the hood door, and a flat-topped trunk door, all features which would suggest a date around 1725. The base panel is crossbanded and rests on a double plinth.

The 12-inch brass breakarch dial has female mask spandrels in the corners and dolphins in the arch. There are fleur de lis half-hour marks, and quartering on the inner aspect of the chapter ring. The dial center is well matted and the center of the seconds ring features an engraved star. Immediately below the name plaque is the regulation for four-or eight-bell chimes.

To the left of the movement may be seen the pin barrel and in front of it, and just visible to the left of the plate, is the lever which pumps it over to change it to the second chime. On the backplate the leaf spring is mounted, which returns it to its original position.

The long back cock provides alternative positions from which to hang the pendulum should the case be leaning too far forward or back. 7′ 3″ (221 cm) high.

## Quarter Chiming and Musical Clocks 167

## 168 Quarter Chiming and Musical Clocks

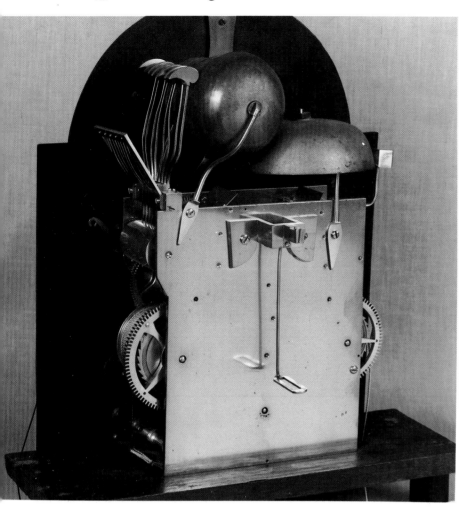

Figure 222 A, B, and C. *George III Quarter-Chiming Longcase Clock by John Taylor, London, 1775.* A particularly slim and elegant quarter-chiming mahogany pagoda-topped longcase clock of classical design with brass capped and strung quarter columns to either side of the trunk, a panelled base and a double plinth.

The fine quality and very substantial three-train movement chimes the quarters on eight bells and is fitted with a repeat. The 12-inch dial has strike/silent regulation in the arch, a recessed seconds ring, and a date aperture above 6 o'clock. It is signed on a semicircular silvered strip across the bottom half of the matted center. The clock has brass cased weights and the pendulum has a flat brass rod and a substantial brass-faced pendulum bob. 8′ 1″ (246 cm) high excluding tip finial.

## Quarter Chiming and Musical Clocks 169

Figure 223 A, B, and C. *Nine-Tube Chiming Mahogany Longcase Clock.* These massive clocks, which were made to the highest standards, must have been extremely expensive to produce, as just about everything was incorporated in them which was felt desirable at the time. Thus they were smothered in carving, much of it of fine quality, and decorative floral panels were put into the base.

The sides and front were glazed to display the nine tubular bells (often nickel plated) for the quarter chime and the hour strike, the brass-capped weights, and in this instance the mercury-compensated pendulum. The movements are very heavy. A large pin barrel runs parallel to the backplate and activates the large felt-covered hammers for the tubular bells. These are by no means easy to set up, but once done correctly, they sound superb.

The shaped brass dial has a silvered chapter ring with raised, individual gilt numerals; spandrels applied over most of the dial, and regulation in the arch for strike/silent, chime/silent, and a choice of three different chimes. 9′ 10″ (274 cm) high.

# 170 Quarter Chiming and Musical Clocks

## MUSICAL CLOCKS

Figure 224 A and B. *Musical Clock by Francis Dorrell, London, Circa 1720*. This relatively early musical movement has a 12-inch brass dial with all the decorative features of a clock made around this period such as wheatear engraving to the border, fine engraving around the subsidiary dials in the arch and the dial aperture, ringed winding holes, and a chapter ring with half-quarter and half-hour marks together with quartering on the inside of the chapter ring. Note the fine, delicate hands.

The dial to the left is inscribed with the names of the four tunes it plays: *God Save the King, Rule Britannia, March Rinaldo,* and *O'Jenny, O'Jenny*. In the center is a year calendar and the signs of the Zodiac. The dial on the right is marked *It Will Not Chime/It Will Chime*, the use of the word *chime* on musical clocks being common at this period.

The substantial movement has 14 bells and 14 hammers with a large pin barrel mounted back to front as was the custom on the earlier clocks. Its open construction is similar to that seen on some organ clocks.

Opposite page:
Figure 225 A, B, and C. *Walnut Musical and Calendar Longcase Clock by Robert Hall, Chichester, Circa 1730*. A quite remarkable clock by the previously unrecorded maker Robert Hall of Chichester. The only reference to a Robert Hall is contained in Baillie (Vol. 1) where he is noted as working in Drury Lane, London in 1743. Perhaps this is the same man and he moved from Chichester to London.

The fine quality case shows the early development of the pagoda top, at this stage quite shallow. The moulds to the top of the hood are well developed, there are sound frets to the front of the pagoda, and the side of the hood and the free standing columns are brass strung. The case is inlaid inside the crossbanding on the trunk door and also on the base.

The musical movement represents the final development of the pin barrel running at right angles to the plates. It is already protruding some 2½ inches beyond the backplate and is almost touching the backboard. Indeed, on some of the early musical clocks the backboard is seen cut away to make room for the pin barrel. Eighteen bells and 18 hammers are used.

Hall imposed on himself an added complication in the provision of a center-sweep seconds hand, presumably because there was no place on the dial to put a seconds ring. The cutout to permit removal of the pallets may be seen in the center of the backplate.

Center-sweep seconds hands were particularly popular on clocks made in the North Country after 1750, but already had appeared on some of the finer London longcase clocks by 1715-1720. Williamson, for instance, made use of them.

The arched dial, somewhat in the Dutch style with the chapter ring moved up into the arch, is both complex and beautifully executed with wheatear engraving to the border and floral engraving around the subsidiary dials. The dial is so filled that there is room for only two half spandrels.

In the matted dial center are subsidiary dials for chime/no chime and music selection, the choice being either *Black Joak, White Joak,* or *Bright Aurilia*. Below is a lunar calendar disc. To the bottom left of the dial is a year calendar, on the right the days of the week, and in the center a beautifully engraved rotating disc giving the days of the week and their ruling deities.

# Quarter Chiming and Musical Clocks 171

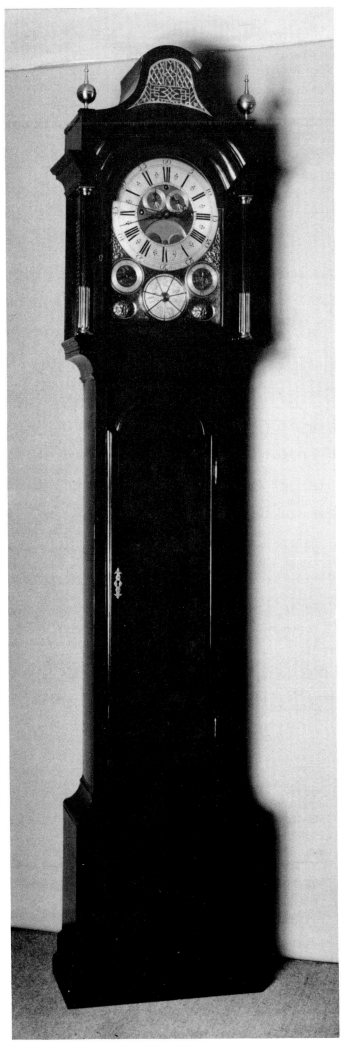

## 172  Quarter Chiming and Musical Clocks

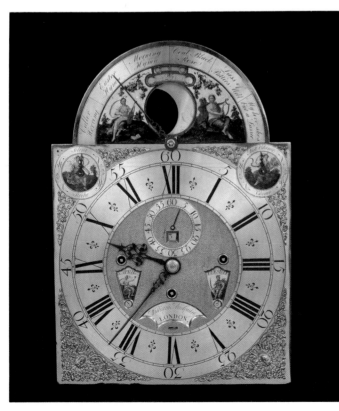

Figure 226 A, B, and C. *Musical Clock by Audouin, London, Circa 1735.* A musical clock of outstanding quality and in remarkably good condition aided no doubt in some degree by the fact that it has been in store for 60 years. However, probably of more relevance is the superb quality of its construction. A good example of this is the way in which the door is made—panelled and then backed to reduce any chance of its distorting to a minimum. It has a very heavy and beautifully shaped deep mould at its edge and crossbanding within. The figuring of the wood is the best that could be obtained.

The case has a breakarch top with frets below and to the sides of the hood, panelling to the sides of the trunk, and a raised and panelled base which rests on a double plinth. A door to one side of the hood gives access to the massive seven-pillar movement which has very heavy and attractively shaped plates and a ten-inch long pin barrel running parallel to it with 25 hammers playing on 13 bells.

Placing the pin barrel parallel to the backplate was a major advance in musical clock construction, permitting a much longer pin barrel with many more pins spread over its greater length. More hammers and bells also could be accommodated. Since two hammers are used on all but one of the bells, a note could be repeated far more rapidly than if a single hammer were used. These two advances improved the music produced and signaled the ultimate development of the musical clock.

The 14-inch brass breakarch dial has a raised chapter ring with hour and half-hour divisions and quartering on its inner aspect. In the bottom corners are beautifully executed mask spandrels. To the top left is a strike/silent dial and to the top right chime/silent. In the center of these rings are well-executed pictures of girls in country scenes. Below 12 o'clock is a seconds ring in the center of which the date is shown. The date can be set without removing the hood by inserting a pin in a slot above 6 o'clock. Segmented apertures flanking the dial center show respectively the day of the week, the ruling deity, the month of the year, and the appropriate zodiacal sign. These are also decorated with attractive paintings. In the arch is a center-sweep tune selector for six different tunes: *Caller Herring, Easter Hymn, Morning Hymn, Coal Black Rose, Lass of Patties Hill,* and *My Love She Is But a Lassie.*

The moon's phases appear in the arch and above are the lunar calendar and the time of high water at London Bridge. The whole of this is surrounded by a beautiful picture showing a boy and girl playing musical instruments. The clock is signed by William Audouin of London and dates circa 1735. 8' (268 cm) high.

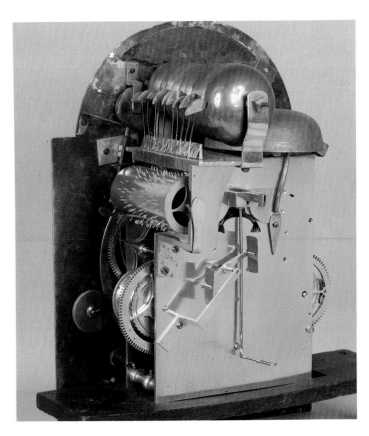

Figure 227 A, B, and C. *Musical Longcase Clock by Farley, London, circa 1780-1785.* This clock has what was undoubtedly considered in its time (and still probably is) the finest of London mahogany cases. The Cuban mahogany veneers are of the best quality and brass-strung quarter columns are used not just on the trunk but also on the base. This has a shaped and raised panel and sits on a beautifully shaped and moulded plinth, a most unusual feature.

The substantial three-train movement was converted sometime in the 19th century to play two verses of *Auld Lang Syne, Annie Laurie, Home Sweet Home* or just chimes at each hour. There are separate repeat cords for the hours and the music. The fine 12-inch dial has strike/silent and tune selection hands in the arch. After removing the ring for the tune selection (which is obviously engraved in the Victorian style), the names of the tunes which the clock played originally can be seen on the back. The barrel has been repinned to play music of the period when it was converted. Most Georgian tunes played on musical clocks are unrecognizable to us today.

Until around 1775-1780 nearly all London-made clocks had matted centers but after this engraving was sometimes used instead. This is a particularly good example of fine floral engraving to all three dial centers. Note that the seconds ring is engraved on the dial plate also.

## 174  Quarter Chiming and Musical Clocks

Figure 228 A, B, and C. *A Musical Clock by Edward Gatton, London.* This clock represents the classic, fully developed musical longcase clock produced in London circa 1760-1780. It has narrow chapter and seconds rings, with no quartering or half-hour marks, finely matted dial centers, floral spandrels, and chime/silent (for the music) and strike/silent regulation in the arch. A long hand sweeping across the arch is used to select the tune: *Easter Hymn, Westermen, March, Hornpipe, Air, Langs Lee,* or *Minuet.*

The fine quality movement has 12 bells and 23 hammers with the pin barrel extending well beyond the plates on both sides. Because of the lack of space due to the bell assembly, the fly has been moved to the lower part of the backplate. The mechanism linked to the center-sweep selection hand, for moving the pin barrel and changing the tune, can be seen to the left.

The case is typical of the best produced at that time. It has reeded and brass strung columns to the hood, quarter columns to the trunk, an attractively shaped trunk door with a well-executed mould around it, a panelled base, and a double plinth. It is possible that the finials predate the clock by about 30 years. 7′ 9″ (234 cm) high.

# Quarter Chiming and Musical Clocks 175

Figure 229 A, B, and C. *Musical and Quarter Chiming Clock by Eardley Norton, London, circa 1780.* A fine clock by the eminent maker Eardley Norton who was one of the best known manufacturers of musical clocks in the second half of the 18th century and was also renowned for his astronomical clocks. In Buckingham Palace is one he made for King George III. He made a beautiful musical clock for the Empress Catherine, and various other fine and complex clocks by him are in museums and private collections.

This clock has a particularly fine and decorative case with a breakarch hood and an embossed brass sound fret immediately below. There are brass frets to either side of the hood and brass-capped fluted columns inlaid with brass. The attractively shaped trunk door and case are made of beautifully faded mahogany and it has reeded and brass inlaid quarter columns with brass capitals. Immediately beneath each of these is a panelled block with satinwood cross banding. The base is panelled and has a fine inlay in the four corners which are repeated on either side at the top of the trunk.

The substantial three-train, six-pillar movement, somewhat simpler than that seen in Figure 226, chimes the first three quarters on two bells, it strikes the hours on one bell, and plays one of six tunes at each hour (or at will) on nine bells with 14 hammers.

The lovely 12-inch dial has chime/silent regulation to the top left and music selection on the other side. Between them is a silvered sector bearing Eardley Norton's signature and the number of the clock, *2450*. There is a narrow chapter ring with inner Roman and outer Arabic numerals.

It has a recessed silvered seconds ring and above 6 o'clock a date aperture. Interestingly, the spandrels are in the Chinese style reflecting the strong Oriental influence at this period as publicized, for instance, in Chippendale's *Director*.

## 176 Quarter Chiming and Musical Clocks

Figure 230 A, B, and C. *Musical Clock by Benniworth, St. Albans.* A fine musical longcase clock contained in a typical George III, "London" mahogany case with double plinth, a panelled base, reeded and brass inlaid quarter columns to either side of the trunk with brass capitals, and a shaped trunk door with an attractive mould around it.

The hood has brass capped columns with brass stringing, glazing to either side, and a side door for access to the movement. The pagoda top which has a fret to the front and finials to either side has been made removable to reduce its height when required.

The very substantial seven pillar movement has a large barrel 11½ inches long and 2 ½ inches in diameter which runs across the backplate of the movement. It plays eight tunes on 12 bells with 24 hammers and has an additional bell and hammer for the hour.

The 13-inch arched brass dial has a raised chapter ring and spandrels, recessed second ring, and a large engraved cartouche in the center enclosing both the date aperture and the signature of Thomas Benniworth. St. Albans is approximately 30 miles north of London. He is recorded as having become a member of the Clockmakers' Company prior to 1773.

There is a rotating moon disc in the arch and around its periphery are the names of the tunes selected by the center-sweep hand. The tunes are, from left to right: *Minuet, Gavot, March, Highland Laddy,* and *The 104th Psalm.* It plays three verses of the chosen tune at every third hour. To the top left of the dial is regulation for strike/silent and to the top right, chime/not chime.

The clock has three contemporary brass-cased weights.

Minor differences from typical London work include the plain, as opposed to matted, dial center and the decorative cartouche around the name. They indicate a clock produced close to London.

## Quarter Chiming and Musical Clocks 177

Figure 231 A, B, and C. *Musical Longcase Clock in the Chinese Chippendale Style by Rayment of Stamford, Circa 1785.* An extremely rare musical longcase clock by Thomas Rayment of Stamford, 1780-1790, in a Chinese Chippendale case similar to an illustration in the *Director*.

The case, made of oak and veneered in mahogany, is of the finest quality and must have been both time consuming and expensive to make. Nothing is straight; everything is tapering, curved or rounded. There are frets on either side of the hood and one of these can be opened to give access to the fly of the musical train and thus adjust its speed. The finials are in the form of pagodas and the hood pillars resemble bamboo with beautifully executed gilt wood capitals.

The eight-day, three-train movement has heavy plates and five ringed pillars. At each hour, besides striking, it plays one of four tunes on 11 bells with 14 hammers. The 12-inch arched brass dial has a raised chapter ring and scroll spandrels with a delicately painted scene of five minstels playing and a cherub flying above them in the arch. 7' 8" (248 cm) high.[2]

---

[2] Tebbutt L. *Stamford Clocks and Watches,* Private Publication, Stamford Lincs, 1975.

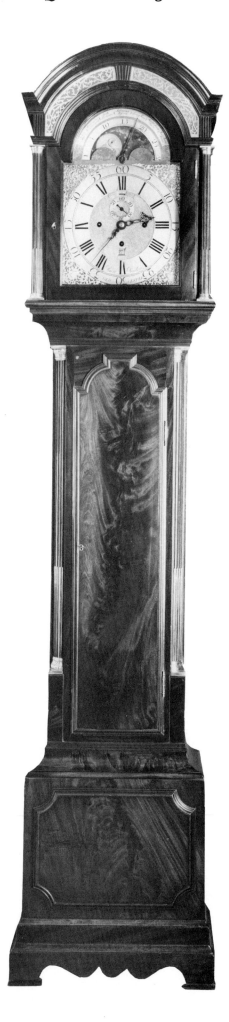

Figure 232 A, B, C. *Musical Clock by Peyton of Gloucester, circa 1765.* A fine musical clock similar to those produced in London at the time but having some marked differences.

The case is of typical London quality with its shaped door and quarter columns and it may well have been made there but the shape of the hood, almost arched at the top, is not quite standard.

The movement, with a long pin barrel with 12 bells and 20 hammers, is different from London practice in that each bell has its individual standard as opposed to all of them being mounted on one rod with spacers between. A further difference is the position of the fly to the left of the barrel.

The 12-inch dial is a little more attractive than contemporary London dials with a beautifully executed moon disc and cloud scenes painted on it and a lunar calendar above. There is a center-sweep hand in the arch for the tune selection which comprises *Emperors Hymn, Jenny Jones, Auld Lang Syne,* and *Home Sweet Home.* There is music/silent regulation below 12 o'clock.

Richard Peyton, an eminent maker, was born at Sandhurst near Gloucester in 1718. He retired and moved to Shepton Mallet in 1774 and it was here where he died eight years later. In Gloucester City Museum is a green lacquer longcase clock of year duration by him that has moonphases in the arch. An extract from the *Gloucester Journal* of 1743 (reproduced in *Gloucestershire Clockmakers*) is given below:

GJ October 25, 1743

**STOLEN** out of the Shop of Richard Peyton, Watch-Maker, in Gloucester, between the Hours of Six at Night on Tuesday the 11th Instant, and Seven the next Morning.

Five Silver WATCHES

viz. One new watch, with a plain Dial-Plate, and plain Middle-Piece; on the Back-Plate, Richard Peyton, Gloucester;' the Verge playing in Diamonds; worth Seven Guineas. One old watch having a scollop'd Dial-Plate; the Name on the Middle-Piece, Dix, London; having on, when lost an old Leathern Strap. One ditto, large-sized; the Name on the Dial-Plate, Washbourn, Gloucester; on the Back-plate, John Washbourn, Gloucester. One ditto, large-sized and old-fashion'd; the Name Andrews, London; with a very small Button sunk in below the Case. And one very small siz'd, Silver-cased, old Ballance Watch, with Chain, very long Hour Figures, and the Quarter Divisions above the Hours; having an old faded Pink-colour'd Riband.

Whoever gives Notice of the said Watches to the aforesaid Richard Peyton, so as they may be had again, shall have Five Guineas Reward: Or whoever will discover the Person or Persons that stole the same, so as he or they may be apprehended and convicted thereof, shall receive Three Guineas Reward. Witness my Hand, Richard Peyton.

## Quarter Chiming and Musical Clocks

N.B. If offer'd to be pawn'd or sold, be pleased to stop them and the person making such Offer, and upon Notice thereof, given as above, you shall likewise be handsomely rewarded.

The said Watches are suspected to be stolen by one John Edwards, a Scotchman, who lately work'd with the said Richard Peyton. The said John Edwards is about five Foot nine Inches high of a fair complexion, with full Grey Eyes, somewhat lean visaged, stoops in his Shoulders, and is about 30 years old; having on, when he went away, a white Wig, a Light-colour'd Surtout Frize Coat, a blue Waistcoat and Leather Breeches.

GJ August 13, 1751

A sober man, that is a good workman at the CLOCK-MAKING business, may have constant Employment at RICHARD PEYTON'S watch maker and goldsmith, in Gloucester.[1]

---
[1] Dowler G., *Gloucestershire Clock and Watch Makers,* Phillimore and Co., Ltd., Chichester, 1984. This extract gives us a little more insight into the life and times of Richard Peyton.

## 180 Quarter Chiming and Musical Clocks

Figures 233 A, B, and C. *Musical Clock by Ferguson, Edinburgh, circa, 1775*. A fine musical longcase clock by Alexander Ferguson of Edinburgh circa 1770, contained in a case equivalent in quality to the best in London at that time and possibly of slightly slimmer and more elegant proportions. However it has two differences: a plain (instead of a panelled) base, and a scalloped border to the pagoda top.

There are further frets to the side of the hood which has fluted and brass strung columns with brass capitals. The trunk door has a breakarch top, a well-executed mould around its periphery, and brass-strung quarter columns to either side. The substantial movement, which has brass-cased weights, chimes the first three quarters on eight bells and at the hour plays on 14 bells with 14 hammers one of the following tunes: *Balance a Straw, Fools Minuet, Lucky in the Kitchen* or *My Love Lies on the Cold Ground*. Most unusually, it may also be set to chime just the four quarters at the hour.

In the arch of the brass dial is a center-sweep hand for the tune selection with rings for music/silent and chime/silent. Within and around these is floral engraving. The solid engraved and silvered chapter ring has Roman hour and Arabic minute numerals with a large seconds ring below 12 o'clock. 8′½″/7′ 8″ (245/234 cm) high with/without top finial.

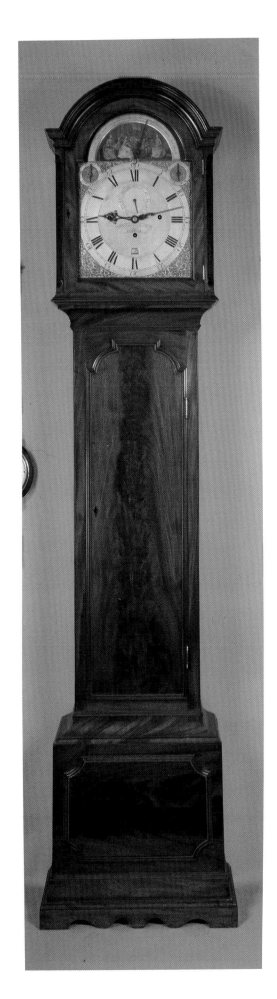

Figure 234 A and B. *Musical Clock with Automata by Samuel Smith, London, Circa, 1770.* The case of this clock is similar to those used by Dutton, Holmes, etc. in that it has canted and reeded corners to either side of the hood instead of pillars. There are glazed apertures to either side, the trunk door is shaped from well-figured wood of good color and patina, and the base is panelled and stands on a double plinth. The six-pillar movement plays one of six tunes at each hour.

The brass breakarch dial has a large seconds ring and a date aperture. To the top left of the dial is regulation for strike/silent and to the right chime/not chime. In the arch is the tune selection lever and a particularly attractive feature is the beautifully painted scene with automata playing the harpsichord, violin and lute.

## 182 Quarter Chiming and Musical Clocks

Figure 235 A, B, and C. *Longcase Clock by Thomas Ashton, Macclesfield, circa 1775.* A fine quality longcase clock typical of those being produced in this part of the country in the last quarter of the 18th century. The swan neck more usually is seen with a boxed top. Because it is a musical clock, fretwork has been used in front instead of the usual painted glass, to let the sound out. There are applied frets to the top of the trunk, quarter columns to either side, and beautifully selected, book-matched veneers on the door and base.

The relatively simple movement, quite different from those made in London, has a selection lever for chime, psalm or dance at the hour. It plays on eight bells with 15 hammers. There is also a music/silent lever which may be seen in the rear view of the movement to the right and appears through the 14-inch dial near 10 o'clock. This has a solid (i.e. filled-in center) chapter ring which is attractively engraved with center-sweep minute, hour and date hands. There are rococo spandrels in the four corners and in the arch is a rolling moon with a scene painted on it. The signature is carried on an engraved silvered strip around the top edge of the dial. Thomas Ashton is recorded as working from 1760-1795. 7′ 11″ (241 cm) high.

Chapter 9.
# Brass-Faced Oak Eight-Day Longcase Clocks

**THE OAK LONGCASE**

Prior to 1700-1710 the vast majority of longcase clocks were made in London and towards the end of this period in other major cities such as Chester, York and Exeter. Nearly all were contained in fine ebonized, walnut, olivewood or marquetry cases and thus an eight-day oak longcase clock made prior to 1700 is a very rare piece.

Oak cases were sold by London makers only when they were supplying a clock to be used "below stairs," i.e. in the servants' part of the house, where a fine case would not be required. This occurred quite often at a later period with longcase regulators.

Occasionally an early eight-day oak longcase clock appears (Figures 236 and 237) which was made in the country. These are usually small, of pleasing proportions, similar to the London clocks, and made of good quality oak. However, it is not easy to give their exact date of manufacture because London styles were slow to spread into the country.

By 1720 oak eight-day longcase clocks, usually with a 12-inch or occasionally an 11-inch square brass dial, were being made in most parts of the country and indeed continued to be made for the next 60-70 years. The majority of the cases were of simple construction and between six feet six inches and seven feet in height. Sometimes a caddy top was added and this style continued for far longer than on the London clocks, well into the 1770s. A lenticle was also featured on some of the early clocks and star parquetry is occasionally seen on the trunk door and occasionally the base.

By the mid-18th century mahogany frequently was used as well as walnut, to decorate the plain case on the hood, and for crossbanding, for instance around the door and base.

Early oak cases tend to be heavy and have an excellent patination, however, later examples are lighter in construction and lack the lovely warm color.

Oak breakarch longcase clocks appeared in the country in the 1720s, but never displaced square-dial clocks as they did in the capital. Breakarch cases were taller than those with square dials and became more elaborate as the century progressed. Some styles, such as the pagoda top, originated in London, but others were peculiar to a particular region.

By the 1760s the all-over engraved and silvered brass dial appeared, although never in large numbers, and primarily in the South. Painted or white dials also appeared around this time and by 1790 had largely superseded brass dials with raised chapter rings and spandrels. However, some of these were still being produced even as late as 1850.

## SQUARE DIALS

Figure 236. *"Edward Wilcox, fecit" (1690-1700)*. The inscription on the dial unfortunately gives no indication of where the clock was made and, as the maker is unrecorded, there is no way in which we can find this out. The carving at the top is most unusual for a longcase clock being reminiscent of certain items of 17th century furniture. The spiral twist columns are attached to the door, there is a convex mould beneath the hood, and a lenticle in the door. The trunk is only 10½ inches wide and the hood originally slid upwards.

The five-pillar movement has count wheel strike and a ten-inch dial with late cherub spandrels. It runs quite happily on a pair of six pound lead weights. 6' 6" (198 cm) high.

## 184  Oak Eight-Day

Figure 237. A slim but very heavy oak eight-day longcase clock with a lenticle in the door and a well-developed mould to the top of the hood, as often seen on early country longcase and wall clocks. The ten-inch dial is similar to those made in London at the time with twin cherub and crown spandrels, engraving between, ringed winding holes, a cartouche around the date aperture, and half-hour marks, circa 1705.

Figure 238. *James Thorn, Colchester.* This clock still has several early features such as the lenticle in the door and the half-hour marks, but would probably dates from circa 1740-1745. The hands are later. 6' 7" (200 cm) high.

Figure 239. *Wm. Anthony, Truro, circa 1750.* This straight-grain walnut longcase clock has star parquetry on the trunk door and base, and stringing around them and the hood. The 12-inch dial is beautifully executed, the center containing fine engraving and showing the moon's phases above 6 o'clock. William Anthony was born in 1688 and died in 1768. 6' 8" (203 cm) high.

Figure 240. A simple country oak eight-day clock, circa 1730. 6' 5" (196 cm) high.

Figure 241. *Bagnell at Talk, circa 1740.* A straight-grain walnut clock with crossbanding to the door and the base. The 12-inch dial has decorative half-hour marks and a ring of engraving inside the chapter ring. 6' 9" (206 cm) high.

Figure 242. A simple oak eight-day longcase clock, circa 1740, with walnut crossbanding.

*186 Oak Eight-Day*

Figure 243. *Nathaniel Hedge, Colchester, circa 1770.* A very pretty pale oak eight-day clock with walnut crossbanding to the door. Note the very late use of ringed winding holes on the otherwise plain 11-inch square brass dial.

Figures 244 and 245. *Left:* Oak eight-day clock by John Oliver, Manchester, who ceased trading in 1749. It has freestanding pillars to the hood, half-hour marks on the dial, and walnut crossbanding. 7′ 2″ (218 cm) high.

*Right: William Nash of Bridge (near Canterbury), circa 1770.* Note the rudimentary pagoda top, now somewhat reduced. 6′ 8″ (203 cm) high.

Oak Eight-Day 187

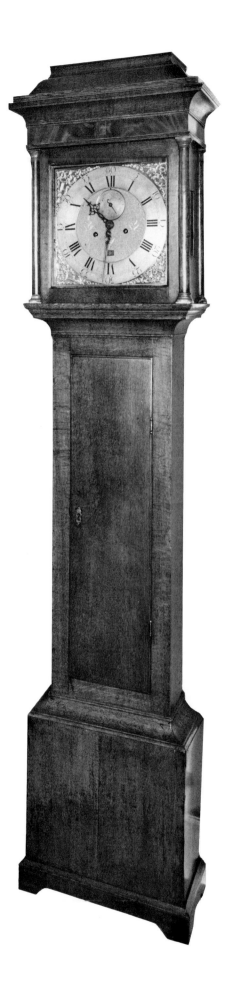

Figure 246. *John Snelling, Alton, circa 1780*. Note the comparatively late use of the caddy top. 7′ 2″ (218 cm) high.

Figure 247. *Massingham, Fakenham, circa 1780-1790*. Towards the end of the 18th century, various different forms of decoration, such as that seen here, were mounted on top of the clock. Although unusually deep plinths are often replacements following damage to the base, this one may be original. Note the simple dial layout with engraving to the center.

188  Oak Eight-Day

Figure 248. *Shapley of Stockport, circa 1750-1760*. This has a 12-inch dial with fine engraving to the center and half-hour marks which continued in the North to circa 1760-1770. Note the delicate hood pillars. 6′ 11″ (211 cm) high.

Figure 249. *Stephen Cranbrook, Dover, circa 1780-1790*. Note the continued use of the caddy top together with the plain dial center with decorative engraving and a recessed seconds ring. The chapter ring, and possibly the dial center, would originally have been silvered. 7′ 0″ (213 cm) high.

Figure 250. *J. Whitern, Abingdon (near Oxford), circa 1790.* A slim, well-proportioned clock with plain dial center except for the signature.

Figure 251. *Samuel Harley, Salop (Shrewsbury), circa 1790.* This clock has a center-sweep date hand, crossbanding to the shaped door, and a raised base panel. 7′ 1″ (216 cm) high.

Figures 252 and 253. *Left: John Christian, Aylesham,* Active 1745-1788.

*Right: Charles Cooper, Whitchurch, circa 1770-1780.*

Figure 254. *George Lambert of Blandford, circa 1780.* This has a fine, pale oak case with an attractively shaped and crossbanded door and a very well executed dial with crosshatched engraving in the center, a recessed plaque for the name, a decorative seconds ring, and center-sweep date hand.

## Oak Eight-Day

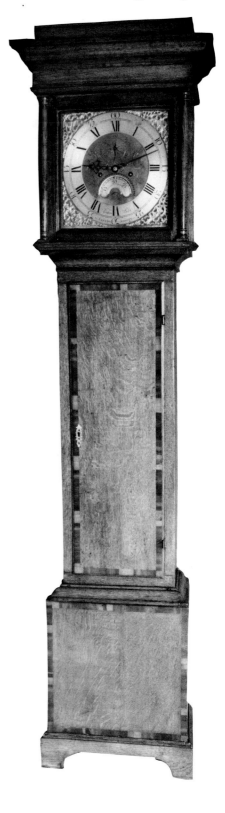

Figure 255. Well-proportioned oak longcase clock by Agar of York, circa 1790-1800. 7′ 1″ (216 cm) high.

Figure 256. Oak eight-day clock, circa 1790-1800. 6′ 8″ (203 cm) high.

Figure 257. *James Pike, Newton Abbott (Devon), circa 1785.* It has an oak case with walnut crossbanding and a large date aperture. Whereas on a London clock applied pillars to the hood door ceased around 1735, on country clocks they continued, as well as freestanding pillars, until the 19th century.

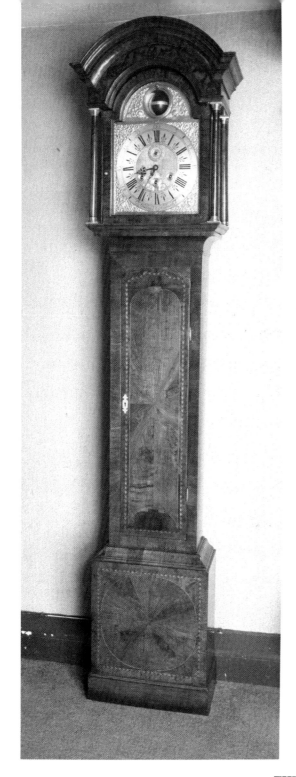

## THE BREAKARCH DIAL

Figure 258 A, B, and C. *Quarter-chiming clock with rotating globe moon, circa 1750.* Although London longcase clocks are almost invariably of fine quality, they tend at any one period to be very stereotyped and thus in some ways uninteresting. However, that cannot be said of the work of the best of the country clockmakers who sometimes produced clocks of excellent quality and great originality. These are invariably much rarer than the standard, London-made longcase clocks.

An excellent example is this superb three-train longcase clock by George Booth of Manchester who is recorded working from 1758 to 1788, although this clock is probably somewhat earlier. The substantial movement, which chimes the quarters on four bells, has a beautifully executed 12-inch brass dial with that rare feature a rotating "Halifax" globe moon recessed in the center of the arch. One side of the globe is silver and the other is deep blue and it is surrounded by a band of wheatear engraving with dolphin spandrels on either side.

The whole dial has wheatear engraving in the border and a further strip below the breakarch. The raised chapter ring has bold fleur-de-lis half-hour marks and quartering on its inner aspect. The dial center is matted and there is a date aperture above 6 o'clock which is surrounded by a well-executed floral cartouche. The winding holes are ringed and mask spandrels used.

The highly individual parquetry case is of very good proportions and excellent quality. All the oak moulds are cross-grained. The top of the hood has a breakarch with a fret below, the pillars are freestanding, and the sides are glazed. The base has a border of cross-grain veneers, a band of wheatear inlay, and a circular panel composed of 16 fan-shaped veneers. The long, slender trunk door is similarly decorated with walnut panels to the top and bottom, a band of herringbone inlay, and cross-banding to the border. The sides of the base and trunk are panelled.

Figure 259. *Sampson of Wrexham, circa 1745-1750.* The oak case is typical of those produced in the Cheshire area at this period. It has lightly ringed winding holes, strike/silent regulation, and walnut cross-banding. 7′ 2″ (218 cm) high without finial.

Figure 260. *Phillip Avenall, Farnham, 1760-1770.* An oak clock case similar to walnut ones which had gone out of fashion in London 30 years earlier. The dial, except for the engraving to the center, is like those produced in the capital in the 1760s. Farnham is 40 miles West of London. 7′ 1″ (216 cm) high.

Figure 261. Pridgen, Hull, circa 1780. The pagoda-top gained favor around Hull, but was usually more pronounced in shape than its London counterpart and the trunk was slimmer. 7′ 1″ (216 cm) high.

## 194  *Oak Eight-Day*

Figures 262, 263, and 264. *Three Oak Pagoda-topped longcase clocks.*

Figure 262. *Thomas Sutton, Maidstone (Kent)*

Figure 263. *Daniel Cornwell, Billericay (Essex)*

Figure 264. *Mathews, Leighton (Essex)*

These three clocks, all made within 40 miles of London, emphasize how strongly the capital influenced clock case design in the Home counties. They are based on the standard London mahogany pagoda-topped style but are simplified by being made of oak, are somewhat smaller and probably all were originally surmounted by three wooden finials as opposed to brass ones used in London. Figure 262 has a double plinth while on the others the second tier has been omitted. Each has a five-pillar movement, one with a strike/silent regulation in the arch, and each has a dial engraved in the center instead of matted as used in London. Two retain the mould around the trunk door, but on the Mathews clock (Figure 264) it has been omitted and the pagoda has at some time been reduced. The clocks date circa 1775-1790.

## Oak Eight-Day 195

Figure 265. *Wm Mayhew, Woodbridge (Suffolk)*
Figure 266. *Joseph Trattle, Newport (Monmouth)*
Figure 267. *Unsigned*

Figures 265, 266, and 267. Three small oak breakarch longcase clocks typical of the southern half of England from circa 1760-1785. One has a brass dial with a raised chapter ring and the other two have flat engraved brass dials. The one by Trattle (Figure 266) is silvered while the other has been left polished brass. The hands on the clocks in Figures 265 and 266 are later replacements and the bases may have been reduced a little on Figures 265 and 267.

## THE SWAN-NECK PEDIMENT

The use of swan-neck pediment was probably first seen in the 1750s, and it soon its use had spread over much of the country. It assumed different forms in different areas, the shapes used in Bristol, Manchester and Edinburgh, for instance, having their own characteristics (see Chapter 15 for regional characteristics).

Although used mostly on clocks with breakarch dials, swan-neck pediments also are found with square dials and on clocks of 30-hour as well as eight-day duration. It continued to be used on white-dial clocks after the brass dial went out of fashion and was still being made in the 1860s.

Figure 268. *R.Bruce, Wigtown, circa 1790.*

Figure 269. *James Smyth, Saxmundham (Suffolk), circa 1800.*

## Oak Eight-Day 197

Figure 270. *James Nicol, 1 Cannongate (Edinburgh), circa 1755.* This example has a long, slim, shaped door and fretwork below the swan neck. 7' 8" (234 cm) high.

Figure 271. *Thomas Elkins, Barslem, circa 1815.* The short trunk door with a panel below was introduced around 1810. The flat, engraved brass dial was originally silvered.

## Oak Eight-Day

Figure 272. *Richard Boyfield, Great Dalby, circa 1765-1770.* This clock has a long, shaped, crossbanded door, brass-capped quarter columns, and a panelled base. 7' 9" (237 cm) high.

Figure 273. *James Berry, Pontefract, circa 1780.*

Figure 274. *John Russell, Falkirk, circa 1800-1810.* Note the extensively inlaid case, a feature which became popular after around 1800.

## Oak Eight-Day 199

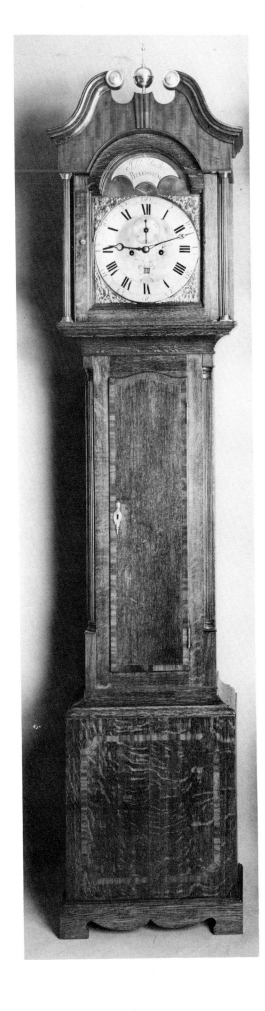

Figure 275. *John Smith, Pittenweem, circa 1790.* There is a solid, silvered chapter ring (i.e. it is not cut out in the center). This feature seems to have appeared soon after the introduction of the painted dial.

Figure 276. A most unusually shaped, engraved and silvered brass dial with center-sweep seconds hand and moon phases at the top. It is a little reminiscent of the oval painted dials used by a few makers such as Noon of Ashby.

200  Oak Eight-Day

Figure 277. *Edward Smith, Newark, circa 1790.*

Figure 278. *Allan Fowlds, Kilmarnock, circa 1800.*

# Chapter 10.
# Country Mahogany Longcase Clocks

The term "country" when used in the context of the longcase clock may be unfamiliar to some readers and thus it might help if we define it a little. The term describes clocks made outside the environs of London as well as the style in which they were constructed. Thus, if a clock looks typical of those being made in London at the period in question, but happens to have been made by a good provincial clockmaker, then it would not usually be called a country clock. This term implies features markedly different from those of the London-made clocks. It may have been designed as a simplification of London designs, for instance by using solid oak instead of finely veneered mahogany, or a reduction in the size of the "London" clock, or with the omission or modification of some of the moulds.

To try to illustrate some of the varying degrees of influence London makers had in the country, often due to distance, clocks are shown which were made both close to the capital and also at a great distance. In the former examples the designs are usually very similar to the London clocks, whereas 300-400 miles from the capital designs were evolved which had nothing to do with it.

Figure 279 A and B. *Jermiah Standring, Bolton, circa 1780.*

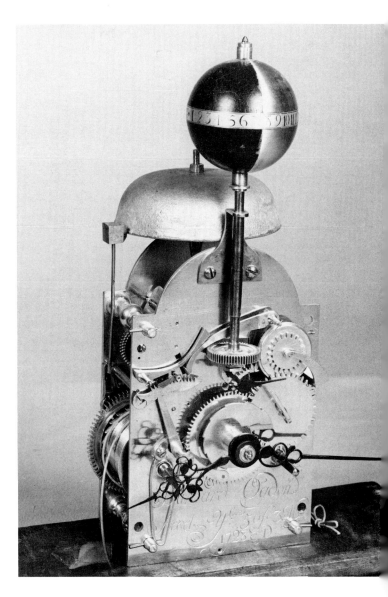
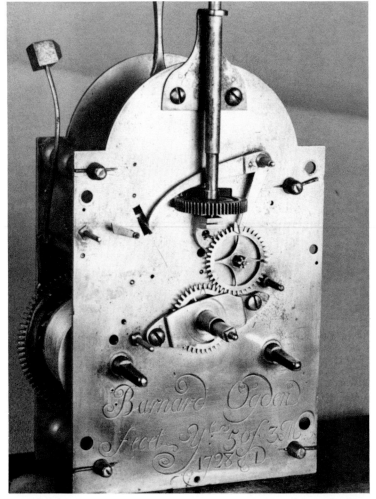

Figure 280 A, B, and C. *John Ogden/Bernard Ogden, Darlington, 1728.* This particular clock is a good example of country clockmaking. The movement, with its rotating globe moon, and breakarch-shaped plates, is full of character, as also is the brass dial with unusual dolphin spandrels flanking the moon. A raised half-round mould surrounds the edge of the dial, a most unusual feature that was used on other clocks by the same family of clockmakers.

It is particularly interesting that this clock is signed by John Ogden on the chapter ring, but all across the frontplate of the movement by his son: *Barnard Ogden Ye 5 of 3M, 1728 AD, FECIT.* Barnard was then 21, so perhaps this was his apprentice piece.

Clocks with rotating globe moons in the arch are rare. A few were made in and around Manchester, Darlington and Halifax (hence the common name "Halifax moon") by families such as the Ogdens and Booths (Figure 258), and some of the better London makers used them also (Figure 112).

*Country Mahogany* 203

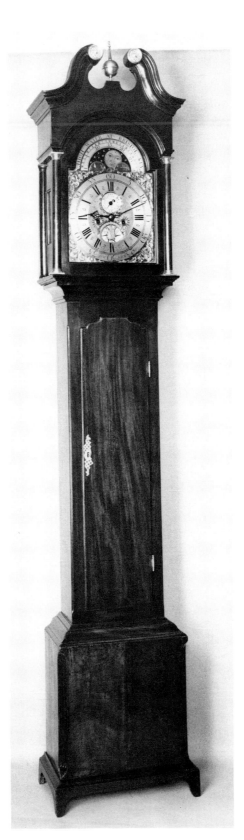

Figures 281, 282, and 283. *Three "Bristol" Clocks.*

Figure 281. *Robert Rogers, Bristol, circa 1780.* 7′ 9″ (236 cm) high.

*Figure 282. Richard Hardwick of Ashwick (Near Bristol).* This is an exceptionally tall and slim clock with a beautifully designed and executed dial with High water at Bristol Key. It has most unusual and decorative treatment of the date aperture and winding holes. 8′ 2″ (249 cm) high.

Figure 283. *Thomas Williams, Chewstoke, circa 1790.* This, like the previous clock, shows high water at Bristol Key. Note the shaped panel applied to the base and the extensive fretwork to the top of the hood. 7′ 5″ (231 cm) high.

Figure 284. *John Gordon, Aberdeen.*

Figure 285. *R. Thomas, Helston (Cornwall), circa 1780.*

Country Mahogany 205

Figure 286. *J.Baker, Hull, circa 1780-1790.* This clock has a solid mahogany case. Note the late retention of the ringed winding holes (small) and quartering on the inside of the chapter ring.

Figure 287 A and B. *Henry Baker, Mallin (now West Malling), 1768-1784.* Although decorative engraving around the date aperture had ceased in London by circa 1730, it was used here 1780. Father Time rocks to and fro in the arch. West Malling is about 35 miles south of London.

*206 Country Mahogany*

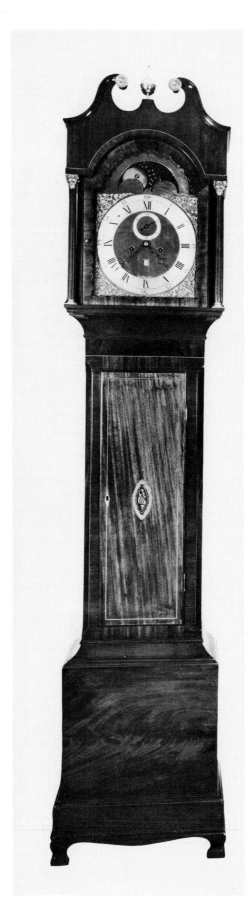

Figure 288. *James Hutton, Edinburgh, circa 1780.* This is a solid mahogany case. 7′ 7″ (237 cm) high.

Figure 289. *James Clarke, Bulcombe, circa 1800.*

Figure 290. *Stephen Meeres, Hampnell (Norfolk), circa 1780-1790.* Note the unusual double swan neck.

# Country Mahogany 207

Figure 291. *John Kay, Liverpool, circa 1780.* This has a Chippendale style case with "brickwork" to the base, quarter columns, and fine book-matched veneers. 7' 10" (339 cm) high with 13-inch dial.

Figure 292. *Robert Darling, Edinburgh, circa 1790.* A piper in the arch rocks to and fro as he marches through the glens.

Figure 293. *Painted Musical Longcase Clock.* This is a highly unusual, painted, mahogany, musical, calendar, longcase clock. The 14-inch 24-hour dial incorporates a year calendar dial and four sectors for the sun and tides. The painted moon disc in the arch is below the inscription *Time and Tide Wait for No Man.* The spandrels are of Rococo form. The three-train movement has a pin-drum operating a rack of ten bells and 20 hammers and two separate bells mounted at the side. The pendulum has a wooden rod. The case has a shaped swan-neck crest surmounted by flaming urn finials. The gothic-arched waist door is entirely crossbanded and inlaid with boxwood stringing. It is well painted with delicately drawn chains and sprays of summer flowers. 8' 10" (270 cm) high.

Photo courtesy Sotheby's, London.

*208 Country Mahogany*

Figure 294. *Louch, St. Albans, circa 1785.* This clock is as fine as those made in London in quality and proportions but has minor differences such as the arched trunk door with no mould around it, the applied fretwork to the hood door, and the smooth, as opposed to matted, dial center. 7′ 8″ (234 cm) high.

Figure 295. (Center) *Edward Burroughs, Fordham, circa 1790-1800.*

Figure 296. *Knight, Chichester, circa 1785.* This small mahogany clock has pleasing proportions. It is a simplified version of those made in London at the time. It has a five-pillar movement with strike/silent regulation. It is common to find clocks made in the Southeast of England having London-style movements and it seems likely that most of these were bought-in, ready-made, from the capital.

*Country Mahogany* 209

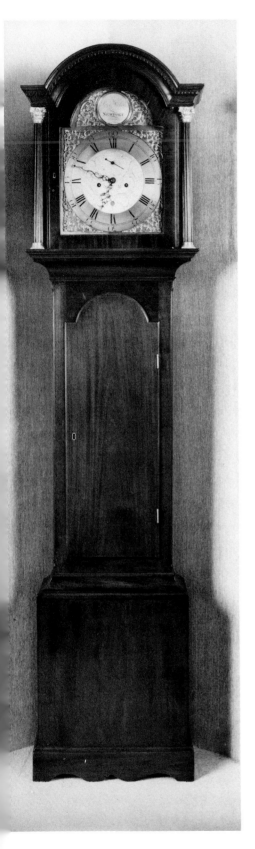
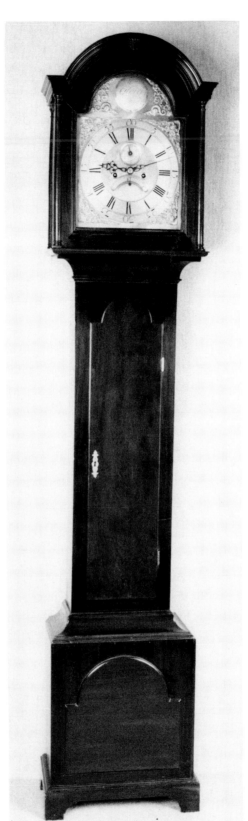

Figure 297. *Peter Roberts, Newport, circa 1780-1790.*

Figure 298. *William Glover, Worcester.* An unusual, solid cuban mahogany longcase clock with a breakarch top to the door which is repeated on the raised base panel. 7′ 3″ (221 cm) high.

Figure 299. *Hugh Pannell, Northallerton, circa 1790.* This large and impressive inlaid mahogany longcase clock has a month-duration movement, center-sweep seconds hand, and an unusual moon disc in the arch. 8′ 1″ (246 cm) high.

*210 Country Mahogany*

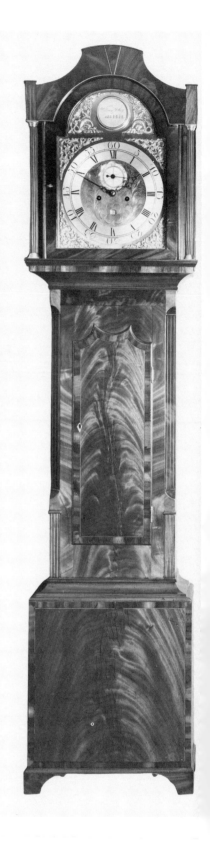

Figure 300. *Richard Allen, Woolwich, circa 1780*. Woolwich, ten miles east of London, is the home of the Royal Naval Training College and The National Maritime Museum. It is thus scarcely surprising that the clock, while London in quality, has a rocking ship in the arch. 7' 0" (213 cm) high.

Figure 301. *Yates Thornton, Romford, circa 1780*. This clock is similar to those made in London but there is no raised base panel, the trunk door has no mould around it, and the treatment of the top of the hood is unusual. 7' 4" (223 cm) high.

Figure 302. *William Hay, Airdrie, circa 1790-1800*. A fine, flame-mahogany, longcase clock, possibly with the caddy top reduced a little. It was probably similar originally to Figure 305. The chapter ring has been almost evenly divided into Roman hour and Arabic minute numerals.

## Country Mahogany 211

Figure 303. *William Whittaker, (town not given), circa 1790.* A fine mahogany case similar in some ways to Figure 305 with an oval panel let into the door and having a center-sweep date hand.

Figure 304. *John Coates, Coventry, circa 1790.* A well-proportioned, solid mahogany, longcase clock with a particularly attractive dial having several features taken from much earlier clocks such as the Roman chapter ring with half hour marks and quartering on its inner edge. However, the bold signature with a plain dial center are fully in keeping with the period the clock was made. 6' 9" (206 cm) high.

## 212 Country Mahogany

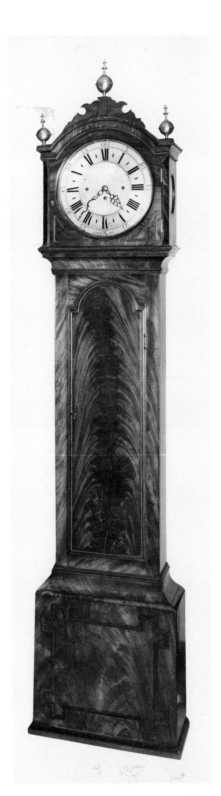

Figure 305. *William Bryson, Kilmarnock, circa 1790-1800.* A fine quality case with beautifully figured veneers, the treatment of the door is particularly interesting. Note the delicate hands and the well-engraved dial center. 7′ 10″ (339 cm) high.

Figure 306. *Quarter-Chiming Clock by Robert Parker, Derby, circa 1800-1810.* The circular dial came into fashion in London around 1790-1800, usually as flat silvered brass. It also was popular in Edinburgh and Derbyshire, however, in these areas it was generally used with a painted dial. In a few instances, as seen here, a brass dial was used with a raised chapter ring. The case is of typical London quality and may well have been made there. The plinth has been reduced.

Figure 307. The vast majority of mahogany, brass-faced, country longcase clocks have a breakarch, however some are found with a square 12-inch dial. This one by Richard Daking of Halstead dates circa 1780-1790.

# Chapter 11.
# Silvered Dials

Several terms have been used to describe *engraved, flat, silvered, brass dials* which are distinct from *brass dials* that have polished brass backplates, a raised chapter ring, and spandrels.

Although silvering had been used on dials on the Continent, and later in England, long before longcase clocks were thought of, its use had been largely confined to chapter rings, name plaques, etc., to contrast the black numerals. The dial plate and decorative pieces, such as spandrels, were polished brass or gilded.

This style of dial continued until about the 1760s when the first "all over" engraved and silvered brass dials started to appear. Only on regulators was the flat, silvered, brass dial used almost from the outset, for instance on those made by Graham and Shelton, because regulators are purely functional instruments for which clarity is the all-important factor and decoration is kept to a minimum.

Brian Loomes, in his book on white-dial clocks, suggests that the all-over engraved dial and the white dial started to be used at the same time, and this could well be true; however, others feel that some of the silvered dials may have been used a little earlier. Loomes argues that with the advent of the painted dial, some engravers were left without work because brass dials were no longer being used in large numbers. To combat this, he contends, they produced flat engraved brass dials, thus mimicking painted dials so far as they were able. If this were indeed the case, it did not succeed, except possibly in certain areas for a limited period. There must be a hundred painted dials today for each engraved and silvered one. Silvered dials must have been more expensive than painted dials, and painted dials were more decorative than silvered ones. Therefore, painted dials became more popular.

Engraving on silvered dials is sometimes of a very high standard but often it can be regarded only as rudimentary. Most silvered brass dials used on London clocks have no decorative engraving at all.

It would have been rare for a clockmaker to carry out his own engraving as this is an entirely different trade. It is more likely that the dials were either "brought in" or made by a travelling engraver when he was in the district. This latter was certainly common in the West country. The author has owned engraved dials from different locations there but obviously made by the same hand.

Silvered dials were rarely used in the North of England, but were in fairly general use throughout the South, including South Wales. They were often used on the later London clocks and seem to have been particularly popular in the Portsmouth/Southampton area.

One of the most attractive forms of silvered dial is that incorporating a painted moon disc which provides an excellent contrast.

Figure 308. *James Scott, Leith, circa 1780.* This clock has a solid mahogany case and a silvered brass dial incorporating a moon disc and a tidal dial in the arch.

## 214 Silvered Dials

Figure 309. *John Spendlove. Thetford, circa 1780-1790.* The fine silvered dial has beautifully executed and carefully balanced engraving to the four corners and the dial center. This has high quality engraving like that seen on the backplates of the better London-made bracket clocks. It has a center-sweep date hand and a painted moon disc in the arch.

Figure 310. *Mitchell and Sons, Gorbals (Scotland), early 19th century.* This has a finely figured mahogany case, a fairly short trunk door with a panel below, and a 13-inch silvered dial. Bold scrolled engraving makes an attractive feature of the maker's name in the arch. 6' 11" (211 cm) high.

*Silvered Dials* 215

Figure 311. *Richard Biggs, Romsey, circa 1785.* This clock has strike/silent regulation in the arch, a large date aperture, and decoration in the four corners similar to that seen on painted dials.

Figure 312. *George Vokins, Newtown, circa 1800.* A fine quality and small mahogany longcase clock with a 12-inch silvered dial which has decoration in the center and four corners depicting flowers, foliage and birds. 6′ 7″ (201 cm) high.

## 216  Silvered Dials

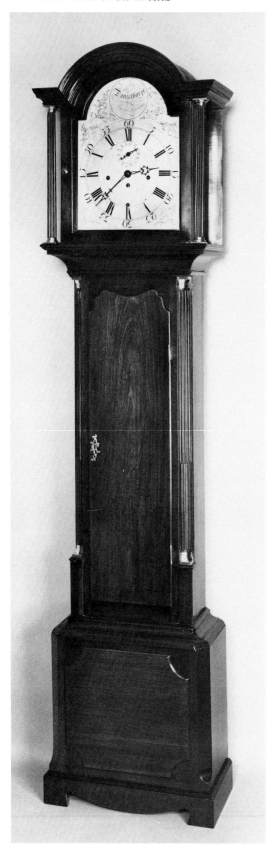

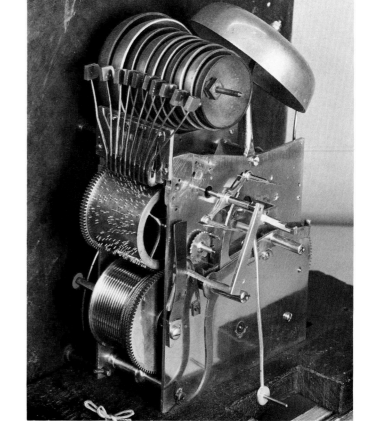

Figure 313 A, B, and C. *A Musical Clock by Donisthorpe, Birmingham, circa 1780.* This interesting musical clock plays one of seven tunes using ten bells and ten hammers every three hours. It automatically changes to a different tune every 12 hours. The dial beneath 12 is marked 1-14, and the hand which indicates the tune travels around once a week. Thus, each tune is played for two 12-hour turns every seven days. The silvered dial shows dragons breathing fire on either side of the arch.

## Silvered Dials

Figure 314. *James Stratton (Minchin), Hampton, circa 1780.* This clock has a silvered dial with a castle and a cottage depicted in the center, and a lion rocking to and fro in the arch in time with the pendulum. The carving on the trunk door is a later addition.

Figure 315. *James Pollard, Plymouth, 1790-1795.* A decorative pagoda topped longcase clock with a silvered brass dial looking very similar to a white dial. It has a large date aperture, floral engraving to the four corners, and shows moon phases and High water at Plymouth in the arch. 7′ 7″ (231 cm) high without finial.

Figure 316. *T.Webster, Piccadilly (London), circa 1820.* Note the simplification of the case relative to those being made in London 30-40 years earlier, for instance the base is not panelled and there is no brass stringing to the front corners of the trunk. The simple silvered brass dial, free of all decorative engraving, as with most London clocks, suites this relatively simple case style well.

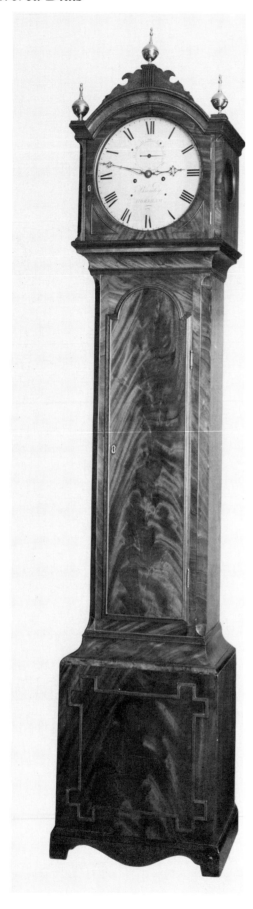

Figure 317. *Bromley, Horsham, circa 1805.* Following the emergence of the circular dial at the end of the 18th century, flat silvered dials became increasingly popular in London and the South East of England. Here you can see the changed style of hands following the move away from the brass dials.

Figure 318. *Leplastrier, Deal, circa 1810.* A typical London case, and indeed probably made there, with canted and brass strung corners to the front of the hood. The silvered dial has unusual, bold Arabic numerals.

# Chapter 12.
# White Dials

We must all be grateful to Brian Loomes who was the first person to research the origins and evolution of the "white dial clock," as he prefers to term it. These clocks were, until fairly recently, much maligned and misunderstood, and yet the movements and cases are exactly the same as brass-faced clocks of a similar period and origin. In certain parts of the country, for instance around Aberdeen, clocks with similar cases and movements were made with either brass, flat engraved silvered, or white dials. Much of the misunderstanding probably arises because white-dial longcase clocks were made in very large numbers for nearly 100 years. Whereas the early clocks are usually of excellent proportions, by 1840-1850 they had become large, heavy, and sometimes of indifferent quality. Many people think only of the late examples when referring somewhat scathingly to North Country painted-dial longcase clocks.

Brian Loomes has unearthed the earliest reference so far known to the white dial which appeared in the *Birmingham Gazette* on 28th September 1772, and is quoted below:

*White Clock Dials*
Osborne and Wilson, manufacturers of white clock dials in imitation of enamel, in a manner entirely new, have opened a warehouse at No. 3, in Colmore Row, Birmingham, where they have an assortment of the above mentioned goods. Those who favor them with their orders may depend upon their being executed with the utmost punctuality and expedition.
N.B. The dial feet will be riveted in the dial, and such methods used as will enable clockmakers to fix them to the movements.[1]

Figure 319. *John Chapman, Loughborough, circa 1790.*

This extract is interesting, not just because it gives us an idea of when white-dial clocks were first made, but also hints at the problem with dial feet. In a brass-faced clock the feet may be riveted on, usually where the chapter ring will hide them, after the dial has been made and in a position suitable for the movement so that they won't foul the pillars or any of the other components. With a painted dial this is impossible, as the feet have to be riveted onto the iron dial plate before it is painted. Obviously, the painting on the dial would be ruined if the feet were fixed after it was finished.

To overcome this problem, false or intermediate plates were often provided with the dials, and it is these which the advertisement is presumably referring to.

The advantage of the additional plate was twofold: 1) It meant that only short dial feet were used and therefore little leverage would be put on them and because of this they were less likely to be moved and the paint on the dial damaged; and 2) The feet going to the movement could be positioned on the falseplate by the clockmaker wherever was most convenient. Frequently, these false plates bear the name of the dial maker.

From 1772 onwards the manufacture of white dials, mainly in Birmingham, increased very rapidly. Wilson and Osborne split up in 1777 and started separate factories and several other firms commenced to make them. Not only were the dials sold virtually all over the British Isles, they were also exported, principally to America.

It might be helpful at this stage if we explain why there is so much confusion so far as the correct name for this type of dial is concerned.

The term "painted dial" is not really suitable as the dials were not painted as we think of it today, but made with white lead ground and varnish applied to the dial and baked hard in an oven at a relatively low temperature. It was then rubbed down with pumice stone etc. until a perfectly flat finish was obtained. Onto this was placed any decoration, the numerals, and usually the name, and finally a protective coat of varnish was applied. Thus, whereas the background is hard and long-lasting, the numerals etc. are far more likely to get rubbed with time and

Figure 320. *Staffordshire moulded and hard-enamelled dial signed Jn Green, St. Martins Court, London.* Note the shaping to the chapter ring, strike/silent dial, and spandrels. The inner and outer aspects of the chapter ring are surrounded by a raised ring. The whole is decorated with flowers and birds.

It has been suggested that the dial may have been made by Anthony Tregent, a well-known enameller of the period whose brother was James Tregent, the eminent clockmaker. A bracket clock by him with a similar dial is in *The Schreiber Collection* by Bernard Rackham, catalogue No. III, No. 408, plate 25. Anthony Theelke published privately in 1983 a small book on these dials entitled *Faces of Mystery*. Photo courtesy Sotheby's, London.

require repainting. The hard, smooth finish obtained on the background of clock dials was similar to that achieved on 17th- and 18th century Japanese lacquer. However, it does slowly craze with age. The term "enamelled dial" is usually confined to the beautiful, hard-enamelled dials which were produced on the continent in the 18th and 19th centuries. In these a superb, very even and hard surface was achieved by firing the enamel onto a copper or occasionally a silver background at high temperatures. These enamels do not craze with age but maintain a perfect surface, unless chipped.

Hard enamels are occasionally seen on London-made longcase and more commonly bracket clocks. However, the enamelling is usually confined to the chapter, strike/silent, and any other subsidiary dials, with the dial plate, spandrels, etc. being of brass. One suspects that most of the true enamelled dials which appear on English clocks were imported from France.

There is one other form of enamelled dial which is seen occasionally, the all-over enamelled dial such as that illustrated in Figure 320 which was undoubtedly produced in England. It is very difficult to make such large enamelled dials and the number produced was very small, partly because of cost, but also because they are not usually so decorative as white dials.

It is for all the foregoing reasons that Brian Loomes quite rightly prefers the term "white dial," the more so as this was the name used in the first instance. However, one suspects that the term "painted dial" will still persist.

The changeover from brass to white dials seems to have been remarkably rapid, although brass dials continued to be made on an ever-reducing scale until the 1820s and even beyond. The changeover occurred at a period of rapidly expanding production of longcase clocks, particularly in the Midlands and the North due to the increasing wealth there following in the footsteps of the industrial revolution.

The reasons for the changeover were probably fourfold, the first being one of economics, a white dial was cheaper to produce than a brass one. The second is that white dials are more decorative. When you buy a white-dial clock you are buying a painting as well. The third is fashion, often the most powerful of all arguments, even if the least justified. The fourth is one of clarity, they are much easier to read.

The majority of white dials were 12-inch breakarch, although 13-inch and later even 14-inch were quite common. The square dial was also used, particularly on the 30-hour clocks, but in much smaller numbers. The round dial, usually without any decoration, was popular in Scotland, particularly in Edinburgh and in London, and other districts such as Derbyshire.

Early white dials sometimes had no decoration or just simple gilt flowers copying brass spandrels. But soon single flowers, such as roses or peonies, or sprays of flowers stared to appear in the corners and the arch. As time went on, the decoration gradually became heavier and spread over more of the dial until it filled the corners, the arch, and finally the dial center. Birds were often featured in the designs and girls in different costumes were used to depict the four seasons. By the mid-19th century religious scenes and parts of the British Empire were depicted on the dials.

Blued steel hands were always employed on brass dials and their use continued for some time on white dials, but gradually a changeover occurred to brass hands which were easier to make and more decorative. Initially they were similar in form to the steel hands but gradually became more ornate and were often gilded. Whereas some of the earlier brass hands were of fine quality and attractively chased and engraved, the later ones were often just punched out.

By the 1840s longcase clock production had almost ceased over much of the country following the flood of imported clocks from Germany, France and America. However, their manufacture was to continue in the North of England for another 20-30 years.

---

[1] Loomes B, *The White Dial Clock*, David & Charles, London, 1974.

## White Dials 221

Figure 321. *Robert Welch, Dalkeith, circa 1780.* An early plain white dial with well-executed, raised-gilt decoration.

Figure 322. *James Upjohn, Brentford (Nr. London), circa 1800-1810.* A simply proportioned mahogany case with flame veneers to the door and base, and the corners to the top of the hood with classical decoration. The 12-inch plain white dial is similar in layout to London-made brass dials. 7′ 0″ (213 cm) high.

Figure 323. *William Payne, Ludlow, circa 1780-1785.* Ludlow is about 30 miles south of Shrewsbury and this oak case with broken-arched top is typical of the style being used in that area. The dial has delicate blued steel hands, raised gilt decoration, and shows moon phases in the arch. 7′ 3″ (221 cm) high.

Figure 324 A, B, C, and D. *B. Parr, Grantham.* This clock is the perfect rebuttal to those who think painted-dial clocks are of poor quality. It is one of the most interesting and original we have seen. It has a five-pillar movement with unusually tall plates and is most ingenious in its design. It has a large, engraved and silvered year-calendar disc which is displayed through a curved aperture above 6 o'clock. Although this disc appears to be perfectly round, it is in fact eccentric and this minor deviation from the round is magnified by a pulley running on its edge and various levers so that the clock can display "Equation of Time" below 12 o'clock. A further interesting feature is a month fly-back calendar which is controlled by a large rack mounted on the backplate and protected by two steel studs.

The 12-inch painted dial shows moon phases in the arch with a gilt moon and a star-studded disc. The moon's age is displayed by Roman numerals. In the upper half of the square part of the dial is a gilt solar hand which indicates *Sun Slower* and *Sun Faster* on a curved sector below 12 o'clock. Running from 4:30 to 7:30 is a further sector graduated 0-31 which gives the day of the month, and below this is a year calendar displayed in an aperture. There is finely executed decoration in the four corners and at the bottom of the arch. It is signed *B. Parr, Grantham*.

The case of highly original design and excellent proportions is a fine piece of Sheraton furniture. It rests on two tapering and reeded shaped square wooden feet and a plinth. The central panel of the base is surrounded by stringing in Greek key design and there is further stringing to the edge of the base. The shaped trunk door has an elliptical panel surrounded by stringing and the sides of the trunk are canted and inlaid. Immediately below the corners, boxwood "fish" are inlaid on a dark ground and this motif is repeated to either side of the pagoda top and below each of the finials. The fish represented are Parr, thus reflecting the clockmaker's name. There is stringing to the reeded hood pillars and also to the hood door. All the veneers used on this clock are of the finest quality. The case bears the label of the cabinet and clock case maker J. Usher who commenced work in 1808 and married in 1809. It was James Ward Usher, a descendant of his, who founded the famous Ward Usher Collection now seen in Lincoln Museum and who produced a beautiful catalog describing and illustrating its contents. 7′ 11″/7′ 3″ (24½/22 cm) high with and without top block finial.

Figure 325 A and B. *George Youl, Edinburgh, circa 1785-1790.* An exceptionally well-proportioned, eight-day longcase clock by George Youl of Edinburgh who was apprenticed to John Skirving in 1773 and became free of the Clockmakers' Company in 1779.

The case is of particularly fine quality and delicately inlaid throughout. There is a swan-neck pediment with a band of inlay immediately above the dial. There are delicate columns to either side of the hood, reeded quarter columns to the sides of the trunk, and canting with inlay to the sides of the base, which is finely panelled with yew wood inlay in the center. The long trunk door is well-figured with banding to the border, inlay in the four corners, and thistle, leaves and flowers are laid into an oval plaque of Harewood in the center.

The early white dial is in excellent order with a finely executed scene of dancers, players and other figures on the Village green. There are delicate, raised-gilt spandrels in the four corners and a center-sweep date hand is provided. 7′ 7″ (231 cm) high.

Figure 326. *Thomas Stewart, Auchterarder, circa 1785.* Note the delicacy of the decoration with rosebuds in the corners and a bird among the foliage in the arch.

Figure 327. *J. Jackson, Boston, crossbanded oak case, circa 1790.*

Figure 328. *James Warren, Canterbury, circa 1795.*

White Dials 225

Figure 329. *Evan Tobias, Llandilo, circa 1800.*
Figure 330. (left) W. Thackall, Banbury, circa 1790.
Figure 331. (right) *Russell Cawston, circa 1805.*

Figures 332 A and B, 333, 334, 335. A series of clocks, circa 1790-1805, from different parts of the country but all with scenes painted in the arch on an oval plaque and with dials possibly by the same maker.

Figure 332 A and B. *John Symonds, Reepham.*

Figure 333. W. Wills, Truro (Cornwall).  Figure 334. *Wm. Robertson, Dunbar.*  *Figure 335. George Deriemer, Portsmouth (Hants).*

Figure 336. *Elliott & Son, Ashford, circa 1790.*

Figure 337. *L. Lambert, Haxey, circa 1795.*

Figure 338. *Thomas Twigg, Shilton, circa 1790.* This clock has a very fine mahogany case with book-matched veneers to the trunk door. The finials and the ogee bracket feet have been removed. There is a lightly painted dial with moon phases in the arch. 7′ 10″ (239 cm) high.

White Dials 229

Figure 339 A and B. *A. Noon of Ashby.* A fine-quality and rare oval-dialled longcase clock by Noon of Ashby. Note the very restrained decoration to the dial.

230 White Dials

Figure 340. *Wm. Fletcher, Gainsborough, circa 1785-1790.*

Figure 341. *Davey, Norwich, circa 1780-1785.* This clock has raised gilt and green spandrels and a fine mahogany case.

Figure 342. *Robson Jnr., N. Shields, circa 1780-1785.* This clock has a very fine mahogany case and an early dial with raised gilt corner decoration, moon phases in the arch and a center-sweep date hand.

White Dials 231

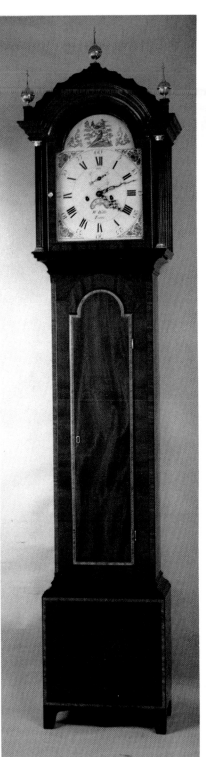

Figure 343. *John Bower, Kirriemuir, circa 1795-1800.*

Figure 344. *William Wills, Truro, circa 1790-1800.* An unusually slim and elegant case with satinwood crossbanding to the trunk door and base, and tulip wood to the canted corners of the trunk.

Figure 345. *Nickisson, Newcastle, circa 1800.*

Figure 346. *John Barron, Aberdeen, circa 1795.*

232   *White Dials*

Figure 347. (left) *Jordan, Great Missenden, circa 1790.*

Figure 348. (center) *Wm. Hunter, Dunfermline, circa 1815.*

Figure 349. (right) *Andrew Rich, Bridgewater, circa 1820.*

Figure 350. *Pierre Poignand, Jersey circa 1795-1800.*

# White Dials 233

Figure 351. *J. 5. Scobell, Ottery St. Mary (Devon), circa 1815-1820.* Note the typical West Country feature of a ship in the arch.

Figure 352. (left) *James Christie, Perth, 1820-1830.* This clock has the shorter trunk door with a panel below, satinwood banding, and a relatively heavily painted dial with a sailing scene in the arch.

Figure 353. (center) *Small tavern clock by Thomas Lee of Potton, circa 1780-1790.*

Figure 354. (right) *James Isaac, Carmarthen, circa 1800-1810.*

234 White Dials

Figures 355 and 319. *John Chapman, Loughborough, circa 1790.*

Figure 356 A and B. *Emmanuel Cohen, Redruth, circa 1815.* A good example of a West Country clock with spiral twist columns to the trunk and showing moon phases and high water at Hayle.

Mr. Cohen (1766-1849) subscribed five shillings for the starving Jews in Tiberias in 1849 and by local tradition he could be seen waiting for sunset on Friday and Saturday evenings to close and open his shop.[2]

---
[2] Brown, H. M. *Cornish Clocks and Clockmakers,* David & Charles, Newton Abbott.

## White Dials 235

Figure 357 A and B. John Sherwood, Hythe, 1807. It is rare for a clock to have been in only one home, yet this particular clock has lived in Hythe all its life and was purchased from Mr. Sherwood of the same town for the princely sum of eight pounds on 24th October, 1807. It is surmounted by three finials with decoration between, has brass-capped fluted columns, solid mahogany sides and front, and an attractively figured and pleasantly faded trunk door. The 12-inch painted breakarch dial has gilt seconds, minute, hour and date hands, a still life painting of fruit in the arch, and further decoration in the corners. It is signed by Sherwood of Hythe. Documented clocks such as this are invaluable when assessing the dates of other pieces. 6' 10" (208 cm) high excluding finial.

Figure 358. *Jn. Heslop, Newcastle, circa 1810.*

236 *White Dials*

Figure 359. *Unsigned, West Country Clock, circa 1815-20.* Note the wavy edge to the inside of the hood door and the sailing ship on the moon disc in the arch. It has a small trunk door typical of this period.

Figure 360. *John Jones, Aberystwyth, circa 1820.*

Figure 361. *George Blackie, Musselburgh, circa 1820.* Interestingly the author of *Old Scottish Clockmakers*, Mr. John Smith is in possession of an eight-day longcase clock by this maker which was made in 1800 for an ancestor. He still has the original receipt for it for seven guineas.

It is also recorded by him that Mr. George Blackie, watchmaker, Mueselborough, presented a handsome clock to the Free Church Congregation of that place. Mr. Blackie was one of the oldest men in the congregation being about 70 years of age. "And that he might live to finish the clock has of late with him been a frequent wish." It is nice to know that he just made it. He died that same year.[3]

[3] *The Witness*, 21-8-1844.

*White Dials* 237

Figures 362 and 363. Two more clocks with automata in the arch.

Figure 362. *John Piggin, Norwich, circa 1820.* A girl dances to and fro and plays with a little dog in the arch.

Figure 363. *Gilbert, Hythe, circa 1805.* A sailing ship rocks to and fro.

Figure 364. (left) *Hutchinson, Retford.*

Figure 365. (right) *Wilmshurst of Burwash.* Small scenes in the four corners and decoration to the dial center, circa 1825.

238  *White Dials*

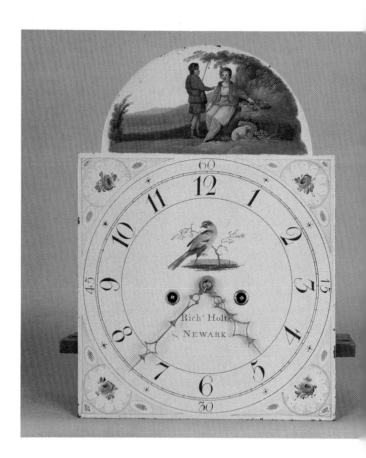

Figure 366 A and B. *Richard Holt of Newark, circa 1820.*

Figure 367 A and B. *Black-dialed clock by Chas. Kinnear, Dundee, circa 1820.* The quality of the case and dial of this clock are of a particularly fine craftsmanship. Only the best mahogany veneers have been used for the case, and the 13-inch dial has a dished (convex) white center with gilt hands and a black surround with a well-executed still life painting of fruit in the arch and flowers in the four corners.

## White Dials 239

Figure 368 A and B. *P. & A. Boffi, Hastings, circa 1825.* A finely figured mahogany case surrounds a well-executed dial showing the four seasons in the corners while in the center of the arch a swan rocks to and fro in time with the pendulum.

Figure 369. A painted dial with religious scenes including the Last Supper, circa 1850-1860.

Figure 370. *S. Kellett, Bredbury, circa 1845.* This fine quality, but very wide, case encloses a clock with a heavily painted 14-inch dial. In the center are parrots and foliage, on the moon disc is a god on his chariot accompanied by a cherub, and below this are two female figures. 8′ 2″ (249 cm) high.

Figure 371 A and B. *A Musical Clock by Bradber of Leyburn, circa 1835.* The case and 13-inch dial of this clock would suggest a date circa 1840-1845 and yet the tune selector in the arch has among its four tunes *God Save the King*, which would imply a date pre-1837. The other tunes are *Jenny's Baubee, Money Must,* and *La Belle Isabelle.* 8′ 3″ (251 cm) high.

# Chapter 13.
# Thirty-hour Longcase Clocks

Most of the very early clocks produced for domestic use, such as the first lantern clocks, were only of 12- to 16-hour duration and thus had to be wound each morning and evening. It was after around 1650-1660 that their running time was increased to thirty hours. However, right from the beginning the majority of longcase clocks produced in London were of eight-day duration and it was only when they started to be made the provinces towards the end of the century that the 30-hour longcase clock appeared. There were, of course, exceptions to this rule in that even the most famous London makers prior to 1700, including Tompion and Gould, produced a few 30-hour clocks. However, when someone wanted an expensive clock, which could be of short duration, it is likely that they chose a lantern clock rather than a longcase. Interestingly, whereas the London makers used plated 30-hour movements, usually with two hands, right from the start most of the early country 30-hour clocks, particularly those who produced in the South, have birdcage movements and only a single hand. Nevertheless, it is dangerous to assume that a clock with a "birdcage" movement is as early as this. The author has seen more than one of these clocks with an original painted dial which could safely be dated post 1790. Similarly, although most 30-hour, single-handed clocks may date prior to 1750 and usually have ten-inch square brass dials, examples were still being produced 50 years later.

Figure 372. *Thos. Moore, Ipswich, circa 1725.* This maker inherited the business of Thos. Moore of London, presumably his father, and it is probably for this reason that this 30-hour dial is so similar to those of the eight-day clocks being produced in London at that time. Note the central alarm disc with the seconds ring carefully cut out to clear it, the cartouche around the date aperture, and the half hour marks.

Nearly all 30-hour clocks are driven by a single weight suspended from a continuous rope or chain and thus require no winding holes in the dial; but very occasionally key-wound examples may be seen.

The 30-hour clock was developed mainly for country cottages with low ceilings which seldom exceeded six feet six inches to seven feet in height. Of necessity they were small clocks. The early clocks were nearly always made of solid oak, were slim and of good proportions, and some are simplified versions of their wealthy London cousins, even down to the caddy tops. However, a few examples may be found (Figure 381) which were heavy and clumsy, made, one suspects, by the local coffin maker.

As the early 18th century progressed, cases tended to become wider and more decorative, maybe with a swan neck pediment or pagoda added on top, and the oak became thinner and of poorer quality. By the 1770s a changeover had commenced from brass to painted dials and at this time also mahogany started to replace oak for the cases. Prior to 1750 the number of 30-hour clocks produced in the country was quite high when compared with eight-day clocks, but after this it gradually declined, with probably only one painted dial 30-hour clock being made for every ten of eight-day's duration.

The whole concept of the 30-hour clock was to produce a small and inexpensive longcase clock which the country folk could afford to buy; thus the production of complex 30-hour clocks is virtually a contradiction in terms, and yet a few were made, mainly by one or two specialist makers such as Lister (Figure 388). The author has seen in the last twenty years three 30-hour musical clocks, one of which had four weights and was also quarter chiming. The mass of chains and weights in the case was bewildering, and the thought of doing battle with them each day a daunting prospect.

Figure 373. *Step Harris, Tonbridge, circa 1720.* This pretty little clock stands only six-feet four-inches high to the top of the caddy with delicate pillars to the hood. It has a single-handed ten-inch dial.

Figures 374 and 375. Two single-handed oak longcase clocks circa 1740-1750; that on the left by Harley of Salop having "Four Seasons" spandrels.

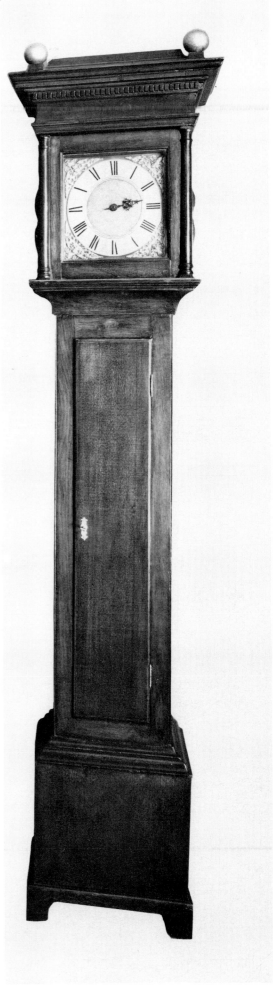

Figure 376. An unusually tall and slim 30-hour clock with ten-inch dial, circa 1740-50. 7′ 1″ (216 cm) high.

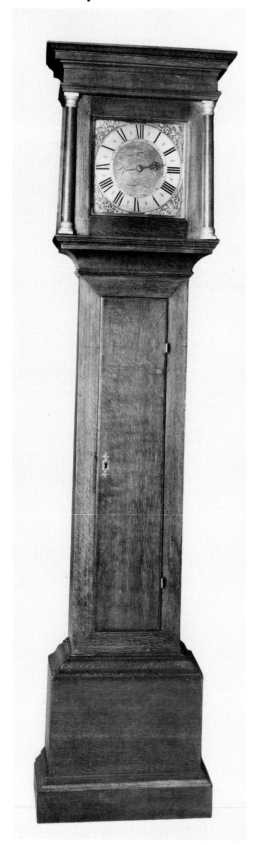

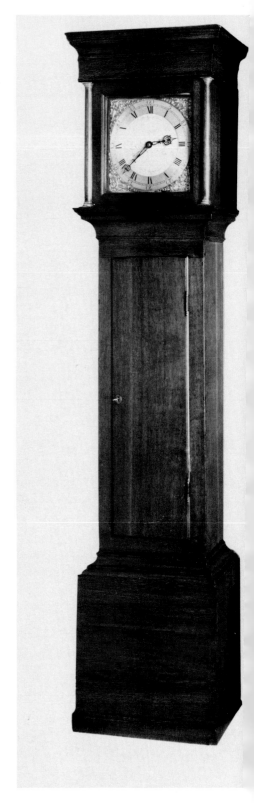

Figure 377. *Obadiah Boddy, Battle, circa 1725.* Note the birds engraved on the dial center, a popular feature at this period.

Figure 378. *William Flint, Charing, circa 1760.* Ten-inch dial, 6′ 3″ (190 cm) high.

Figure 379. *James Irish, Lewes, circa 1790.* Ten-inch dial, 6′ 3″ (190 cm) high.

# Thirty-hour 245

Figures 380, 381, and 382.

Figure 381. (center) *William Hoadley, Rotherfield, circa 1760.* Note the ringing to the center of the dial and the pegged construction of the heavy oak case. The base may have been reduced, but with an unconventional case such as this it is difficult to be certain. 6′ 9″ (206 cm) high.

Figures 380 and 382. (left and right) Two 30-hour clocks, that on the left with 11-inch dial while the one on the right is 12 inches. Both have two hands and there are now minute marks on the outsides of the chapter rings but no quarter hour divisions. However, that on the right by William Lawson of Keighley still has half-hour marks although it was probably not made until 1780-1790. The dial centers are engraved and both have apertures for the date.

Figure 383. *Charles Stevenson, Congleton, circa 1770*. Note the dummy winding holes which would make the owners' friends think that he had not a 30-hour clock but one of eight-days duration. 6′ 8″ (208 cm) high.

Figure 384. *Thomas Sharly of Leighton, circa 1780-1790*. The use of a breakarch door on a square-dial clock was not unusual for country clocks.

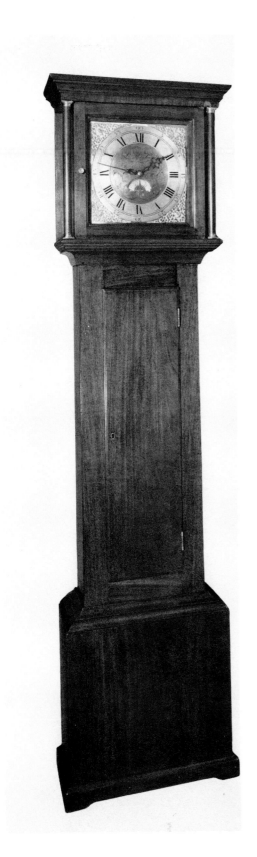

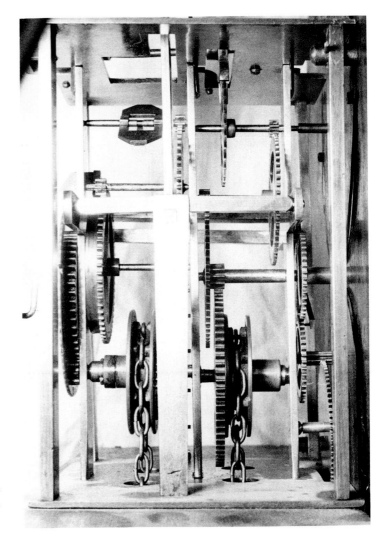

Figure 385. *P. Thackwell, Ross, circa 1780-1790.* This is a solid mahogany case.

Figure 386 A and B. Flat engraved and silvered brass dials are occasionally found on 30-hour clocks and here we see a particularly fine example by James Newman of Lewes. Several interesting makers, such as Richard Comber, worked in Lewes. It has a birdcage movement and is laid out like a regulator with a center-sweep minute hand, a large seconds ring to the top of the dial, and hours below.

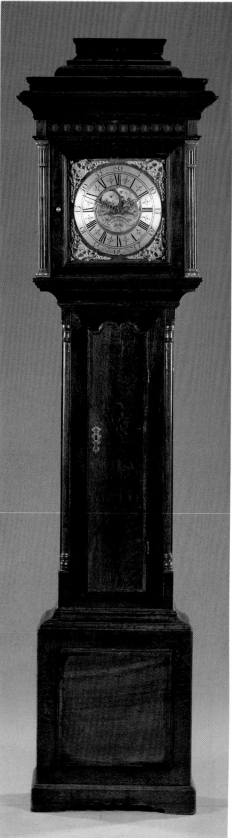
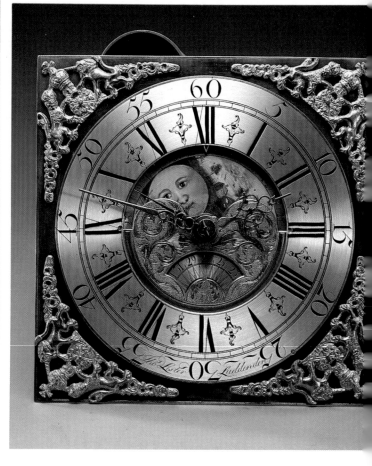

Figure 387. *Donisthorpe, (no town given), circa 1790.*

Figure 388 A and B. *Thomas Lister, Oak, Musical Longcase, circa 1740.* A rare George II oak musical longcase clock, circa 1740, the 12-inch dial signed *Thos. Lister, Luddenden.* It has unusual lion, unicorn and crown spandrels. The center has a painted moon disc, a calendar sector, and a subsidiary blued steel hand. The three-train, 30-hour plated movement with rope drive and outside countwheels plays a tune on six large bells at every fourth hour. The oak case is crossbanded with walnut and has a stepped caddy cresting, hood pillars, serpentine-topped trunk door, flanking pilasters, and a panelled plinth. 7' 6" (229 cm) high.
Photos courtesy Sotheby's, London

Thomas Lister the elder was born in 1718 and died in 1779. The tune played by this clock is a hymn called *York.* The central blued steel hand is thought to originally have been part of the silencing mechanism. This clock is illustrated in detail and discussed by Brian Loomes' *Grandfather Clocks and their Cases,* 1985, pages 310 to 312. Also see *Clock Magazine* November, 1983, pages 25-27.

## THE WHITE DIAL

The white or painted dial first appeared on clocks in the 1770s (see Chapter 12) and was probably introduced on 30-hour clocks and eight-day clocks at nearly the same time. The 30-hour movement was far easier to add than an eight-day clock, since if there were no seconds hand there was little to line up. With an eight-day clock, four holes in the dial have to be in their correct positions if it is to fit (i.e. for minutes and hours, the seconds hand, and the two winding holes).

By the time the painted dial had been introduced, a large number of cases were being made in mahogany, not oak, and the popularity of the 30-hour clocks had declined considerably. You will thus see more painted dial clocks with eight-day than 30-hour movements.

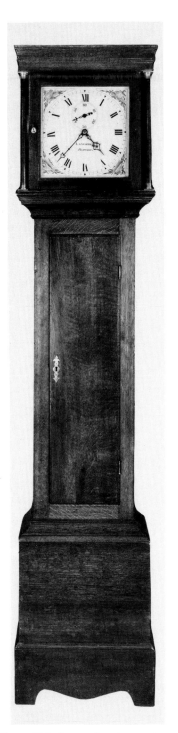

Figure 390. *Longridge, Burwash, circa 1800.*

Figure 391. *Hosmer, Tonbridge, circa 1800.* A mahogany case on which the hood pillars have been replaced by canted corners.

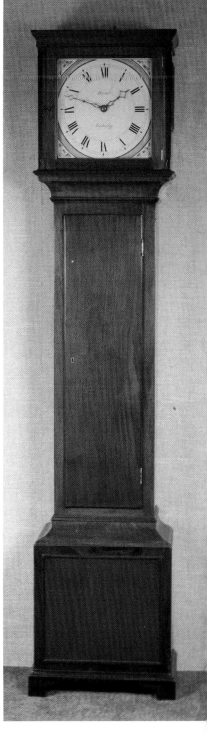

Figure 389. *N. Kemp, Place unknown, circa 1800.*

## 250  Thirty-hour

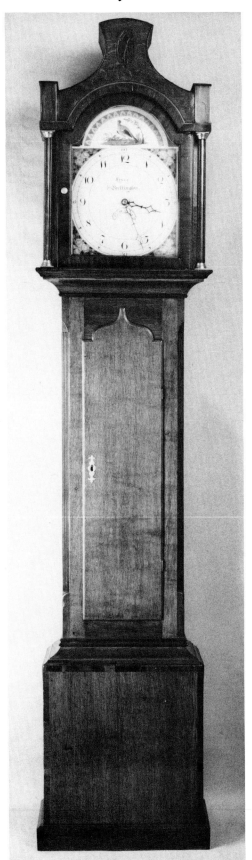

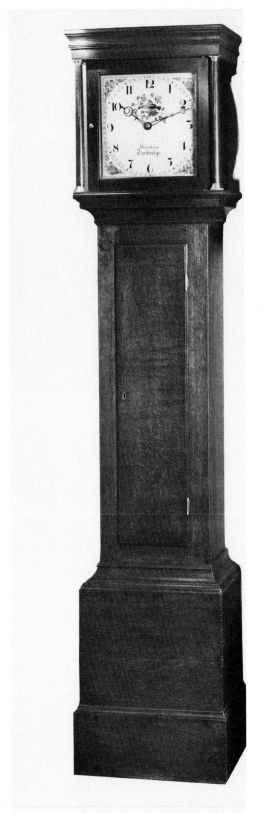

Figure 392. *Fryer, Pocklington, circa 1810-1820.* Pocklington is about 30 miles Northwest of Hull. This clock has a case typical of those being made in that area with a tall pagoda and crossbanding to three sides of the base.

Figure 393 A and B. *Barcham, Tonbridge, circa 1830.*

Figure 394. *B. Rice, Neath, circa 1790-1800.*
Figure 395. *H. Evans, Llangadock, circa 1820.*
Figure 396. *W. Cawdle, Torquay, circa 1830.*

## 252  Thirty-hour

Figure 397 A, B, and C. *Regulator with Gravity Escapement by Brock.* Although longcase clock manufacture declined rapidly after 1830, the production of longcase regulators increased dramatically. This was in large measure due to the rapidly increasing requirement from clock restorers and jewelers to have a master clock by which to set their watches and ordinary clocks. The demand for these latter items soared as the wealth of the nation increased and mass production made them readily available.

The regulator seen here is typical of the massive engineering of many of the best Victorian clocks, an approach commonly taken at that time. It was made by Brock who was the foreman overseeing the construction of the clock commonly known as "Big Ben" and thus it is scarcely surprising that it uses Dennison's gravity escapement. The movement is of immense strength with a pulley offset at the top to give maximum fall for the weight of 34 pounds as the gravity escapement absorbs a great deal of power. Moreover, the zinc/steel compensated pendulum weighs 44 pounds.

The clock is an odd combination of fine functional precision timekeeping and decorative features such as the cherub spandrels, raised chapter ring on the dial, and elaborate finial to the top of the case, which is made of teak. 7' 5" (226 cm) high.

# Chapter 14.
# The Victorian and Edwardian Eras

By 1785-1790 the production of longcase clocks in London had started to decline and by 1820-1830 had almost ceased. This was brought about largely by changing fashion, there being a strong swing away from the longcase to the bracket clock and also because of the rising tide of imports, initially from France and later from Austria, Germany and America. In contrast the demand in the country districts and particularly the North of England, for painted dial longcase clocks increased from 1780-1810 and continued quite strongly until 1850-1860.

A few distinctive clock styles originated in the Victorian era, thus there was the Gothic revival which started in the 1840s which strongly influenced some clock case designs. The dome topped regulator (Figure 466) was entirely Victorian in concept being produced from 1845-1900, but probably the most spectacular creations were the five- and nine-tube chiming late Victorian/Edwardian longcase clocks (Figures 408 and 409). These were nearly always large clocks, from seven feet nine inches to nine feet six inches tall, usually with glazed doors and sometimes sides, and either inlaid or carved cases—occasionally even both at the same time. These clocks must have been extremely expensive to produce and may be regarded as the last titanic effort of the British horological industry to survive before it was submerged under a flood of cheap imports.

Grandmother clocks (between five feet and six feet six inches high) are seen occasionally from Victorian times but rarely encountered in the 18th century (Figures 400 and 401). Some were made especially but others were made from modified earlier movements.

The latter half of the 19th century probably marked the start of reproduction clock manufacture with styles copied generally from the 17th and 18th centuries (Figure 403).

Figure 398. *Late 19th Century Mahogany Longcase Clock.* This clock, basically of 18th-century style, has brass capped hood pillars and canting to either side of the shaped and well-figured door.

The 12-inch square dial has half-hour markings and cherub spandrels, similar to those used around 1700. The weight-driven movement has strike/silent regulation and brass-cased weights. 6' 4" ( 193 cm) high.

## 254 Victorian and Edwardian

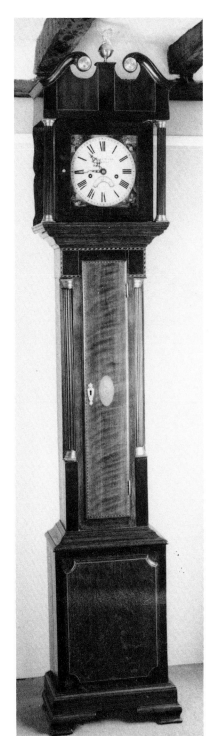
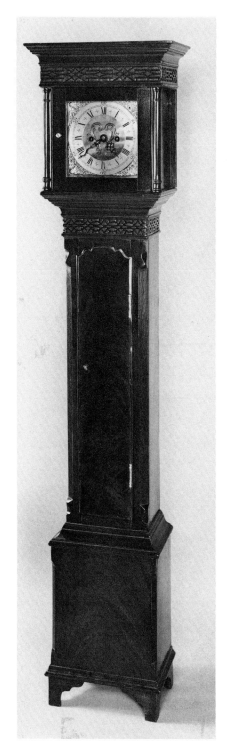

Figure 399. *Joseph Fannell, Poole, circa 1840.* This clock, which has well-figured veneers, might be regarded as a regulator as it has dead beat escapement, maintaining power, and a wood rod pendulum with heavy brass cased bob seen through the glazed door. However, it has a conventional longcase dial and a striking train, which some would say excludes it as a regulator, although many fine French regulators strike.

The shape of the trunk door and hood give a hint of the Gothic style which became more pronounced on other clocks.

Figure 400. *Mid-19th Century Grandmother clock.* This attractive little weight-driven clock standing just 6′ 4″ high is particularly well proportioned for a clock of this period. It has an eight-inch square painted dial with the four seasons featured in the corners and a circular chapter ring with a recessed date ring in the center. It is signed by Charles Allen of Bewdley.

The case is of fine quality, the major part being of oak. The hood is veneered in mahogany, there are freestanding columns to either side of the trunk, and the base is panelled.

A problem with these clocks is the small space in which the pendulum can swing, all too often it hits the sides.

Figure 401. *Spring-Driven Grandmother Clock.* To overcome the lack of space within the trunks of Grandmother clocks (this one has an internal width of only 6½ inches) spring-driven, as opposed to weight-driven, movements were sometimes used thus eliminating the need to provide space for the weights to fall in the trunk. The well-proportioned case has been made to a high standard. 5′ 7″ (170 cm) high.

## Victorian and Edwardian

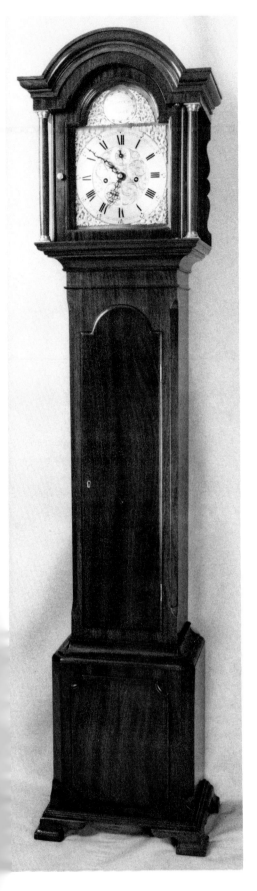

Figure 402. *Late 19th century mahogany longcase clock.* A good quality eight-day, weight-driven longcase clock striking on a gong with solid mahogany sides and the front made of oak veneered with mahogany. The 10-inch dial has *Tempus Fugit,* beloved of the makers at this time, and a well-engraved dial center. 6' 2" (188 cm) high.

Figure 403 A and B. *Late Victorian reproduction pagoda-topped longcase clock.* This clock is almost a straight copy of a George III pagoda-topped longcase clock, although the case is a little wider than the originals. The 13-inch brass dial has an unusually small seconds ring and regulation for Whittington/Westminster chimes in the arch. The movement is of good quality but many features of its construction, for instance the dial feet, immediately give an indication of its age. As with many Victorian clocks the maker's name is omitted, often so that the retailer could, if he wished, insert his name instead.

## 256 Victorian and Edwardian

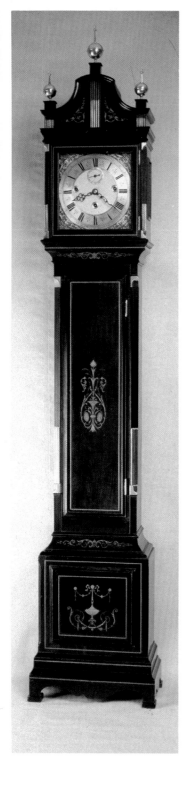

Figure 404. *Late 19th century chiming longcase clock.* Although much relatively poor cabinet work was produced towards the end of the Victorian and in Edwardian times to satisfy the enormous demand, the quality of the finest pieces by eminent cabinetmakers such as Edwards and Roberts was superb. Moreover, they often abandoned heavy decoration and carving in favor of the classic furniture designs by Sheraton of over 100 years earlier. This clock is a perfect example.

The case is of excellent proportions with a long trunk door, canted, reeded and brass strung corners on either side which are repeated on the hood, a panelled base, and a double plinth. The inlay is beautifully executed and used in a restrained manner. Indeed, although this clock has obviously been inspired by many different features from the Georgian era, it comes over as a superbly conceived original design in its own right. For instance, the pagoda top, as modified in this instance, was never used in Georgian times with a square dial. The design of this dates from the 1700-1710 era with late cherub spandrels, half-hour marks, and ringed winding holes.

The movement matches the case in quality. It chimes the quarters on either four or eight bells, strikes the hour on a gong, has strike/silent regulation, and is provided with brass-cased weights. 7′ 6″ (229 cm) high.

Figure 405. *Edwardian Longcase Clock.* The case style of this clock is typical of those being produced in the Edwardian era, with solid or glazed doors. The inlay, generally used in a restrained way with classic motifs, is usually well executed. The movements were either two or three train, still weight driven at this period. Many different forms of chime were provided, for instance on four or eight bells, just four bells, eight bells and four gongs, just four gongs or on tubular bells.

This particular example, signed *Mirgan, London,* chimes the quarters on either eight bells or four gongs and strikes the hours on a gong.

### THE TUBE CHIMING LONGCASE CLOCK

Towards the end of the 19th century clocks started to be produced in England with tubular bells. Usually they were nickel plated and were possibly inspired by those used in the temples in the Far East for many hundreds of years which create such a pleasant sound. They were used in sets of four or eight with a ninth much-larger bell for the hour. The sizes of the bells varied with the height of the clock, but the range was usually from two to five feet.

Simpler clocks played *Westminster* chimes on four bells, but the best would play two or three different chimes such as *Whittington, St. Michael* and *Westminster* with up to eight bells. These clocks, when correctly set up, sound superb, coming very close to the chimes produced by church bells.

The large and possibly slightly earlier clocks seem to have been produced in two forms, a typically inlaid Edwardian case, nearly always with a glazed door and about eight feet high, and a much larger, deeper and wider clock around nine feet high. This was heavily decorated with carving and fretwork and usually had inlaid panels (Figure 409). Detailed pictures of one can be seen in Figures 223 A, B, and C.

When tube-chiming clocks were produced only a handful of manufacturers (such as Elliott, Smiths of Clerkenwell, and Gillet and Johnson in Croydon) remained in the market so the designs were standardized. It is interesting to note, for instance, that the inlaid panel seen on the base of the clock in Figure 408 is virtually identical with that in Figure 223 A.

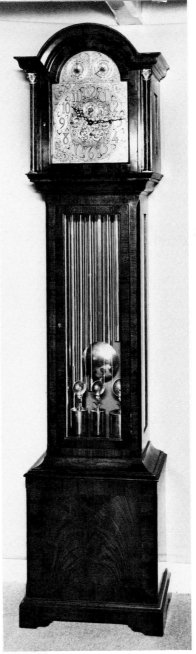
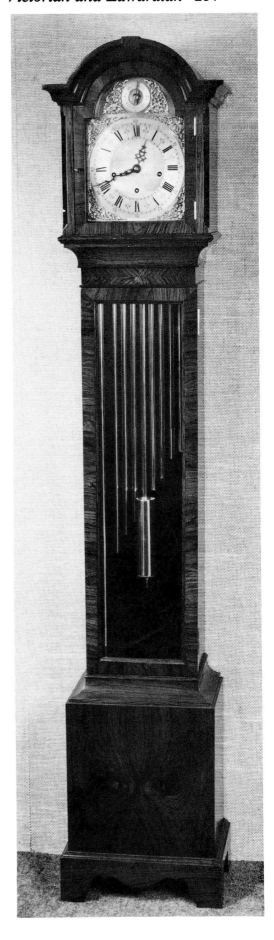

Figure 406 A and B. *Tube-Chiming Clock Signed Terry, Manchester.* The introduction of large tube-chiming clocks stimulated demand from the general public for similar clocks. These were scaled down and simplified to fit into average homes at affordable prices. The typical clock shown here, standing about 6' 10" high, has a quality mahogany case in late Georgian style, but with a glazed door to show off the tubes. Although many clocks were being made entirely in England, continuing demand for inexpensive examples led to the import of cheap movements from Germany for English cases.

Figure 407. *A Spring-Driven, Tube-Chiming Grandmother Clock.* This is a small (5' 10") and comparatively rare rosewood-veneered, spring-driven, eight-day, tube-chiming clock. The 7 ½-inch breakarch brass dial has regulation for chime/silent and either *Westminster* or *Whittington* chimes, circa 1910-1920.

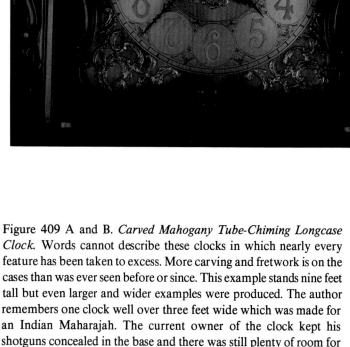

Figure 409 A.

Figure 408. *Edwardian Nine Tube Chiming longcase clock.* A typical Edwardian nine-tube-chiming longcase clock with an oval glass panel in the door displaying the brass cased weights and tubes. The brass dials used at this time were more decorative than anything previously produced. Individual gilt arabic numerals were laid into decorative engraved plaques on silvered chapter rings. Raised gilt decoration was also used extensively, for instance in the dial center and arch where you can see the regulation for strike/silent and fouror eight-bell chime.

Figure 409 A and B. *Carved Mahogany Tube-Chiming Longcase Clock.* Words cannot describe these clocks in which nearly every feature has been taken to excess. More carving and fretwork is on the cases than was ever seen before or since. This example stands nine feet tall but even larger and wider examples were produced. The author remembers one clock well over three feet wide which was made for an Indian Maharajah. The current owner of the clock kept his shotguns concealed in the base and there was still plenty of room for the weights. Many years ago we had one in which my wife could comfortably stand without her head coming up to the top of the trunk door.

These immensely expensive clocks were exported to wealthy customers all over the world who wished to make their wealth obvious.

The clocks represent the final fling of English horology, one final effort to produce the ultimate clock before retiring, murmuring "see if you can beat that."

The massive movement of this clock has dead beat escapement, maintaining power, and a wooden rod pendulum with pewter bob. Some featured mercurial compensation. See also Figures 223 A, B, and C. 9′ 0″ (274 cm) high.

*Victorian and Edwardian* 259

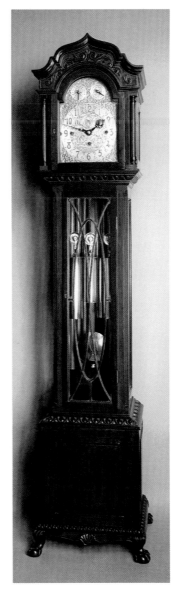

Figure 410 A and B. *Tube Chiming Clock by Elliott, London.* Elliott was probably the leading maker in England of chiming clocks. They were well made and this is a typical example. It has a more restrained style than the one seen in Figure 409, and is considerably smaller (7' 6"). It has chime/silent regulation and plays either *Westminster, Whittington* or *St. Michaels* chimes. The raised, individual gilt numerals are visible in the view of the movement, where the bells can be seen suspended from a cast iron bracket attached to the movement. The clocks in Figures 408 and 409 have bells suspended by cords from the backboard.

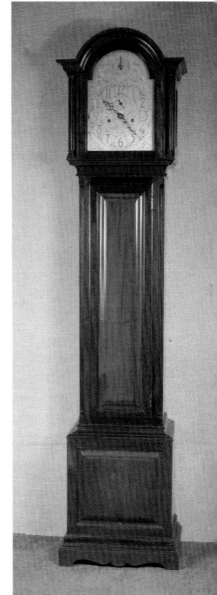

Figure 411. *Elliott, London.* Around the turn of the century, Elliott devised a system whereby a two-train clock could be produced with full chime (i.e. on all four quarters). Prior to this, two-train, quarter-chiming clocks had a system known as "pump-action quarter chime" whereby they chimed the first three quarters, and at the hour the pin barrel was "pumped over" so that it only engaged the hammer for the hour bell.

This example has a very good quality weight-driven movement with dead beat escapement, maintaining power, and Elliott's own form of compensated pendulum. It chimes the quarters on either four or eight straight rod gongs in contrast to the coiled gongs used in the 19th century. 6' 10" (208 cm) high.

260  *Victorian and Edwardian*

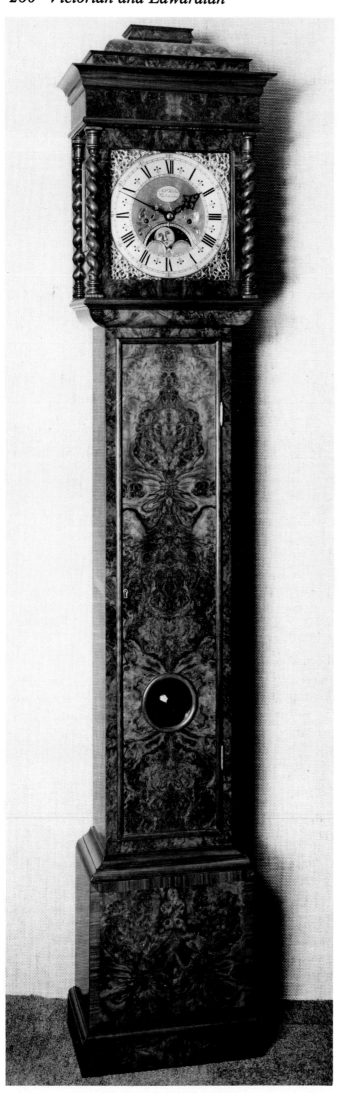

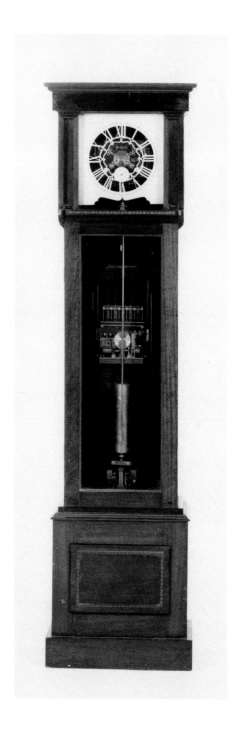

### AFTER 1945
Since the last war a small number of longcase clocks have been produced in England, usually of good quality and nearly always of 17th and 18th century designs. Figures 414 and 412 show two typical examples. Nearly all of the mass-produced grandfather/grandmother clocks have been made outside England.

Figure 412. *Walnut Longcase Clock in the Queen Anne Style.* This clock of excellent proportions and good quality is typical of those which have been produced in the last 30-40 years. Clocks are being made in very limited numbers (often in batches of 4, 6 or maybe 12) by different craftsmen in various parts of the country.

Figure 413. *Electrical Horology.* Although electrical clocks started to be made in the 1840s following Bains work in this field the number of electric longcases produced was very small. Just one is shown here for completeness. It employs a Hipp toggle impulsing the invar pendulum and is provided with quarter chime, a relatively unusual feature. 6′ 5″ (196 cm) high.
Photo courtesy Sotheby's, London.[4]

---
[4] *Antiquarian Horology*, Vol. 9, No. 7, June 1976, pages 794-799.

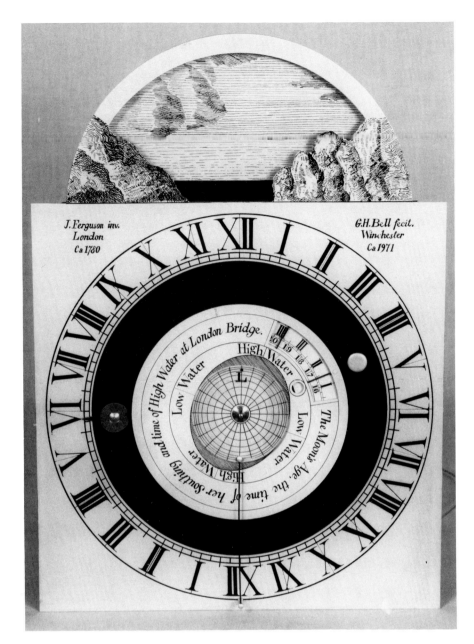
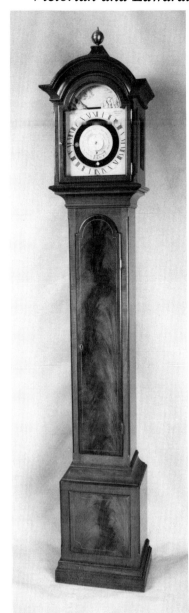

Figure 414 A, B, and C. *C. Ferguson's Tidal Clock, made by Geoffrey Bell, 1971.* This is one of a small series of longcase clocks made by Geoffrey Bell in 1971 to Ferguson's design, which is described in detail in Ree's *Clocks, Watches & Chronometers,* pages 81, 82, 83 & 132. It is most ingenious, giving the relationship between the sun and moon, and the Time of High Water at London Bridge. The sun, or hour, hand rotates around the main chapter ring, graduated I XII twice, once a day and the moon hand, which is attached to a central disc, rotates anti-clockwise once every 29 ½ days.

An aperture in the disc gives both the times of High Water at London Bridge and the age of the moon, and the moon's phases are displayed through a small circular cutout. Around the edge of the rotating disc "High Water" is engraved twice and in the center an ellipse represents the state of the tides so that this may be seen for the hour in question. A shutter in the arch rises and falls to represent the sea rising and falling in the estuary.

The substantial movement has tapering plates and an ingenious system of chain wind for the wire line on which the rolling weight rests. The good quality case has excellent proportions and glazing to either side of the hood. 6′ 8″ (203 cm) high including finial.

262  *Victorian and Edwardian*

Figure 415 A and B. Aberdeen was a particularly fine area for this type of clock, and this lovely example comes from Stonehaven (15 miles south). It has a delicate swan-neck pediment with brass paterae and finial, and shaped satinwood panels below. The fluted hood pillars have brass Corinthian capitals, and there are brass-capped quarter columns to either side of the trunk. The door has beautifully figured mahogany, with a satinwood line and crossbanded border. The base is similarly finished and rests on a shaped plinth.

The dial, as befits a fishing town, has a maritime theme and is beautifully executed with a seascape in the arch and brown and gilt anchors in the four corners. It has light Roman hour and delicate Arabic minute numerals with subsidiaries for seconds and date. It has good, blued steel hands and is signed *William Buchan, Stonehaven*, circa 1790. 6′ 8½″ (205 cm) high excluding finial.

Figure 416. A mahogany longcase clock by William Spark of Aberdeen, probably made 20 to 30 years later than those seen in Figures 415 and 417. Although still of good proportions, the quality of the case is not so fine and the dial is more heavily decorated.

# Chapter 15.
# Regional Characteristics

Throughout most of the period of longcase clock production in London, the designs tended to be stereotyped because of strong social pressures which dictated what was fashionable in clothing, houses, and furniture. A further factor was the small number of cabinetmakers producing clock cases mostly using similar patterns. Therefore, although the basic design and execution of cases and movements was excellent, the repetition practically stifled all individuality and ingenuity. There were, of course, some exceptions, particularly prior to 1725. Many of Tompion's clocks had great individuality; Delander produced his fine clocks with unusual flat tops or shallow breakarches; Williamson made a series of clocks showing equation of time or solar time; makers such as Pinchbeck and Kipling evolved some complex astronomical clocks; and Clay, among others, made lovely organ clocks.

The further from London the less the influence. Thus in towns within 50 miles, such as Rochester, Uxbridge or St. Albans, and towns and cities with a close relationship to London such as Oxford and Cambridge, designs followed those of London quite closely, although usually with some simplifications. For instance, moulds tended to be lighter and less complex, and on mahogany clocks the raised base panel was often omitted. In the second half of the 18th and first half of the 19th centuries, much more individual designs evolved in areas such as the southwest of England and Scotland. The simplified pagoda top probably originated in London, whereas the swan-neck top (which was popular in the provinces in the latter part of the 18th century), was never used there.

While it is impossible to describe all the regional characteristics in this work, the range of this subject may be gleaned by referring to the books on clockmaking in the British Isles which are listed at the end of this chapter. Here we endeavor only to highlight some of the principal characteristics of various areas. For instance, the wavy inside borders to the hood door of many West Country longcase clocks and a highly individual form of pagoda top made in and around Hull.

To demonstrate regional styles and their influences on nearby makers, maps are included that indicate most of the places where clocks were produced in the United Kingdom in the late 17th, 18th, and 19th centuries.

Figure 417. The best of the longcase clocks made in Scotland in the last quarter of the 18th century are among the finest produced anywhere. They are always of excellent proportions, slim and relatively small. This longcase clock by Pringle of Dalkeith is a classic example. It has a swan-neck pediment with a finial between and frets below; brass-capped columns to the hood; and quarter columns to either side of the trunk, which has an attractively shaped door with boxwood stringing and panelled base. The movement is far more substantial than most and has a 12-inch brass dial with a raised chapter ring; Indian head spandrels in the corners; dolphin spandrels and name plaque in the arch; a large second hand, date aperture, and strike/silent regulation above 12 o'clock. Circa 1775. 7′ 2″/6′ 11½″ (219/212 cm) high with/without finial.

## Scottish Clockmaking

Although clockmaking began in Scotland at the end of the 15th century, it was to be another 150 years before the number of clockmakers started to increase appreciably. They then became recognized as a branch of the locksmiths' trade, which made them members of the Hammermen Incorporations.

The early longcase clocks produced in Scotland mirrored those being made in London, but often with cases of simplified design. It was only from 1770-1780 onward that Scotland really came into her own, producing clocks of fine quality and great technical merit. Some of the clocks had excellent proportions and style, the quality of which has seldom been exceeded anywhere.

One of the reasons for this change was the academic excellence that prevailed, particularly in Edinburgh at that time. This gave rise to brilliant clockmakers, particularly in the field of precision timekeeping. An example that springs to mind is Reid, who wrote an excellent *Treatise on Clock and Watchmaking* and who in conjunction with his partner, Auld, made a fine astronomical regulator for Lord Grey's private observatory at Kinfauns Castle in 1811 (Figure 464). This clock had a 45-day duration and at the time was probably the most technically advanced regulator in existence. It employed spring pallets to detach the pendulum from the direct influence of the movement and give it a constant impulse (i).

William Hardy was another fine maker. In 1820 he was awarded 50 guineas and a gold medal for designing another form of regulator with spring pallets. His workmanship was of the highest order, as was that of James Ritchie (who devised an interesting compensated pendulum), and Robert and Alexander Bryson. During this period the precision timekeeping work and ingenuity of the finer Edinburgh makers outshone London.

The clock cases these and many other Scottish makers employed both for their ordinary clocks and their regulators are of simple, almost austere lines and are usually relatively small. They are made of solid mahogany and either have a circular engraved and silvered (or occasionally a painted) dial. One particular feature often seen at this time is the tapering trunk.

The clock cases produced from 1770 to 1810 on Scotland's East Coast, mostly around Aberdeen, are in direct contrast. A study has revealed so many identical details in their design and construction that it would seem more likely that they are all the products of one firm of cabinetmakers.

The earlier examples of these clocks have brass dials with a raised chapter ring and spandrels or occasionally all-over engraved and silvered dials. The later ones have especially well-executed painted dials. However, there is a considerable overlap between the use of all three types of dial.

Strike/silent regulation was often provided on the brass dials, which were either 12 inches or 13 inches whereas the painted dials tended to be 13 inches or on the later clocks even 14 inches.

The cases are of the finest quality, with beautifully chosen veneers, a shallow swan-neck pediment with fretwork below, quarter columns to the slim trunk which has an attractively shaped door and crossbanding to both this and the base. On the later clocks a shorter door was employed with a panel below.

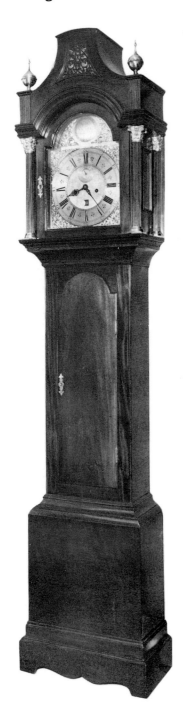

Figure 418. Mid-18th-century longcase clock by James Nicoll, Canongate, Edinburgh, who was apprenticed to Alexander Brande in 1721, gained his freedom as a clock and watchmaker on 2 July 1729, and continued working until 1760.

Although this clock is similar to the classic London mahogany longcase clock, it is of simpler construction and lacks fine veneers. There is no mould around the door or panel on the base and the style of dial would have gone out of fashion in London by the time this clock was made. 7' 8" (234 cm) high.

A much finer and more complex Scottish mahogany pagoda-topped longcase clock by Alexander Ferguson is to be seen in Figure 233 A, B, and C. This is equal to a London longcase in quality but has distinct differences, such as a plain base and a scalloped border to the pagoda.

References:

Roberts, D., *Precision Pendulum Clocks.* Pp. 30, 31, 32, 76-78. Derek Roberts Antiques, June 1986.
Smith, J., *Old Scottish Clockmakers Up to 1900.* David Watts and Sons Ltd., 1982.
Hudson, Felix, *Scottish Clockmakers.* Felix Hudson, 1984.

Figure 419. A Scottish mahogany longcase clock circa 1830-1840. The white painted dial with relatively heavy Roman numerals shows seconds below 12 o'clock, date above 6 o'clock, and *Tempus Fugit* in the arch.

Note the relatively small rectangular trunk door and the spirally reeded columns to the hood and sides of the trunk. By this period the beautiful proportions achieved with the earlier clocks had been lost. 7' 1" (216 cm) high. Photo courtesy of Phillips, Scotland.

Figure 420. An early 19th-century Scottish mahogany longcase clock quarter chiming on eight bells and showing the moon's phases in the arch. Note the typically Scottish feature of columns to the hood and trunk. The plain and rather crude swan-neck applied to the flat top is a relatively late feature and is quite frequently missing on these clocks. 7' 9" (236 cm) high, with a 14-inch dial.

Figure 421. The beginning of the 19th century was to usher in, particularly around Edinburgh, an era of plain, functional case styles of which this is a typical example. Circular dials between 12 and 14 inches of either engraved and silvered brass or painted were employed. Solid mahogany was the timber of choice for the cases.

This eight-day striking clock has a 13-inch circular painted dial with subsidiary rings for seconds and date, to which access may be gained by a hinged bezel. There are small raised panels below, reeded quarter columns to either side of the trunk, and a panelled base. The plinth has been reduced. It is signed by J. Durward of Edinburgh, who was admitted a freeman clockmaker to the Canongate Hammermen on 20 October 1775. On 26 May 1819 he handed over his business to James Ritchie. This clock would date circa 1800. 6' 7" (201 cm) high. Photo courtesy of Phillips, Scotland.

266 *Regional*

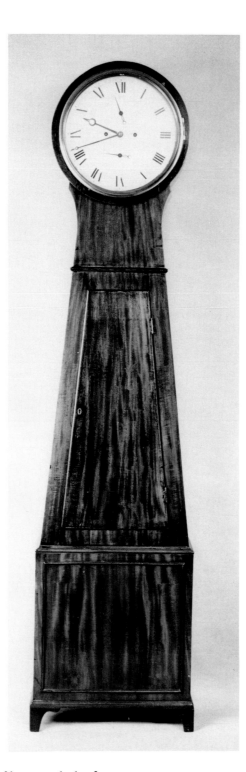
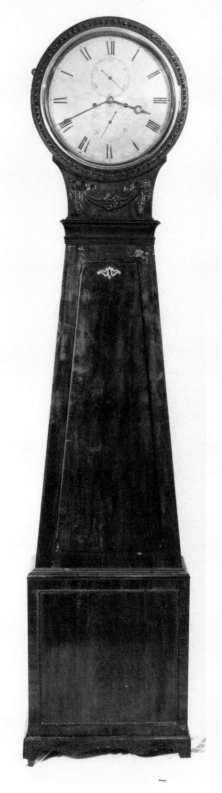

Figure 422. The use of the "drumhead" style of longcase clock, often of fine quality, is confined almost entirely to Scotland and most often Edinburgh, although London clocks also occasionally employ this form of case. Like most other clocks being made in Edinburgh at the time, it draws maximum attention to the engraved and silvered brass dial. The trunk has fluted columns to either side and there is a recessed panel below the door and to the base. It is signed by Christopher Lawson, who worked at 19 North Bridge, Edinburgh, from 1820 to 1837. 7′ 4″ (214 cm) high. Photo courtesy of Phillips, Edinburgh.

Figure 423. The use of a tapering trunk on longcase clocks was virtually confined to Scottish clockmakers; the design possibly being chosen by men such as Ritchie because of the great stability it obviously provides, particularly for regulators. In nearly all cases solid mahogany was used.

This striking clock has an engraved and silvered brass dial with subsidiaries for seconds and date and a panelled base. 6′ 4″ (193 cm) high.

Figure 424. A somewhat similar but possibly better-proportioned longcase clock than that seen in Figure 423. It also has a silvered dial with subsidiaries for seconds and date. There is a raised, carved husk border outside the brass bezel and further carving below the dial.

Because of the difficulties caused by having side hinges to a tapering door, this example can be completely removed by undoing the top lock and lifting the door vertically away from the two pins or posts which retain it at the bottom.

It strikes on a bell, is provided with dead beat escapement, and is signed by Whitelaw of Edinburgh (which could be either David Whitelaw, who worked at 16 Princes Street from 1815 to 1835; or James Whitelaw, who was active from 1820 to 1846). 6′ 7″ (200 cm) high. Photo courtesy of Phillips, Edinburgh.

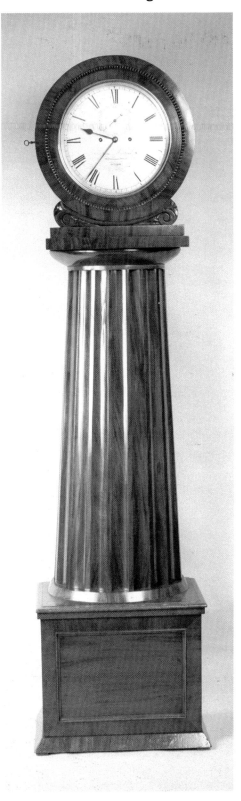

Figure 425. A much more substantial longcase than that seen in Figures 423 and 424. The front of the case is decorated with fine mahogany veneers, there are panels above and below the flush trunk door, and the front of the hood is like a large crossbanded wooden bezel. The fluted brass dial has center-sweep minute and hour hands and seconds below 12 o'clock. The massive movement, which strikes on a gong, is built to full regulator standards. It is signed by Blaylock of Carlisle and dates circa 1850. 6′ 10″ (208 cm) high.

Figure 426. A mid-19th-century longcase clock signed by J.D. Reid of Airdrie with features typical of the period, including a gold-lacquered circular painted dial with carving below and above the hood and a shaped and moulded solid mahogany door. It strikes on a gong. 7′ 6″ (230 cm) high. Photo courtesy of Phillips, Edinburgh.

Figure 427. A mahogany longcase clock signed on the engraved and silvered brass dial by James Muirhead, 90 Buchanan Street, Glasgow, where he worked from 1817 to 1841. The fine quality drumhead case, a little reminiscent of the one which Tompion produced for the pump room in Bath over 100 years earlier, has a substantial fluted and tapering trunk with rings to the top and bottom. It rests on a panelled base. 6′ 11″ (211 cm) high.

Figure 428. An early Victorian mahogany longcase clock signed on the painted dial by George Lumsden of Pittenweem (1818 to 1849). It has features typical of the 1840s, including substantial barley twist columns to the hood and trunk, carving above the hood, and a short shaped and moulded trunk door.

Smith in *Old Scottish Clockmakers* comments on Lumsden: This maker was an apprentice of his celebrated townsman John Smith (who made one exceptional musical clock). He carried on an extensive trade, and his productions are to be found all along the coast fishing towns and villages in the East Neuk of Fife. Suiting the design of his clock dials to the tastes of his customers, many of them have painted on them ships, fishing boats, and the like, which were in a large number of the cases the production of the pencil of James Brown, joiner, Pittenween. 7′ 5″ (225 cm) high. Photo courtesy of Phillips, Edinburgh.

Figure 429. A mahogany longcase clock by Thomas McGregor of Ayton circa 1835-1840. Note the short shaped door surmounted by a finial. The split columns to either side and the curved panels above and below are relatively late features, as also is the heavily decorated dial with girls in suitable costume representing the four seasons in the corners and a tavern scene in the arch. 7′ 3″ (220 cm) high. Photo courtesy of Phillips, Edinburgh.

Figure 430. An unsigned Scottish longcase clock circa 1840. The flat-topped hood is surmounted by a swan neck. The well-figured rectangular trunk door is surrounded by crossbanding and has quarter columns to either side. The 14-inch painted dial is heavily decorated with Figures depicting the four seasons and "Peace and Plenty." The clock strikes on a gong. 7′ 5″ (229 cm) high. Photo courtesy of Phillips, Edinburgh.

Figure 431. A late 19th-century mahogany longcase clock with a brass dial with a raised chapter ring and spandrels. It has regulation in the arch for chiming on eight bells or four gongs and is signed by William Alexander and Son, Glasgow.

It has a carved and pierced swan-neck pediment, a bombe base with a carved panel, and is extensively decorated with blind fretwork. 7′ 6½″ (230 cm) high. Photo courtesy of Phillips, Edinburgh.

*Ireland*

In the 18th century Dublin was one of the most important cities in the British Isles, and the quality of the products of its numerous craftsmen second to none. Fine silver was being produced, along with some excellent furniture which had its own original character. The number of longcase clocks produced was relatively small, but nearly all those made were of very good quality. Whereas in London the square dial had largely gone out of favor by 1720, in Dublin its use continued until the 1780s. With the earlier clocks the dial was sometimes 12 inches square, but more often it was 13 inches and quite frequently 14 inches, which gave rise to clocks around eight feet tall.

The cases, a little heavy in style, are always well and substantially made, usually out of solid mahogany (sometimes veneered). The earlier examples occasionally are finished in walnut or burr elm and these generally have flat tops.

A swan-neck pediment is employed with either brass or, frequently on the later clocks, wooden paterae and finials. The hood columns are usually brass capped. Matching quarter columns are employed for the trunk, the door of which is often attractively shaped. In most instances the base is panelled and the clock rests on substantial ogee feet.

The dials are relatively conventional, but on some of the earlier clocks a band of wheatear engraving is visible on the inside of the chapter ring, and ringed winding holes may be employed. Large cherub spandrels are frequently found as seen on the clock in Figure 432.

Reference: Fennell, G. *A List of Irish Watch and Clockmakers*. National Museum of Ireland, 1963.

Figure 432. A typical Dublin-made eight-day longcase clock circa 1770 signed by William Raymen(t) of Dublin. Note the lightly ringed winding holes, large cherub spandrels, and panelled base. 7′ 7″ (231 cm) high, excluding the top finial. Dial is 13 inches square.

270 Regional

## Lancashire Clocks

When we speak of Lancashire clocks, we have to qualify this a bit because of the changes which took place in the county boundaries after the last war. Here we are obviously referring to Lancashire as it was prior to these changes and are also including neighboring districts such as Halifax, which produced similar clocks.

From as early as 1640, Liverpool was manufacturing components for watches. In the ensuing years it became the major supplier in England for watch and clock components and the tools required to make them. Their watch trade continued until the 19th century.

Few clocks were produced in Lancashire prior to 1700. Prior to 1750, their numbers were still relatively low, many of the clocks being 30-hour oak clocks with plated single-handed movements. After 1750 their numbers grew quickly due to the rapidly increasing wealth and rising population of the district coupled with the great importance of Liverpool as a port for shipments all over the world (and in particular to America).

This gave rise to what is now generally known as the Lancashire longcase clock, which first appeared circa 1760. They have a distinctive style of their own, possibly owing a little to Chippendale, and are almost always of very fine quality, being made by such eminent cabinetmakers as Gillows of Lancashire. They were usually decorated with beautifully figured mahogany veneers.

Characteristic features are the box top seen behind the swan neck; the panels below this, frequently of painted glass but occasionally fretted; the quarter, half, and even full columns to either side of the trunk; the characteristic shape to the top of the door; and the "brickwork" base often seen on the better clocks. On the larger and more important pieces such as that seen in Figure 121 A, B, C, and D, double columns were used to either side of the hood.

They are generally large clocks (averaging about eight feet) and are somewhat wider than their London counterparts, although they still are well proportioned when compared with many clocks made in the North after 1820. The earlier clocks were provided with brass breakarch dials, usually either 13 inches or 14 inches, and frequently had a rolling moon above which a bold signature commonly was featured. After circa 1780 painted dials were more commonly used; this was usually accompanied by some simplification in the case design.

A case design which appeared in the Northeast prior to the typical Lancashire style was that seen in Figure 438. It was produced in walnut, oak, and mahogany and dates back to the 1720s.

Several fine makers such as Lister of Halifax, Ogden, and Booth (Figure 258 A and B) seem to have favored this style and often incorporated revolving ball moons, sometimes known as Halifax moons, in the arch. This type of case continued on, often in simple oak form, until the end of the century. Reference: Loomes, B., *Lancashire Clocks and Clockmakers*. David and Charles, London, 1975.

Figure 432 i. A fine quality unsigned Lancashire longcase clock quarter chiming on eight bells, circa 1785. It features a painted rolling moon in the arch of the 14-inch dial and a typical boxed swan-neck top with *verre eglomisee* panels and well-carved wooden paterae with further draped carving between them. The hood and trunk doors and the base are all picked out with satinwood to either side of the hood and below the quarter columns flanking the trunk. The clock would originally have had three wooden finials. Note the beautifully executed cube inlay to the top of the trunk which gives a three-dimensional effect.

Figure 433 A, B, and C. A most interesting longcase clock by Green of Liverpool circa 1830 with center-sweep second hand. It is included here not because of the unusual escapement or the case style but to display the fine quality of the engraved dial center on some Liverpool clocks. This particular example is most unusual because a date dial is set into the center. 6′ 8″ (203 cm) high.

Figure 434. An eight-day longcase clock by John Kay of Liverpool, who was working in the last quarter of the 18th century. Note the beautifully figured book-matched veneers to the door and base. The solid panels below the swan neck may be replacements for glass panels, and the hood could originally have had rear quarter columns. The brickwork base was employed on many of these clocks.

The 13-inch brass dial has a raised chapter ring and spandrels, a rolling moon in the arch, date aperture and seconds hand, and a beautifully engraved dial center with crosshatching. This was a fine feature of some of the best makers in and around Liverpool at that period and can be seen more clearly in the close-up of the dial in Figure 433 B. It was presumably the product of one family of engravers.

This clock was bought in the 1920s by a friend, then an impecunious house surgeon at the local hospital in Liverpool, after much scrimping and saving, for the princely sum of £5 when all other longcase clocks he was offered were only £2. 7′ 9″ (236 cm) high.

Figure 435. A typical Liverpool longcase clock circa 1775 made by John Wyke, who was a famous watchmaker and clock toolmaker. He worked at Wykes Court, Dale Street, and died in 1787.

As usual it is decorated with finely figured book-matched veneers and has *verre eglomisee* panels (painted glass) below the swan neck. There is a "brickwork" base with applied frets to the corners of the panel. These also are featured below the hood and at the base of the columns on either side of the trunk. It has a 14-inch dial. 8′ 10″ (269 cm) high. Photo courtesy of Sotheby's, Bond Street.

Figure 436. A George III oak longcase clock signed on the 14-inch brass dial which shows the moon's phases *Burton & Pattison, Halifax*. It has a center-sweep second hand (not seen here) together with dead beat escapement and a date hand above 6 o'clock. There is applied fretwork below the swan neck and at the top of the trunk. The trunk door and base are crossbanded with mahogany. 7′ 6½″ (230 cm) high. Photo courtesy of Sotheby's, Bond Street.

Figure 437. A late-18th-century Lancashire-style longcase clock by Thomas Twigg of Sheldon (Derbyshire). The clock, while still of fine quality with beautifully chosen book-matched veneers to the door, is somewhat simpler than those being produced some 20 to 30 years earlier. Note that there is no applied fretwork or carving and that by this time the "brickwork" was no longer used to decorate the base.

The 14-inch painted dial has roses in the four corners and shows moon phases in the arch. There is a second hand; date aperture and brass hands also are used. The clock now lacks its ogee bracket feet and finials to the top, probably to reduce its height. 7' 8" (234 cm) high.

Figure 438. A simple oak eight-day longcase clock with a 12-inch white dial delicately decorated in the four corners with raised giltwork and moon phases in the arch. Note the shaping of the door similar to that of the Lancashire clocks of the period. It is signed by William Payne of Ludlow (a town 35 miles south of Shrewsbury in Shropshire).

This case style with breakarch top originated in the Northwest in the late 1720s and was still in use until the end of the century.

274 Regional

**The Northeast**

Just two case styles are illustrated here, one of which was popular around Newcastle-on-Tyne (Figure 439) and the other around Hull. Although these (with those of Lancashire) might all be called North Country clocks, it will be seen that their characteristics are entirely different, which illustrates how dangerous it is to group clock case styles over too wide an area.

When speaking of North Country clocks, people too often think of the massive and excessively wide clocks, often of indifferent quality, produced mainly in the industrial regions of the North after 1830. Nothing could be more dissimilar than the superbly proportioned clocks made on the East Coast mainly around Hull but also in Lincolnshire in the last half of the 18th century.

Figure 439. This style is usually characterized by a well-made case, frequently of solid mahogany; a much heavier than usual swan-neck pediment (generally with carved wooden paterae with little space between for a finial); a long slim trunk and door flanked by brass-capped and fluted quarter columns which have beading down either edge and a panelled base with canted corners. The design probably originated in the 1750s and lasted for some 30 years. The 12" brass breakarch dial of this clock is signed William Tickle, Newcastle and would date circa 1765. 7' 7" (231 cm) high.

Figure 440. An early 19th-century oak eight-day longcase clock by William Blanchard of Hull. It has an excellently proportioned case with canting to either side of the trunk. There is crossbanding around the hood and trunk doors and inlay let into the center of the fine pagoda top, which is outlined by a delicate mould. Note that the base is crossbanded on three sides only, a characteristic feature of these clocks. The 12-inch painted dial has roses in the four corners, and as befits a clock made at an important port, it has sea shells in the arch.

Thirty-hour longcase clocks are to be seen with similar cases. One of the most delicately designed and executed longcase clocks which the writer has had the pleasure of owning is illustrated in Figure 324. The case of this clock is signed by J. Usher, cabinet and clock case maker, who made it for B. Parr of Grantham the clockmaker.

Figure 441. The pagoda-topped longcase clocks made around Hull and on the East coast of Lincolnshire from circa 1760 to 1820 are among the best-proportioned clocks ever produced. Many prefer them to their wealthy London cousins because they are slimmer and have a more delicate pagoda top, although, of course, they are not usually made to such a high standard. However, the clock seen here signed by Branson of Hull is similar to a London clock complete with brass-strung quarter columns but has a slimmer trunk, a more compact base, and a relatively narrow hood which makes the dial look particularly full. 7′ 4″ (224 cm) high excluding top finial.

Reference
Walker, J.E.S. *Hull and East Riding Clocks*. Hornsea Museum Publication, 1982.

### Norfolk and Suffolk

Unfortunately, relatively little has been written about clocks originating in East Anglia. However, one case style produced here from toward the end of the 18th century until roughly 1825 is worth mentioning because, although it is similar to clocks produced in other counties, it does have some distinct differences. In addition, it is of a size (usually less that seven feet) that makes it eminently suitable for most homes.

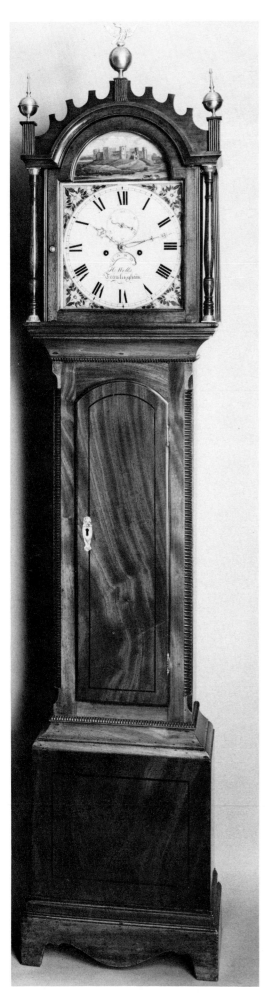

Figure 442. This clock, signed *H. Wells, Framlingham*, circa 1815, is typical of those produced in Norfolk and Suffolk at this time. It has decorative turned pillars, a scalloped design to the top of the hood which is surmounted by three finials, an arched top to the door which (like the base) has an ebony line let into it, and beading around the top and bottom of the trunk and delicate quarter columns to match. Like most of these clocks, it has a pine carcass. The 12-inch painted dial has a second ring; date aperture; brass hands; floral decoration in the four corners; and in the arch a picture of Framlingham Castle, a 12th-century building of which only the gatehouse and external wall remain. Reference: Hagar and Miller, *Suffolk Clocks and Clockmakers.* Hagar and Miller, 1974.

### The West Country (Devon and Cornwall)

As befits a seafaring breed who have supplied many of our greatest seamen, this is strongly reflected in the design of the clocks that were produced in this part of England. The moon is depicted frequently on their clocks because it had such a strong bearing on their lives, providing light to assist them on journeys and aid them in fishing. It was also important because of its influence on the tides and the subsequent scheduling of sea trips. For this reason the "time of high water" was indicated on many dials. Another feature frequently seen on West Country clocks is a rocking ship in the arch. Sometimes this will be named and will be one of historic significance, but in most cases it will be abstract.

The decoration applied to the case usually had a nautical theme. Thus a wavy outline was seen on the inner edge of the hood door and rope twist columns were often employed on the hood columns and those to either side of the trunk. The earlier clocks had full-length trunk doors, and this continued with the early painted dial clocks. By the 1820s, however, the clocks began to get wider and a much shorter door was employed.

References:

Bellchambers, J.H., *Devonshire Clockmakers.* The Devonshire Press, 1962.
Brown-Miles, *Cornish Clocks and Clockmakers.* David and Charles, 1961.
Ponsford, Clive, *Devon Clocks and Clockmakers.* David and Charles, 1985.

Figure 443. A fairly small, typically West Country, well-figured mahogany eight-day longcase clock. It has relatively heavy rope twist columns to either side of the hood and trunk and a scalloped surround to the hood door. Both the short trunk door and base are crossbanded. The painted dial shows moon phases in the arch with birds depicted flying about the world on either side. There is a date aperture, various buildings in the four corners, and in the center a windmill with sails that actually rotate. It is signed by Thomas Ellory of St. Austle, circa 1830. 6′ 9½″ (207 cm) high.

Figure 444. An eight-day mahogany longcase clock with crossbanding to the trunk door, which is flanked by quarter columns and also the base. The 13-inch painted dial depicts sea shells in the four corners and sailing ships in the arch. It is signed by Guilliam of Falmouth and dates circa 1815. 6′ 10″ (208 cm) high.

278 *Regional*

Figure 445. A solid mahogany eight-day longcase clock with wavy inner border to the hood door, canted and fluted corners to either side of the trunk, and an unusual boxwood-strung cloverleaf-shaped panel applied to the base. The 12-inch brass dial has a second hand and a large date aperture. In the arch a galleon is seen rocking to and fro against the background of a painted sea. The curved and engraved silvered brass strip above bears the name of the maker, Moses Jacob of Redruth (Cornwall), circa 1785. 6′ 10″ (208 cm) excluding finial.

Figure 446 A, and B. An attractively figured mahogany eight-day longcase clock with a wavy outline to the top of the hood, rope twist columns to either side of the trunk, and boxwood stringing to the door and base. The finials are replacements. The 12-inch painted dial has honeysuckle in the four corners and a moon dial incorporating a galleon in the arch. Besides the lunar calendar, the time of high water at Hayle is shown. The earth's hemispheres are depicted below the moon and it has decorative brass hands. It is signed *Em Cohen Redruth* who was born in 1766 and died in 1849. This clock is dated circa 1810. 6′ 11″ (217 cm) high excluding finial.

## Bristol Longcase Clocks

Although Bristol might be considered by many to be part of the West Country, it is treated separately here, in part because the type of clocks being made there in the second half of the 18th century are entirely different from those being made in Devon and Cornwall. Bristol was a wealthy port through which passed much of the nation's trade. The houses of many of the prosperous merchants and others were large and spacious, with good ceiling heights, and they wanted clocks to match them. The clocks made in Bristol were among the most elegant produced anywhere.

Like the West Country clocks, they frequently show the moon's phases in the arch. This often is combined with a display of high water at Bristol Key. Indeed, on some of the more sophisticated clocks, a universal tidal dial is provided so that the time of high water may be known at any other port, allowing the owner to plan his voyages accordingly.

The engraving on the dials is usually of a very high standard (Figure 282) and a large decorative date aperture often is used. The trunks of the cases are unusually long and slim and this is accentuated by the reeded and canted corners to either side. The bases are generally but by no means always panelled. A characteristic feature of these clocks is the basically flat top of the hood with a high swan-neck pediment mounted on top. The swan necks are often spaced widely apart and have beautiful wooden paterae with foliate carving, frequently gilt, which trails down from them. A clock that epitomizes all the characteristics of a Bristol longcase clock (to an almost exaggerated degree) is seen in Figure 282.

## The Southeast

The towns and cities around London were so greatly influenced by the capital that they did not develop the strong regional characteristics seen elsewhere in the country and particularly in the North and Southwest. Instead, they tended to produce London designs modified or simplified to suit their needs, and this was done to varying degrees. Undoubtedly, in some of the major towns and cities they merely ordered the cases from London cabinetmakers, possibly with minor alterations to the design. However, the further one strays from London, the greater the likelihood of finding cases of local manufacture. These cases typically were smaller than their London counterparts. They were often of solid mahogany instead of finely figured veneers; the moulds were simplified and in some cases omitted, around the edge of the trunk door, for instance. Likewise the raised base panel was often omitted and a single instead of a double plinth was employed (Figure 287 A and B). Typical of the minor modifications employed was the addition of a rocking ship (Figure 300) or occasionally a rolling moon to clocks made on the coast or more specifically the Thames estuary.

280 *Regional*

Figure 447. A considerably later clock than those seen in Figures 448 and 449. It still has the very slim trunk and tall base together with the high and widely spaced swan neck, but there is no longer a flat top for this to sit on and there are panels above and below the shortened trunk door. The veneer applied to the trunk and base, which are shaped and crossbanded, are of the finest quality. The dial has lost its characteristic Bristol features. Circa 1815. 8′ 7″ (246 cm) high.

Figure 448. A well-proportioned mahogany longcase clock by Thomas Williams of Chewstoke, a village seven miles south of Bristol. The widely spaced swan-neck pediment has applied fretwork below it and this continues down to the arch above the dial. The long, shaped trunk door is flanked by canted and reeded corners and the base has a shaped and raised panel typical of many West Country clocks. The 12-inch brass dial has a second hand; a large date aperture above 6 o'clock; and an engraved and painted rolling moon in the arch above which are indicated the moon's phases and the time of high water at Bristol Key, graduated V11-V1 twice. 7′ 9″ (236 cm) high.

Figure 449. A solid mahogany eight-day longcase clock signed within the recessed second ring by Thomas Nevitt, who became a burgess (gained the freedom) of the City of Bristol in April 1754 and took an apprentice circa 1772. This clock would probably date circa 1765, and in some ways is less decorative than the slightly later Bristol clocks due to its simple rectangular door and plain plinth. The dial is similar to that seen in Figure 448 showing moon's phases and the time of high water at Bristol Key. Note the fine and deep chasing to the dial center and the beautifully executed floral paterae to the ends of the swan necks with trailing foliage. 7′ 10″ (239 cm) high.

References:

Bates, K. *Clockmakers of Northumberland and Durham*. Pendulum Publications, 1980.
Beeson, C.F.C., *Clockmaking in Oxfordshire 1400-1850*. Museum of the History of Science, 1967.
Bellchambers, J.K., *Somerset Clockmakers*. Antiquarian Horological Society, 1968.
_____, *Devonshire Clockmakers*. The Devonshire Press, 1962.
Miles-Brown, H., *Cornish Clocks and Clockmakers*. David and Charles, 1961.
Darnell, J.A., The Making of Clocks and Watches in Leicestershire and Rutland. Leicestershire Archaeological Society, 1952.
Dinsdale, W.V., *The Old Clockmakers of Yorkshire*. Dalesman Publishing Company, 1946.
Dowler, Graham, *Gloucestershire Clock and Watchmakers*. Phillimore, 1984.
Elliott, D.J., *Shropshire Clock and Watchmakers*. Phillimore, 1979.
Fennell, G., *A List of Irish Watch and Clockmakers*. National Museum of Ireland, 1963.
Hagar and Miller, *Suffolk Clocks and Clockmakers*. Hagar and Miller, 1974.
Hodson, Felix, *Scottish Clockmakers*. Felix Hudson, 1984.
Hughes, R.A., *Derbyshire Clock and Watchmakers*. Derby Museums, 1976.
Legg, Edward, *The Clock and Watchmakers of Buckinghamshire*. Bradwell Abbey Field Study Centre, 1976.
Loomes, Brian, *Yorkshire Clockmakers*. Dalesman, 1972.
_____, *Lancashire Clocks and Clockmakers*. David and Charles, 1974.
_____, *Westmoreland Clocks & Clockmakers*. David & Charles, 1974.
Mason, B., *Clock and Watchmaking in Colchester*. Country Life, 1969.
Mather, H.H., *Clock and Watchmakers of Nottinghamshire*. Friends of Nottinghamshire Museums, 1979.
McKenna, J., *Watch and Clockmakers of the British Isles, Birmingham*. Pendulum Press.
_____, *Watch and Clockmakers of the British Isles, Warwickshire*. Pendulum Press.
Moore, Nicholas, *Chester Clocks and Clockmakers*. Grosvenor Museum, Chester.
Norgate and Hudson, *Dunfermline Clockmakers Up to 1900*. David Watts and Sons Ltd., 1982.
Peate, I. C., *Clockmakers in Wales*. National Museum of Wales, 1960.
Penfold, J.B., *The Clockmakers of Cumberland*. Brant Wright Associates, 1977.
Ponsford, CLive, *Devon Clocks and Clockmakers*. David and Charles, 1985.
Ponsford, C.N., *Exeter Clocks and Clockmakers*.
Ponsford, C.N.; Scott, J.M.; Auther, W.P., *Clock and Watchmakers of Tiverton*. Ponsford, Scott, and Auther, 1977.
Reid, C.L., *North Country Clockmakers*. Pendulum Publications, 1925.
Seaby, W., *Clockmakers of Warwickshire and Leamington*.
Smith, J., *Old Scottish Clockmakers from 1453-1850*. E.P. Publishing Ltd., 1975.
Snell, Michael, *Clocks and Clockmakers of Salisbury*. Hobnob Press, 1986.
Trekerne, A.A., *Nantwich Clockmakers*. Nantwich Museum.
Trible and Whatmore. *Dorset Clock and Watchmakers*. Tanat Books, 1981.
Tyler, J., *Sussex Clockmakers*. The Watch and Clock Book Society Ltd.
Walker, J.E.S., Hull and East Riding Clocks. Hornsea Museum, 1981.
Waters, I., *Chepstow Clock and Watchmakers*.

advanced design produced. The eight-day movement is of the highest quality with heavy plates and shaped and screwed pillars. It has fine, six-spoke wheelwork, high-count pinions and maintaining power. The pallet and the 'scape wheel arbors are jewelled. The four legs of the gravity escapement, which are different from Denison's original design, project from a steel ring.

The movement is driven by two heavy weights which are on one continuous line. They are arranged to drop on either side of the movement. The power, absorbed by the gravity escapement with its fly, is far greater than on a conventional regulator and thus a much heavier driving weight is needed.

The 12-inch, 24-hour dial is surrounded by a silvered bezel and is of classic regulator layout. All numerals are Arabic and the hands are blued steel. The dial is signed *Chas. Frodsham & Co. Clockmakers to the Queen. 27 South Moulton Street, late of 84 Strand, London, No. 1562.*

There are four medallions surmounted by crowns on the dial: "By APPOINTMENT to the Queen"; "Gold Medal of Honour Paris Exhibition"; "Grand Gold Medal 1860, Russian Survey"; and "1st Class Gold Medal 1871, Naples Exhibition." The case is of gilded brass top and bottom with gilded rods framing the corners. Heavy bevelled glasses form the sides and front and winding is through the front glass, the hole being covered by a brass dust cover. The case slides on runners off the base of the backboard to allow access.

Figure 450 A and B. *Charles Frodsham's Gravity Regulator, No. 1652.* A most-impressive wall regulator by Charles Frodsham & Co. with four-legged gravity escapement. The massive six-pillar movement is mounted on a substantial brass bracket which is itself screwed to a heavy and attractive panelled rosewood backboard. The aluminum pendulum rod, which is suspended from the bracket, has two long metal tubes containing mercury and would seem to be the most-

# Chapter 16.
# Precision Timekeeping

It is likely that man has been striving to measure time more accurately since the day he first started recording it. In Roman times Clepsydra (water clocks) were in common use, and in at least one instance the hole through which the water ran out was jewelled to improve its accuracy. However, if a Roman senator, whose speech the Clepsydra was timing and thus limiting, wanted to speak for a longer time, he merely sprinkled sand into the bottom to partially block the hole and slow down the clock!

The first major advance in timekeeping occurred toward the end of the 13th century with the invention of the mechanical clock using a verge escapement with a foliot. The ultimate development of the verge was the "Cross Beat" (Figure 452 A and B), all knowledge of which had been virtually lost until comparatively recently when Professor Hans von Bertele of Vienna rediscovered it. The "Cross Beat" was invented by Jost Burgi circa 1584. As a clockmaker, his ingenuity and workmanship were of the highest order. It was gradually refined by Burgi and his pupils, but despite its excellent performance it does not appear to have come into general use, possibly in part because of the relatively precise nature of the work required and also because the pendulum rendered it obsolete. The "Cross Beat" is a precision escapement, each pallet being carried on a separate arbor, which permits very accurate adjustment at the point of contact of the pallet with the vertically mounted 'scape wheel. The two arbors are geared together and to the end of each is attached a counterbalanced arm. As the clock beats these arms cross over each other, hence the name "cross beat." It is a fascinating action to watch.

The next advance in accurate timekeeping was the construction of the first pendulum clock by Salomon Coster to Huygen's design in 1657. This reduced their error from around five to ten minutes a day to two to three minutes a week.

## CIRCULAR ERROR

When Huygens analyzed Galileo's work on the isochronicity of the pendulum he found that Galileo's observation was accurate only if the swing passed through a somewhat steeper path than an arc known as a "cycloidal curve." To overcome this error, which on the verge escapement with its relatively large arc of swing could be appreciable, he suggested the use of "cycloidal" cheeks on either side of the pendulum's suspension to modify its arc. However, these were relatively seldom applied since the error was not then considered to be significant. The importance of circular error was decreased greatly by the invention, some 12 to 13 years later, of the anchor escapement, which reduced the arc of the pendulum's swing and thus the errors involved.

Figures 451 A (Left). *Longcase Regulator with Nine-Rod Gridiron Pendulum by Walsh, Newbury, 1837* and (right) (Figure 452) *Frodsham Wall Regulator.* For description see page 285.

Figure 451 B and C. (left) This regulator has a very substantial mahogany case with a breakarch hood with canted corners and good mouldings. The trunk has a glazed door through which one sees the mahogany backboard with boxwood and ebony stringing to match the shape of the door. A nice feature is the inlaid flower pattern around the holes through which the screws pass to fix it to the wall.

The 12-inch silvered dial has an unusual layout with the hour ring at 3 o'clock and the seconds ring at 9 o'clock, minutes being from the center. The moon-shaped hands are of blued steel. The substantial eight-day movement has five screwed pillars and is mounted on a massive mahogany seatboard 1½-inches thick with brass plates for aligning the movement. The choice of a pinwheel for the escapement is quite rare for an English regulator and the counterpoised pallet arm of this contains jewelled pallets. The nine-rod, brass and steel, gridiron pendulum is suspended from a heavy 2¼-inch block of mahogany screwed to the backboard and positioned over the seatboard making one solid unit. The pendulum is fixed by locking screws for accurate alignment and has a heavy brass bob with graduated rating nut below.

Henry Walsh was apprenticed to Robert Howard Fish of Hanover Square, London, in 1827. He subsequently moved to a house on the north side of Newbury Bridge, probably now the Tudor Cafe. In 1836 he made his famous siderial and mean time regulator to the design of Joseph Vines which is now in the Science Museum, London. This clock, which also employs a gridiron pendulum but adjusted to siderial time, has two dials, the left one showing mean time and the right one siderial time and the age of the moon. 6′ 9″ (206 cm) high.

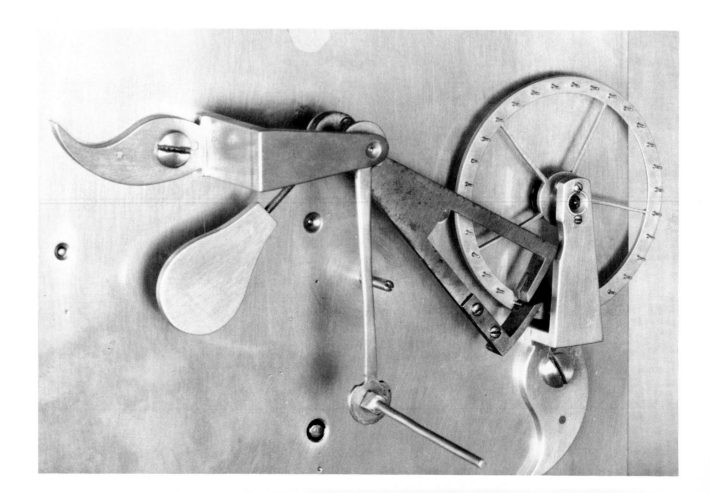

Figure 452. (See page 283, right.) *Frodsham Wall Regulator, No 1097*. A superb quality and unusually small mahogany wall regulator, one of a group of six made in the mid-19th century by *Charles Frodsham, Clockmaker to the Queen; 84 Strand, London* and numbered 1097. The case has an arched top, a delicate fret below the dial, and an attractive half-round, cross-grain mould around the full-length glazed trunk door. There is an ogee mould at the base.

The well-executed, mercury pendulum has a finely finished cast-iron jar with a T-piece fitted to the rod so that it may be held when regulating the clock. Running down the front of this is the pointer for the regulation scale at the top of the jar. The boxwood beat scale is beautifully made and signed by Frodsham. It is graduated in both inches and degrees of a circle, the radius of which is 36 inches. The key, typical of English practice at that time, has a brass handle and steel square.

The substantial movement has five pillars, five-spoke wheelwork, chain fuzee with maintaining power, jewelled pallets, and jewelling to the ¢scape wheel and pallet arbors. It is held down to the seatboard by three relatively heavy brackets. Beat regulation is provided. 43″ (109 cm) long.

Figure 452 A and B. *Radeloff Rolling Ball Clock With Cross Beat Escapement, circa 1660.* This clock, originally the property of Professor Hans von Bertele, is powered by balls rolling down a spiral. It utilizes the cross beat escapement, of which a close up is shown here. This was invented and developed by Jost Burgi of whom Radeloff was a pupil. The basic principle of the cross beat is the use of a single 'scape wheel and two separate pallets, each of which is carried by its own arbor, these being interlinked by two pinions so that as the clock "beats" the arms attached to the ends of the arbor's cross and recross, hence the name "cross beat." This escapement can be adjusted far more precisely than an ordinary verge and therefore keeps better time.

## THE ANCHOR ESCAPEMENT

Because the anchor escapement permitted the pendulum to have a much narrower arc, it was possible to increase the pendulum's length. The "Royal" pendulum was thus evolved, a little over 39 inches long and taking exactly one second to complete each beat, which made it a simple matter to provide a seconds hand on clocks. The accuracy of timekeeping now improved from variations of a few minutes a week to possibly no more than a few seconds. Even at this stage clockmakers became increasingly aware of the movement, and in particular any variation in the power transmitted had on the speed of vibration of the pendulum and thus its isochronicity. Two factors predominate here, both of which are greatly effected by the mechanical condition of the clock's components: (a) the power transmitted to the pendulum via the 'scape wheel, pallets and crutch, and (b) the power absorbed by locking and unlocking the 'scape wheel. In this respect the evolution from the verge escapement, with the pendulum linked directly to the pallets, to the anchor escapement was a marked improvement. A further advance took place with the perfection of the dead beat escapement, which until recently has been attributed to George Graham circa 1715. However, this is now a matter of some debate and it may well be that it was conceived circa 1709 in Tompion's workshops. (Ref.: Roberts, Derek, "Precision Pendulum Clocks" *Report of the Research Laboratory for Archaeology and the History of Art of Oxford University,* 1986, page 53.) This escapement avoids train reversal with the inevitable errors inherent therein. It reduces the effect of the movement on the pendulum and by employing a much narrower arc of swing minimizes circular errors.

## TOMPION'S CLOCKS AT GREENWICH

In 1676 Tompion made two of the first and certainly the most important precision clocks to be constructed in England by that date. They were specially ordered and paid for by Sir Jonas Moore for use by **Flamsteed** in the new observatory commissioned by the King in 1675.

The prime task of the clocks was to ascertain whether the Earth rotated at a constant speed. No clocks until that time had been accurate or reliable enough to determine this. Tompion's clocks subsequently achieved this goal and thus enabled objective work to go forward on the determination of longitude.

The specifications for the clocks were interesting in that both were of year duration, both had maintaining power, possibly to prevent train reversal and damage to the escapement during winding, and both had 13-foot pendulums beating two seconds mounted above the movement. Unfortunately no details of the construction of the pendulums remain. The hands and dial were exquisite (Figure 93).

The escapement was a form of dead beat initially designed by Towneley and subsequently modified by Tompion (5, 14). The movement of the pendulum at its tip was only ⅝ of an inch, an extremely small amount, did much to eliminate circular error.

Unfortunately, in their first two years the observatory experienced considerable difficulty with the clocks, possibly due to dust and dirt affecting the movement and also to problems with the escapement. During this period it would appear that they held time to within about five seconds a day. Although it is likely that their performance subsequently improved considerably, no record exists of this.

## EQUATION OF TIME

In the early days the easiest method of checking on the accuracy of a clock was by using a sundial, and because clocks were relatively inaccurate the difference between "solar" and "mean solar" (our) time was of little consequence. However, with the advent of the long pendulum clock and its greatly enhanced accuracy, the difference became increasingly significant. Clockmakers thus published tables (which sometimes are still seen fixed to the insides of doors of 17th- and early 18th-century longcase clocks (Figure 453)) which show the difference between solar and mean solar time throughout the year.

So that owners might be able to know both apparent and mean solar time without having to refer to tables, Tompion and others made clocks which showed, by means of an indicator in the arch (Figure 125), the equation of time, i.e. the difference between the sun's time and that shown on the clock, which varied by up

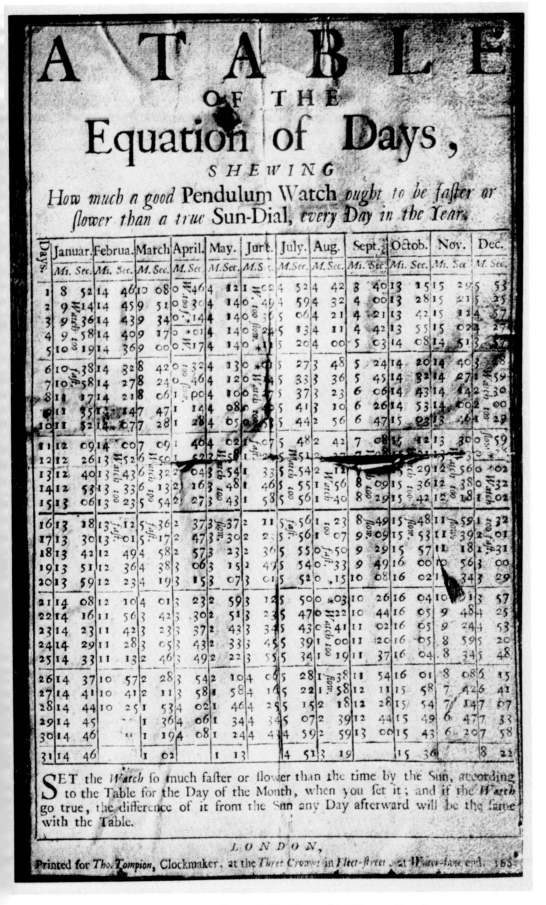

Figure 453. *Equation of Time Tables, Printed for Thomas Tompion.*

to 16 minutes during the year. The days on which the clock and the sun should usually agree are the 15th of April, 15th of June, 31st of August and the 24th of December.

Several different methods of showing solar time (or the equation of time) as well as mean time have been evolved over the years. One of the earliest was

# 288 Precision Timekeeping

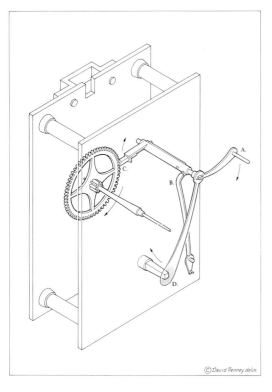

devised by Williamson, who made several equation clocks for Quare, including one of year duration which was supplied to Charles II of Spain. Some of his clocks were also made to show solar time by lengthening or shortening the effective length of the pendulum (refer to Figure 126).

## MAINTAINING POWER

As clocks improved time lost during winding, when the power was removed from the train, became increasingly significant. This obviously did not apply to the 30-hour clocks fitted with Huygens's continuous rope, but it did apply to virtually all clocks of eight-days and more duration. Thus a form of maintaining power known as *bolt and shutter* was devised (Figure 454). With this a cord is pulled or a lever depressed that actuates a spring-loaded bolt which engages on a wheel in the train and continues to supply power when the force of the driving weight is removed during winding. To make certain that this was always used, shutters were placed immediately behind the winding holes which were removed automatically when the maintaining power was actuated, thus permitting the clock to be wound.

This particular form of maintaining power was used extensively on longcase clocks until around 1690-1700 when it gradually fell out of favor. However, it continued to be used on regulators until the end of the 18th century, although its popularity decreased rapidly with the advent of Harrison's maintaining power, known as the "going ratchet" (Figure 455).

Figure 454 A, B, and C. *Bolt and Shutter maintaining power.* When the lever (a) to the right is depressed, the spring (b) is deflected and the bolt (c), which is lightly spring loaded so that it can move in and out, is raised and engages a tooth on the center wheel. The spring (b) now depresses the bolt and thus applies pressure to the center wheel, keeping the clock going. At the same time, the shutter (d) is moved away from the winding square. As the center wheel rotates, everything gradually returns to its original position.

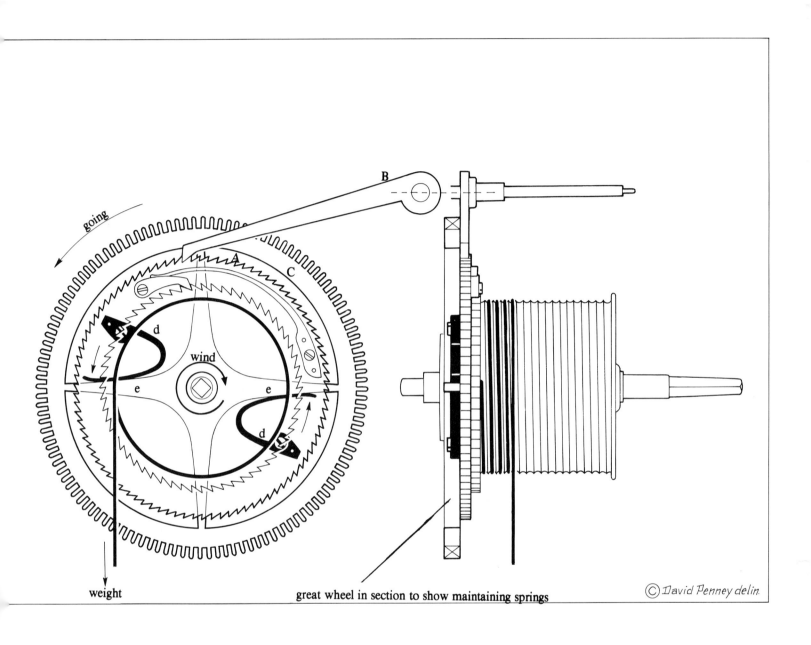

Figure 455. *Harrison's maintaining power.* In this, an additional wheel (a) with fine teeth and its own maintaining detent (b) is used next to the great wheel (c), to which it is connected by two springs (d) & (d) which, when the power of the weight is removed from the train during winding, still apply pressure to it via the spokes of the great wheel and keep the clock going. The springs gain their power by being compressed by the weight during normal running. A section through the great wheel on the right shows the maintaining springs.

# 290 Precision Timekeeping

## THERMAL COMPENSATION

Clockmakers became increasingly aware of deviations in timekeeping due to temperature changes. Probably the earliest method to try and overcome this was the use of wood instead of brass or iron for the pendulum rod. Whereas wood contracts and expands appreciably in cross grain with changes in temperature and humidity, most timbers remain quite stable in long grain. The small amount by which they do change in long-grain can be approximately compensated for by using a relatively large, brass-cased, lead bob which expands up or down from the rating nut with changes in temperature (Figure 456). It has been calculated that a cylindrical bob 14 inches long will compensate for the expansion of a 46-inch deal rod[1]. The efficiency of this form of compensation is borne out by the Westminster clock which has a wood-rod pendulum and is subject to dramatic changes in temperature; it varies on average by only yet one second a week. By the end of the first quarter of the 18th century two more precise means of temperature compensation had been devised, one by Harrison and the other by Graham.

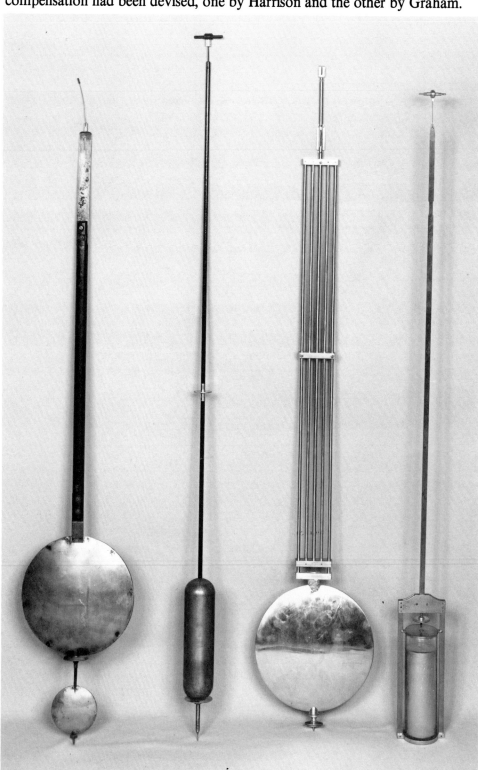

Figure 456. *Four different types of compensated pendulum (left to right):* (a) Wooden rod pendulum with small auxiliary bob for fine regulation. (b) Pendulum from a regulator by Dent with an invar rod fitted with a tray for the addition of fine regulation weights. The zinc bob is supported at its midpoint by a glass tube to compensate for the small amount of expansion of the rod. (c) Five rod gridiron pendulum. (d) Mercury compensated pendulum.

## THE "MERCURIAL" PENDULUM

It was known by the mid-17th century that the expansion and contraction, with changes of temperature, of different metals varied, but not by how much. In 1715 George Graham carried out experiments to measure these variations. His purpose was to find two metals which, when used in opposition to each other for the construction of a pendulum, have rates of expansion which "compensate" or cancel each other out. He found that the difference between the expansions of brass and iron (the metals most readily available then) was so small that it was useless to work with them. He gave up this line of investigation and subsequently turned his attention to mercury, which has a much greater expansion relative to other metals. He wrote up his earlier findings (*Philosophical Transactions*, London, 1726) and from the figures contained in these Harrison developed the gridiron pendulum.

In 1722 Graham produced a mercurial compensation pendulum (Figure 457), the principle of which is very simple. Mercury contained in a jar is used as the bob. It has a thermal coefficient roughly six times as great as steel. If, for instance, a column of mercury six inches long is used in conjunction with a 36-inch steel rod (i.e., a ratio of 1:6), then the thermal coefficients will exactly cancel each other out. In practice it has been found that a column of mercury 6.8 inches long is usually required.

This form of thermal compensation has the advantage that it can be finely regulated by the addition or removal of small quantities of mercury in the jar. A disadvantage, particularly of the earlier mercurial pendulums, is the use of glass to contain the mercury, since this is a very poor thermal conductor. Moreover, since the temperature differential is usually relatively small, it takes a considerable amount of time for changes to be transmitted to the mercury and thus for compensation to occur. It seems likely that thermal compensation achieved over a short period, for instance a hot summer day with a variation between midday and midnight of 25 degrees Fahrenheit, would be particularly poor, which could be important as timekeeping on a day-to-day basis is important in astronomy and, for instance, shipping.

However, it is likely that over a longer period the errors produced by poor thermal conductivity would tend to cancel themselves out and certainly the compensation for the changes in temperature throughout the seasons may well be relatively satisfactory.

Whereas on the Continent the gridiron became the most popular compensated pendulum during the 18th and 19th centuries, in England the mercurial pendulum was generally used. To improve its performance by speeding its response to thermal change, two variations were developed during the 19th century. The first was the use of several (from two to nine) separate glass jars, thus reducing the thickness of the glass required, greatly increasing the surface area, and decreasing the mass of mercury in any one container, all factors which speed its thermal compensation.

The second alternative was substitution of cast iron for glass, the iron greatly speeding heat transfer. One of the finest pendulums produced in England using mercury compensation is seen in Figure 450 A and B which uses two long, metal containers. Possibly the most accurate pendulums using mercurial compensation were designed by S. Riefler immediately prior to the invention of Invar. They are recorded as having a maximum error of approximately 0.005 second per day per degree centigrade.

## THE GRIDIRON PENDULUM

The relative expansions of steel and brass are approximately 3:5. Thus, if a five-foot steel rod in a pendulum expanding down were opposed by a three-foot brass rod expanding up, the length of the pendulum should remain unchanged. In practice a simple way to achieve this, and indeed the way described by Harrison, is to join several short sections of steel and brass in the same ratio. Thus, the five- (Figure 456) and nine-rod (Figure 451) gridiron pendulum was evolved, of which for technical reasons the latter was preferred.

The classic regulator produced with a gridiron pendulum was made by Harrison, the performance of which was almost legendary. He claimed that it kept time to within one second a month over a period of ten years. However, this claim is subject to some doubt as even barometric errors for which it was not

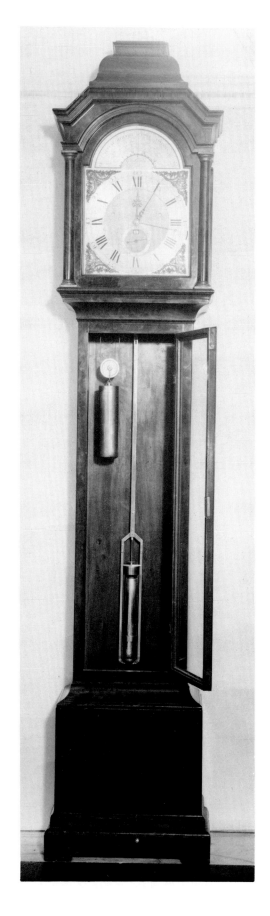

Figure 457. Graham longcase with his mercury compensated pendulum. Photo courtesy of the Trustees of the British Museum.

compensated would have caused a greater error than one second a month. He also claimed that its total aggregate of variations from mean time over a period of 14 years did not exceed 30 seconds, a performance which was probably not bettered for the next 100 years.

Harrison's regulator, as well as some of his clocks, was self-lubricating. Therefore, its performance was unlikely to vary from year to year, unlike conventional regulators in which friction increased as the oil gradually dried up.

A disadvantage of variations in power coming through the train due to increased friction is that the pendulum's arc of swing varies causing circular errors. Obviously if the pendulum's arc of swing remains constant, circular errors are constant and can be eliminated by regulation.

Many other makers produced compensated pendulums during the 18th and the first half of the 19th centuries, but it is doubtful if they performed any better than those already mentioned. Possibly the best was patented by Thomas Buckney in 1885, which made use of concentric tubes of zinc and steel and was used on several observatory regulators.

## INVAR

Invar is a nickel/iron alloy with small quantities of carbon and manganese. Its extremely low coefficient of expansion (only one-twelfth that of steel) varies with its exact composition, meaning that any pendulum rod made of invar can be almost discounted (Figure 456).

On a high-grade regulator where even a small degree of expansion is important, this can be corrected by supporting the bob on a tube(s) of brass, aluminum, glass or other suitable material depending on the degree of compensation required.

The length of the tube(s) needed can be calculated by finding (from the manufacturers) the coefficient of expansion of the invar rod or having it tested in a laboratory. To keep the center of oscillation constant the mass of the pendulum rod and bob must be taken into consideration.

## THERMOSTATIC CONTROL OF TEMPERATURE

The ultimate way of preventing thermal changes in the length of the pendulum is to keep it at a constant temperature by using a thermostat. Either the room in which the clock is situated can be kept at a constant temperature or the clock can be contained in a sealed jar so that the temperature of the air within that jar can be carefully controlled.

## BAROMETRIC PRESSURE

Although barometric compensation was never used on ordinary longcase clocks, it is mentioned here to complete the story of man's quest for accuracy with mechanical clocks.

As the performance of regulators continued to improve towards the end of the 19th century, and such problems as thermal compensation and the "detachment" of the pendulum from the influence of the movement had been largely overcome, attention was turned to the resistance of the air to the passage of the pendulum through it and the way in which this resistance varied with barometric pressure. It has been calculated that a drop of 0.6 inches in barometric pressure can produce a change in the rate of a clock of 0.25 of a second a day; however, this will be varied by such factors as the size and shape of the pendulum bob and its arc of swing.

To prevent any variation in barometric pressure, Leroy, Riefler and Shortt enclosed the clocks in airtight cases of copper, steel and glass with a glass dome on top. Leroy and Riefler maintained pressure at 25-26 inches mercury, while Shortt reduced it to about one inch.

When the regulator could not be contained in a sealed container at constant pressure, Riefler (on No. 341, Figure 458) used a vacuum tank with a carefully calculated weight on top which rose or fell in inverse proportion to the barometric pressure. Other makers, such as Dent, also produced barometric compensation units.

## MECHANICAL IMPROVEMENTS

As the precision pendulum clock evolved, the quality of the movement also improved. Higher train counts were used to reduce friction. Six- and eight-leaf pinions were abandoned in favor of those of ten to 14 and sometimes even higher

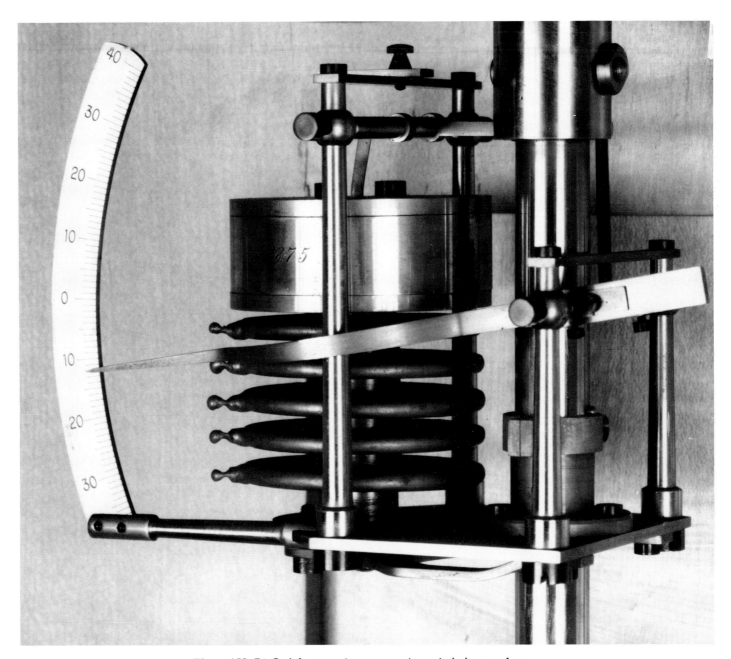

Figure 458. *Riefler's barometric compensation unit.* As it expands up, the unit carries the compensating weight resting on top. These units had a very high degree of accuracy.

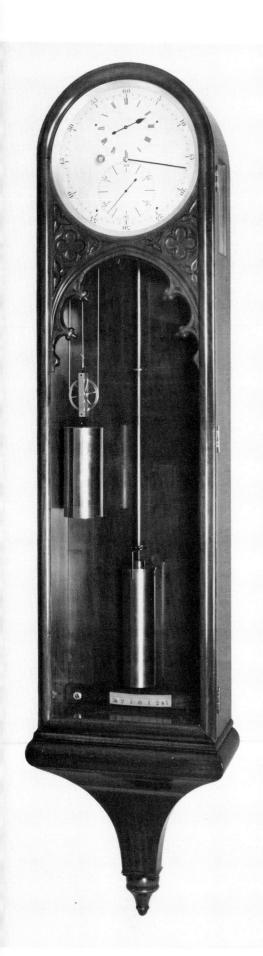

counts. The counts of wheels tended to increase also, as did their diameter and the precision with which they were cut. Six- or sometimes five-spoke wheelwork was used to produce a lighter wheel with greater rigidity.

## WEAR

Wear on a clock was an important factor, particularly in the 18th century when the quality of lubricants was relatively poor. Apart from the actual damage caused to the movement, wear affected the clock's performance by increasing friction which caused variations in depthing on the pivot holes, between wheels and pinions, and more particularly between the pallets and the 'scape wheel. As the efficiency of the movement decreased, the power transmitted to the pendulum, and thus its arc of swing, was altered and therefore the circular error varied.

## JEWELLING

To reduce wear hard jewels were introduced to one of the contacting surfaces. They were first used on the pallets and later to the pivot holes. The latter were normally only used on the 'scape wheel and pallet arbors where movement was rapid relative to the rest of the train. Jewels were also used as end stones to control the position of the arbor between the plates and prevent the shoulder of the pivot from butting against them. Other refinements were introduced such as the enclosure of the movement within a wooden or a metal dust cover, and the provision of some form of beat regulation.

Many of the regulators made in the 19th century were so efficient that they run for a week, providing impulse to a pendulum weighing 12 to 16 pounds on only a three- or four-pound weight. Indeed, many Austrian regulators will run for a year on only a 20-pound weight, which is the equivalent of a little over six ounces of motive power per week. However, there is some debate as to whether a light or heavy driving weight is most favorable for accurate timekeeping.

## THE SEARCH FOR THE "FREE" PENDULUM

This is a complex matter and will not be dealt with in detail here. For those who would like to delve deeper into it, there are suitable references mentioned at the end of this chapter.

One of the most famous persons to address the search for a "free" pendulum was Lord Grimthorpe who, in the 1850s, devised an escapement known as a "gravity escapement" (Figure 459 A, B, and C) which was fitted to the famous Westminster Clock popularly known as "Big Ben." A much more significant introduction, as far as accuracy is concerned, was the escapement devised by Sigmund Reifler 30 to 40 years later, in which impulse is transmitted to the pendulum by flexion of the suspension spring. The final solution was devised by C. O. Bartrum and W. H. Shortt circa 1917 when they propounded the theory of using two pendulums, one as a "master" to maintain perfect time and the other the "slave" to do all the work and yet remain under the control of the master.

The master pendulum was kept in motion by a gravity lever fitted with a roller which acted on the inclined face of a pallet at the bottom of the master pendulum. The gravity lever was released once every minute by a slave. After impulse transmission, the lever fell onto a contact which established a circuit which restored the lever to its original position. The establishment of the contact was also used to reveal whether the pendulum was isochronous. If it was not, then the slave was automatically regulated to bring it back to time by varying the tension of a helical spring attached to it. Shortt was undoubtedly the first to produce clocks in which the pendulum was truly "free" and their performance was almost legendary, their daily rate only appearing to vary by one or two thousandth of a second from the smoothed rate.[2]

Figure 459 A and B. *Wall Regulator by Brock, No 1209.* Mahogany wall regulator with Denison's four-legged gravity escapement, signed on the backplate *Brock, 21 George St., Portman Sq, London, No. 1209.* It has the usual layout for a gravity regulator with seconds and hour rings reversed because of the position of the escapement at the bottom of the movement. Points to note are the large anti-friction

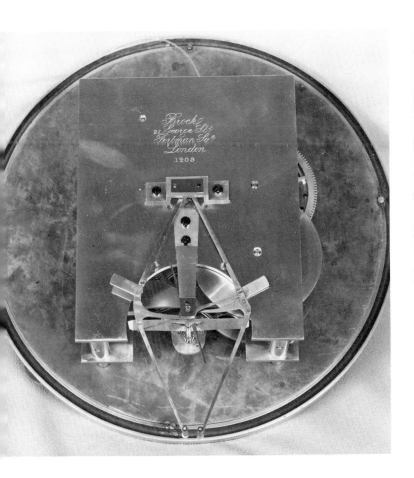

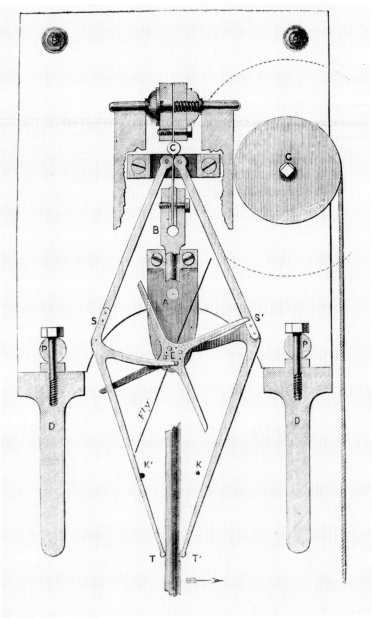

rollers supporting the great wheel arbor which are required because of the heavy driving weight and the massive mercury-compensated pendulum with a cast-iron jar. The large fly may be seen in the side elevation and the ivory-covered pins which give impulse to the pendulum. The additional weights fitted to the outside of the gravity arms are most unusual and may be a later addition.

James Brock appears to be a somewhat enigmatic character. despite his eminence as a clockmaker remarkably little is known about him. No reference is made to his activities in Mercer's excellent book on Dent, despite the fact that he was their foreman and supervised the construction of "Big Ben." Probably soon after this he left Dent and set up his own business in George St., both Nos. 18 and 21 appear as his address.

The similarity of the gravity regulators signed by both Dent and Brock are so great that there can be little doubt that Brock was responsible for making all of them. It is interesting that he had a very good relationship with Denison who had a great respect for his workmanship. He died in 1893 at the age of 67.

Figure 459 C. Denison's four-legged gravity escapement as illustrated in his book *Clocks, Watches & Bells*. The gravity arms which are pivoted at their upper extremities are lifted by two sets of pins near the center, pointing alternately backward and forward as the pallets are in different planes (i.e. one in front of and one behind the 'scape wheel). Impulse is given to the pendulum by the pins at "T" which are usually made of ivory to reduce friction.

## THE EFFECT OF GRAVITY ON THE SPEED OF OSCILLATION OF A PENDULUM

It is interesting to note that before supplying one of his regulators, Reifler[11] always checked on the latitude and altitude above sea level of the place where the clock was going to be installed so that the pendulum length could be adjusted to allow for local gravitational force. The length of a pendulum beating exact seconds at Greenwich would have to be shortened a little more than three millimeters at the Equator and lengthened two millimeters at the Pole if it were still to beat seconds[3].

## REGULATOR CASES

From a comparatively early period, regulator cases assumed a style all their own, although they were influenced to some extent by changes in fashion during the period over which they were made. Because they are essentially functional instruments often being used for scientific purposes, the dial is usually placed at eye level to make it as easy to read as possible. This inevitably resulted in a small clock, usually no more than six feet to six feet six inches high, which is one of their most attractive features. The cases are relatively free of ornamentation. The dials were in most instances engraved and silvered brass or occasionally, on examples after 1790, painted. Normally a square dial was used on the early clocks, but after 1800 circular dials are most commonly seen. The flat-topped breakarch was sometimes used on regulators made between 1770 and 1810. When regulators were used in an observatory, a 24-hour layout was generally favored and this often showed sidereal and not mean time, but in most other instances a 12-hour dial was chosen.

With virtually all 18th- and most 19th-century regulators made in England, a center-sweep minute hand was used in conjunction with a large seconds dial at the top and a balancing hour dial below. However, on the Continent the standard dial layout used in longcase clocks was generally also used for their regulators. During the 18th century a full-width, front-opening hood door was used, but by circa 1820 it was common for just the glazed front bezel to open. Although 18th-century regulators are seen occasionally with glazed trunk doors, it was not until the Regency period that this habit became common, usually to display a beautifully executed mercury-compensated pendulum. Around the turnover period, regulators are seen with framed doors which were sometimes glazed and sometimes fitted with a well-figured, wooden panel.

George Graham (Figure 460) was probably the first to produce a series of regulators using a distinctive style of case and the classic simplicity and proportion of these have never been improved upon. Although the majority of cases were of mahogany, when they were provided "below stairs," as in the servants' quarters, they were often made of oak.

John Shelton was the principal person employed by Graham to make his longcase regulators and it was thus natural that when Shelton started working on his own, his regulators (Figure 461 A and B) were very similar to those he made for Graham. Elliott also produced a small series of regulators (Figure 462 A and B), nearly always of very fine quality and often incorporating one of his forms of compensated pendulum. Although the cases were very much more complex than those used by Graham and Shelton, they were also comparatively small.

The number of regulators made before 1780-1790 was small, but as the production of chronometers rapidly increased after this period, so did the number of regulators. Some of these had wooden-rod pendulums but all those used in observatories and for regulating chronometers would have had fully compensated pendulums, usually mercurial.

Figure 460. *Oak longcase regulator by George Graham, London.* Note the aperture for the date, a somewhat unusual feature for a regulator. Photo courtesy R. A. Lee Fine Art.

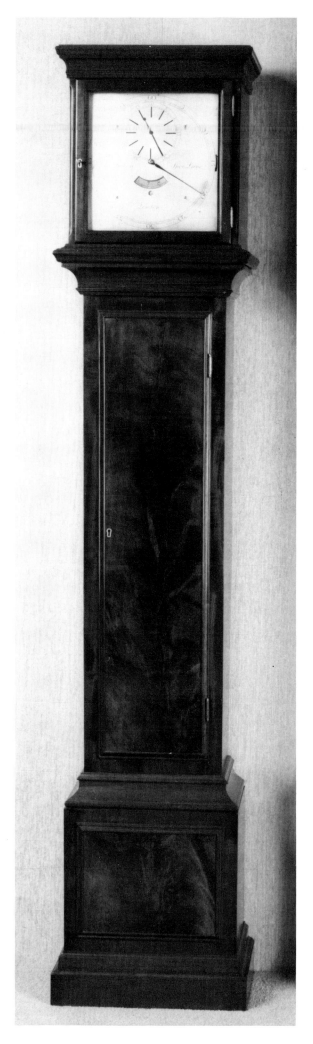

Figure 461 A and B. *Longcase Regulator by John Shelton.* A fine, month-duration, longcase regulator by John Shelton typical of those produced by this maker. The substantial movement has plates which are chamfered at the top corners and held together by six heavy latched pillars. The dial is fixed to the movement by four screwed feet. Bolt and shutter maintaining power is provided and this is actuated by depressing a lever immediately outside 3 o'clock on the dial. Stopwork is also fitted.

The wheelwork has four spokes except for the 'scape wheel which has six. The pallets for this span 10 ½ teeth and have a long anchor. Beat regulation is provided on the crutch. The movement is bolted to the seatboard by four brass brackets and a metal dust cover is provided which slots onto the back of the dial.

The 10-inch square silvered brass dial which is signed *John Shelton, Shoe Lane, London* has a seconds ring below 12 o'clock and shows hours through an aperture immediately above the winding hole.

John Shelton, a fine maker who specialized in precision clocks, was the principal person employed by George Graham to make his longcase regulators. This immediately becomes apparent when one compares this regulator with those signed by Graham. He supplied regulators for various expeditions, usually housed in functional, rather than beautiful, cases.

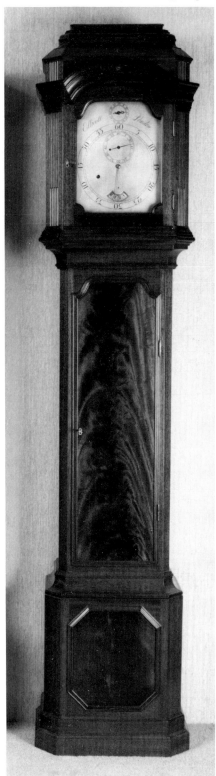
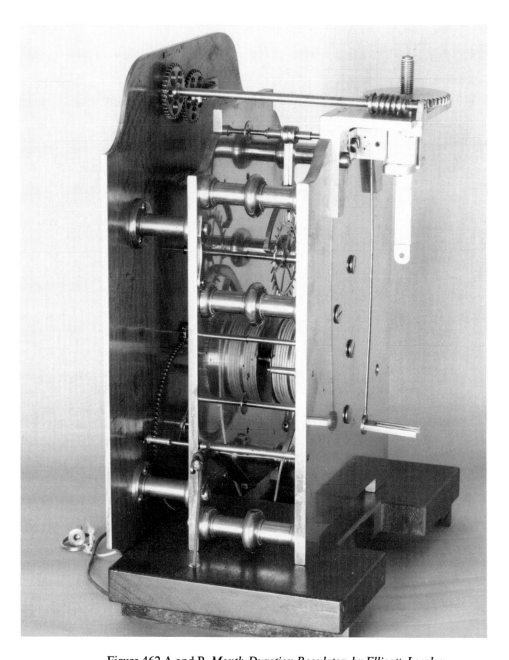

Figure 462 A and B. *Month-Duration Regulator, by Ellicott, London 1706-72.* A month-duration longcase regulator by John Ellicott contained in a reproduction mahogany case of exactly the style favored by him. This movement, which is typical of Ellicott's work, has substantial shaped plates with six latched pillars and four latched dial feet. The wheelwork is six spoke throughout and the pinions are 12 leaf except for that meshing with the great wheel which has 12. The maintaining power is bolt and shutter. The very delicate six spoke 'scape wheel has pallets which span 11 ½ teeth. The long anchor is split to provide beat regulation.

The flat-topped, breakarch dial, a design which was also favored by Delander, Mudge and one or two other makers, measures 11½ inches by nine inches. It has a rise and fall aperture for the hours just above 6 o'clock. An unusual feature is the concave cutouts in the two bottom corners. John Ellicott, an eminent maker of both clocks and watches, devoted much of his time to the development of precision clocks producing at least two different types of compensating pendulum.

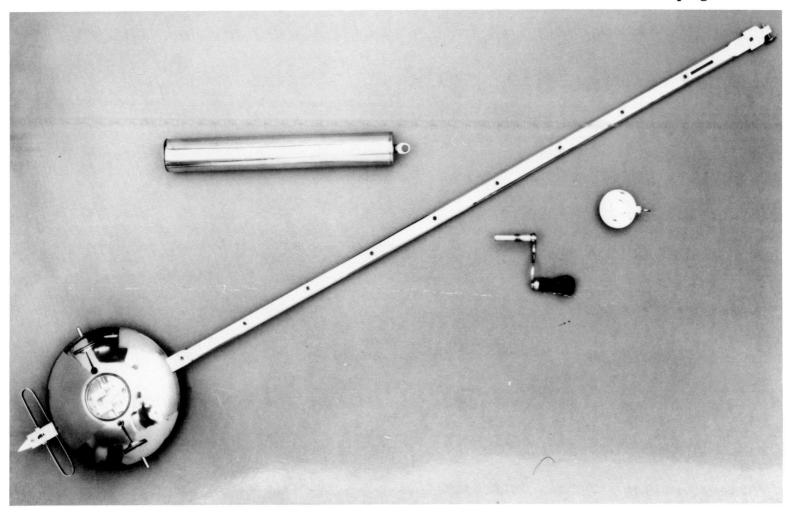

Figure 462 C. *Ellicotts' compensated pendulum* employing pivoted levers which automatically raise the bob as the rod lengthens. The degree of compensation being effected may be adjusted by the recessed knurled nuts on either side.

Photo courtesy of Gerald E Marsh.

Production of the classic Regency style longcase regulator with slide off hood (Figures 463 A and B and 465 A and B) lasted from circa 1800-1850 but towards the end of this period the Victorian arch-topped regulator started to appear (Figure 466). Initially they were usually only around five feet eleven inches to six feet two inches high and of excellent proportions with restrained fretted out carvings below the dial and delicately framed fully glazed doors. However, as the century progressed the cases were made larger and wider and in the process lost their excellent proportions. These regulators, which almost invariably have fine movements, were produced in larger numbers than any others and are still seen in many jewelers' and clockmakers' shops where they have been used for regulating clocks and watches for over 100 years.

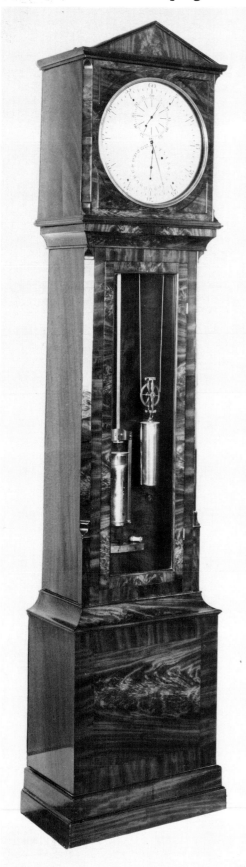

Figure 463 A and B. *High Grade Maritime Regulator.* A typical high grade "working" regulator produced by Thwaites & Reed during the first part of the 19th century, this particular example was used by a shipping line throughout its life, mainly to set and regulate their chronometers. The quality of the movement and the 24-hour dial tend to differentiate this from the somewhat simpler, domestic variety which generally have 12-hour dials. It has a massive mahogany backboard which was bolted to the wall in six places.

The wheelwork is six spoke and the escapement has jewelled dead beat pallets. The pivot holes to the upper part of the train are jewelled and end stones are used. This probably reduces friction by preventing the shoulder of the pivot from butting onto the plates.

The movement has very heavy stepped plates with six pillars and is bolted to the seatboard with three substantial brackets. The pendulum has mercurial compensation.

The quality of the hands and the engraving on the dial are particularly fine and typical of this period when some of the best and most aesthetically pleasing regulators were produced by well known makers such as Molyneux, Frodsham, Dent and others. This regulator, without even being fixed to the wall, keeps time, on average, to three to four seconds a month and thus presumably in its original home would have performed even better.

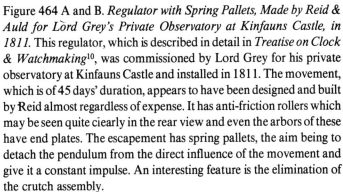

Figure 464 A and B. *Regulator with Spring Pallets, Made by Reid & Auld for Lord Grey's Private Observatory at Kinfauns Castle, in 1811.* This regulator, which is described in detail in *Treatise on Clock & Watchmaking*[10], was commissioned by Lord Grey for his private observatory at Kinfauns Castle and installed in 1811. The movement, which is of 45 days' duration, appears to have been designed and built by Reid almost regardless of expense. It has anti-friction rollers which may be seen quite clearly in the rear view and even the arbors of these have end plates. The escapement has spring pallets, the aim being to detach the pendulum from the direct influence of the movement and give it a constant impulse. An interesting feature is the elimination of the crutch assembly.

The well laid-out, shallow breakarch dial is signed in the arch *Reid & Auld, Edinburgh.* The case is a particularly attractive faded mahogany.

The provenance of this regulator is known from its making. It stayed with Lord Grey and his descendants until 1934 when, sadly, it was converted to dead beat escapement. At that time Mr. Howard of Howard Bros. (Veneers) London acquired it.

In 1958 it was purchased by Professor Hans von Bertele of Vienna who, with the help of various people, restored it to its original design. It remained in his possession until his death in 1984 when we acquired it.

Figure 464 C. Reid's original drawings for his spring pallet escapement as included in his book *Treatise on Clock and Watchmaking.*

Figure 465 A and B. *Lecomber Regulator*. A fine Regency mahogany longcase regulator with a relatively small case (six feet two inches) of excellent color and patination which is fully glazed on the front and sides of the hood and the trunk. The base is panelled and there are canted corners to the front of the hood which has a full width door incorporating a silvered bezel.

The movement can only be described as of exceptional quality with massive, slightly tapering plates held together by six heavy screwed pillars and fixed to the seat board by three substantial brackets. There is no underdial work and the seconds, minute and hour arbors are all supported by very large and beautifully finished bridges. The well executed wheelwork is six spoke; the dead beat 'scape wheel has jewelled pallets, and a lovely feature is the way in which the pallets may be depthed both to front and rear by screw regulation, a feature seldom seen on English regulators. Beat regulation is provided and there are finely turned end stops with screw regulation throughout the train. The entire movement is protected by brass dust covers held in place by knurled brass screws and a pulley offset is provided.

The substantial mercurial pendulum has screw regulation for tolerance at the crutch and is provided with a silvered beat scale. The attractive pulley for the brass cased weight is six spoke and the 12-inch circular engraved and silvered brass dial is of typical English regulator layout, being signed *Lecomber, Liverpool*. However, one suspects that it may well have been made for him by Condliff.

1. Becket, E. *Clocks, Watches & Bells, 7th Edition* Crosby and Lockwood and Co., London, 1883.
2. Boucheron, P. H. *Just How Good Was The Shortt Clock?* Horological Journal, December 1985.
3. Boucheron, P.H. *Effects of the gravitational attractions of the Sun & Moon on the period of a Pendulum*, Antiquarian Horology, March 1986.
4. *Christie's Catalogue of Important Clocks and Fine Watches.* 16th, December 1982, pp.62-65.
5. Howse, D. *The Tompion Clocks at Greenwich & the Dead Beat Escapement*, Antiquarian Horology, December 1970 and March 1971.
6. Neilson, M. *Important Siderial Regulator by Thomas Tompion & Edward Banger, No. 483 c 1709*, Antiquarian Horology Vol. X No. 2 pp 214-216, Spring 1977.
7. Rawlings, A.L. *The Science of Clocks & Watches, 2nd Edition*, Pitman Publishing Corporation, 1948.
8. Rees, *Clocks, Watches & Chronometers, 1819-1820* David & Charles Ltd. Reprinted 1970.
9. Reid, T. *Clock & Watchmaking, Theoretical & Practical*, Blackie & Son, 1826.
10. Reid, T. *Treatise on Clock & Watchmaking, 4th Edition*, Blackie & Son, 1849.
11. Riefler, D. *Riefler-Prazisionspendeluhren, 1890-1965* Callwey Verlag, Munchen, 1981.
12. Roberts, D. *Precision Pendulum Clocks*, Derek Roberts, Antiques, 1986.
13. Saunier, C. *Treatise on Modern Horology*, W & G Foyle Ltd., page 692. Reprinted 1975.
14. Symons, R.W. *Thomas Tompion, His Life and Work* B. T. Batsford Ltd., London and New York 1951.
15. Todd, W. *Communication in Antiquarian Horology*, Vol. X No.3, Summer 1977. p 366.

Figure 466. A particularly good and well-figured Victorian arched top, mahogany longcase regulator only 6' 1½" high. It has a delicately framed, fully glazed door with a fine fret below the dial, glazed sides, and recessed panelled base.

The substantial movement has six-spoke wheelwork, jewelled pallets to the dead beat escapement, stop pins, maintaining power, beat regulation, a well-executed mercury compensated pendulum, and a very decorative six-spoke pulley. There is a beat plate mounted on the board behind the glazed door. The 12-inch circular engraved and silvered brass dial is well laid out with very "full" seconds and date rings. It is signed by *Josh Moses of Bishop Auckland.* 6' 1½" (187 cm) high.

# Chapter 17.
# The Makers

*A brief account of "The Clockmakers' Company" and the clockmakers who made significant contributions to the design of the longcase clock.*

## THE WORSHIPFUL COMPANY OF CLOCKMAKERS

The origin of the "guilds" or "mysteries" of London goes back to old Saxon law which exacted sureties for good behavior from every freeman over 14. This money was then used to meet claims against any of the members of the craft.

Initially, there were two types of guild: Ecclesiastical and Mercantile. But during the 14th century they gradually became amalgamated.

The oldest of the guilds is the weavers', which was formed by Royal Charter in 1184. By the 13th century apprenticeships had been widely adopted and in the 1400s the system by which each had to produce his "masterpiece," which satisfied the master and wardens of his ability, prior to gaining his freedom of the company, was adopted. This gives some idea of the strict control the Companies exercised over their members and apprentices at that time.

It was not until the reign of Edward I that the guilds assumed a distinguishing dress known as the "Livery" or "Clothing" of that particular company.

It is interesting to note that whereas public clocks appeared in England and other parts of Europe at roughly the same time (circa 1350), that England was far slower in introducing domestic clocks. Indeed it was not until around 1600 that they first started to be made in any quantity.

At first, because of their small number, clockmakers were enrolled in one of the other London companies, usually the blacksmiths and indeed it was they who were responsible for the construction of most of the larger early clocks. However, by 1627, due to increasing competition from immigrant French clockmakers, the advance in mechanical science, and the greater refinement and accuracy of clocks which was becoming so important, arrangements were made for clockmakers to form their own company. In 1630 a committee was assembled to petition Charles I and on the 22nd August 1831 a Royal Charter was granted.

This Charter gave the company widespread powers of control over the construction of clocks within the City of London. To this end they could regulate the number of apprentices any clockmaker might have and had the right to search out and destroy any clocks being made which they considered to be of too low a standard.

The first master of the Clockmakers' Company was David Ramsay. The first wardens were Henry Archer, John Wellowe and Sampson Shelton, and the first assistants of the art or mystery of clockmaking were James Vautrollier, John Smith, Francis Forman, John Harris, Richard Morgan, Samuel Lynaker, John Charlton, John Midnall, Simon Bartram and Edward East. Of all these founders of the Clockmakers' Company, it was only Edward East who lived long enough to see the evolution of the longcase clock and the rise of British clockmaking to the pre-eminent position in Europe which it retained until the mid-18th century.

## AHASUERUS FROMANTEEL (THE ELDER)

The Fromanteel Family was one of the most prominent in the field of horology during the 17th century. Although of Dutch stock, in 1607 Ahasuerus Fromanteel was born in Norwich, and it was here that he was apprenticed, probably to Jacques van Barton. In 1631 he came to London when, on production of his Certificate of Apprenticeship, he was admitted to the Blacksmiths' Company and settled in East Smithfield, becoming known as a maker of steeple clocks. In 1632, only a year after its formation, he joined the Clockmakers' Company as a Brother. However, it was not until 1655-1656 that he was appointed a full Freeman.

According to Brian Loomes, it is doubtful whether any work of his made prior to 1655 and bearing his signature is known to still exist, possibly because his earlier years were spent as a journeyman. But in 1658 his name was to rise to a pre-eminent position in the horological world when he first introduced the

# WORSHIPFUL COMPANY OF CLOCKMAKERS
# ARMORIAL BEARINGS

pendulum clock into England. This was proclaimed by him in the following advertisements which appeared in the *Mercurius Politicus* for the 27th October, 1658, and in the *Commonwealth Mercury* of Thursday the 25th November, 1658.

> There is lately a way found out for making Clocks that go exact and keep equaller time than any now made without this Regulater (examined and proved before His Highness the Lord Protector) by such Doctors whose knowledge and learning is without exception, and are not subject to alter by change of weather, as others are, and may be made to go a week, a month, or a year, with once winding up, as well as those that are wound up every day, and keep time as well, and is very excellent for all House Clocks that go either with springs or weights; and also steeple clocks that are most subject to a change of weather. Made by Ahasuerus Fromanteel, who made the first that were in England. You may have them at his house on the Bankside, in Mosses Alley, Southwark, and at the sign of the Mermaid, in Lothbury, near Bartholomew Lane End, London.

Further evidence of his rise to fame is contained in the following extracts. Under date 1st November 1660, Evelyn, in his *Diary* writes:

> I went with some of my relations to Court to shew them his Majestys cabinet and closset of rarities... Here I saw... amongst the clocks one that shew'd the rising and setting of the sun in Ye Zodiaq, the sun represented by a face and raies of gold upon an azure skie, observing Ye diurnal and annual motion rising and setting behind, and landscape of hills, the work of our famous Fromanteel.

Again, under date 1st April 1661, Evelyn records that he "dined with that great mathematician and virtuoso, Monsieur Zulichem (Huygens), inventor of the pendule clock;" and on 3rd May, "I return'd by Fromanteel's, the famous clockmaker, to see some pendules, Monsieur Zulichem being with us."

It is thought that Fromanteel's introduction of the pendulum came about in 1657, when he sent his son John to Holland to learn the art of making pendulum clocks following their invention by Huygens and manufacture by Coster. Apprentices he is recorded as having taken on are: December 1637—Jacob Hulst; August 1646—Robert Collins; December 1646—John Mountage; May 1651—Stephen Smith; April 1652—his son John; April 1654—his son Abraham; September 1663—his son Daniel; June 1664—his stepson Joshua Winnock. By 1668 he had set up a business in Amsterdam, and by that time the output of the family business must have been quite prolific. His earlier clocks were often signed with his full name but later ones were signed just *Fromanteel* in either London or Amsterdam. He died in England in 1693.

Other members of the Fromanteel family are as follows:

*John Fromanteel,* son of Ahasuerus the elder. Born in 1638. Apprenticed to his father in 1654. Sent to the Hague where he worked on the pendulum clocks with Salomon Coster.
*Ahasuerus Fromanteel,* the younger son of Ahasuerus the elder. Born in 1640. Apprenticed to Simon Bartram. Freed in July 1663. Worked in London and Holland. Finally settling in Amsterdam where he died in 1703.
*Abraham Fromanteel,* son of Ahasuerus the elder. Born circa 1646. Apprenticed to father 1663-1669. Freed in 1680. Spent much time in Holland. In London until 1711 when he moved to Newcastle on Tyne where he died in 1730 at the age of 84.
*Daniel Fromanteel,* son of Ahasuerus the elder. Born circa 1651. Apprenticed to his father in 1663.
*Ahasuerus Fromanteel III,* born in 1666. Son of John Fromanteel to whom he was apprenticed.
*Fromanteel and Clark.* This Partnership was formed in Amsterdam between Christopher Clark who married Ahasuerus Fromanteel's youngest daughter and either Ahasuerus the younger or Abraham Fromanteel or both.

It is likely that all the Fromanteels worked for the Family firm and that because of this comparatively few pieces bear their individual signatures. Their total output must have been prolific, judging by the relatively large number of pieces bearing their name which have survived from the period 1655-1690. Many of these are illustrated in Dawson, Drover and Parkes' *Early English Clocks.*

Besides the references at the end of this chapter relating to all the makers described, the following refer in particular to the Fromanteels:

Loomes, B. "Country Clocks and Their London Origins," *Antiquarian Horology,* March, 1975.
Loomes, B. "Complete British Clocks" *Clocks Magazine,* May, June, and July, 1980.
Lee, R. A. "The First Twelve Years of the English Pendulum Clock or the Fromanteel Family and Their Contemporaries. 1658-1670." Catalogue of loan exhibition at the galleries of Ronald A. Lee, London, Feb., 1969.

Figure 468 A and B. An early ebony veneered longcase clock signed *A Fromanteel Londini* with verge escapement, shaped plates, bolt and shutter maintaining power and fine Corinthian capitals. Photos courtesy of R. A. Lee.

# EDWARD EAST

Edward East, the son of John East, was born in Southill, Bedfordshire, in 1602, and was baptized on the 22nd August of that year. He was apprenticed to Richard Rogers of the Goldsmiths Company (there being no Clockmakers' Company at that time) on the 27th March, 1618, and was freed to work on his own account in 1627. As already mentioned Edward East was one of the ten original assistants at the time of the incorporation of the Charter of the Clockmakers' Company. From the start, he took an active interest in its proceedings and in 1645 was elected its master, an honor which was bestowed on him again in 1652. He is recorded by Loomes[5] as supposedly working in Pall Mall in 1632 and at the "The Musical Clock" in Fleet Street in the 1640s. It was here that the meetings of the Clockmakers' Company were held while he was master. This was because the company never had a Hall of their own. In their early days they used to meet at various taverns and later were given hospitality by some of the other companies. In November 1660 East was appointed Chief Clockmaker to the King and in 1671 was described as the only existing founder member of the Clockmakers' Company.

In his will dated 1688 and proved on the 3rd February 1697, presumably shortly after he died, he gave his residence as Hampton, Middlesex. A little prior to this in October 1692 he gave £100 to the Clockmakers' Company for the assistance of poor members.

It is thought[5] that the East family's business probably had a large work force including the following masters: Peter Bellon, David Moody, Benjamin Hill, Michael Cornish, Jeremy East and other members of the East family; John Matchett, William Partridge and Evan Jones together with their apprentices. It is probably because of this extensive work force and thus the large number of clocks and watches manufactured that so many examples of East's work survive from this early period of English clockmaking. An interesting example of Edward East's association with Charles I is contained in *Some Account of the Worshipful Company of Clockmakers' of the City of London* published by them privately in 1887.

Another night His Majesty (Charles I) appointed Mr. Herbert to come into his Bed Chamber an hour sooner than usual in the morning; but it so happened that he overslept his time, and awaken'd not until the King's Silver Bell hasten'd him in. 'Herbert, said the King, 'you have not observ'd the command I gave last night.' He acknowledged his fault. 'Well,' said the King, 'I will order you for the future; you shall have a Gold Alarm-Watch, which, as there may be cause, shall awake you; write to the Earl of Pembroke to send me such a one presently.' The Earl immediately sent to Mr. East, his Watchmaker in Fleet Street, about it, of which more will be said at his Majesty's coming to St. James's.... The Earl delivered a Military Officer who was going to St. James's the Gold Watch that had the Alarm desiring him to give it to Mr. Herbert, to present it to the King.

Edward East's premises in Pall Mall were near the Royal Tennis Court where The Prince of Wales, subsequently Charles II, often played. It is reputed that the stakes for the game were often an "Edwardus East" which was His Royal Highness's way of referring to a watch made by Edward East.

East made watches and clocks of all types including a very large silver alarm clock-watch for Charles I, night clocks, lanterns and fine bracket and longcase clocks. The simple classical proportions of the latter have seldom been equalled or improved upon.

Probably the best account of Edward East's life and work is contained in the Sixth Edition of F. J. Britten's *Old Clocks and Watches*[2] and numerous examples of his work are illustrated in *Early English Clocks*[3].

Figure 469. *Edward East.*

## WILLIAM CLEMENT

The story of William Clement, a most important maker, is something of an enigma. It seems almost certain that he, probably in collaboration with Hooke, a brilliant scientist and mathematician who may have been the inventor, made the first clocks to be fitted with an anchor escapement. The evidence for this is twofold:

a) William Clement made the first turret clock with an anchor escapement for Kings College, Cambridge. Now in the Science Museum, London it is stamped *Gulielmus Clement Londini Fecit 1671*.

b) John Smith, in his *Horological Disquisitions* published in 1694, wrote that several makers were engaged in eliminating the shortcoming of Huygen's verge clock but "that eminent and well-known artist Mr. William Clement had at least the good fortune to give it the finishing stroke, he being indeed the real contriver of the curious kind of pendulum which is at this day so universally in use among us."

The importance of the anchor escapement, and with it the Royal or seconds beating pendulum which came into almost universal use, cannot be overestimated so far as the performance of the longcase clock is concerned. The improvement in timekeeping was so dramatic that the error of a clock over a week was suddenly reduced from minutes to seconds.

The basic difference between the verge and anchor escapements is that with the latter the teeth are in the same plane as the wheel, as opposed to being at 90 degrees, which automatically eliminates the need for a contrate wheel. Furthermore, far less movement of the pallets engaging the 'scape wheel teeth is required to lock and unlock them and thus the arc of swing of the pendulum is greatly reduced enabling a much longer pendulum to be used. Usually, this is approximately one meter long and seconds beating, but occasionally longcase clocks are seen with 1¼ seconds beating pendulums. Indeed the author has owned one such clock by Francis Clement, probably the brother of William, and William himself is known to have made a few examples.

Loomes, in his excellent book on the early clockmakers, tried to unravel William Clement's history but obviously finds it more than a little confusing and postulates that there may well have been two William Clements (Elder and Younger). Because of the doubts which exist, no dates have been quoted as regarding William Clement's life and work. Those who wish to delve further are referred to Loomes and also to John B. Penfold's paper in *Antiquarian Horology* 12th September, 1962.

Figure 470. *Turret Clock by William Clement*. This is the first clock known to exist with an anchor escapement. It is stamped *Gulielmius Clement, London Fecit 1671* and was installed in Kings College, Cambridge until 1817 when it was transferred to St. Giles Church. It is now in the Science Museum, London. It has 1¼ seconds pendulum bolt and shutter maintaining power. Photos courtesy of The Science Museum, London

## THOMAS TOMPION

Thomas Tompion is thought of as the greatest of English clockmakers. He has been called both "The Father of English Clockmaking" and "The Father of English Watchmaking," and when one looks at the immense diversity and on occasions complexity of his work, coupled with the extremely high standards which he kept, it will be appreciated just how justified that opinion is. It would not be too much to say that he raised the quality of English clockmaking to levels never previously attained, and it was he and a handful of other makers who gave England her pre-eminent position in the horological world which was to last for 70 to 80 years.

Tompion, the son of a blacksmith, was born at Ickwell Green near Northill, Bedfordshire in July 1639. A bell cast for St. Laurence Church at Willington near Northill is signed *Thomas Tompion Fecit 1671*. It may be that he served his apprenticeship as a clockmaker and worked in that part of the country prior to moving to London. However, it is more likely that he spent his early years working with his father as a blacksmith, and indeed the family forge still exists and bears a plaque commemorating its historic connection with Thomas Tompion. It must also be remembered that at that time many of the turret clocks, then called "Great Clocks," were still being made by blacksmiths and that when Tompion became a Free Brother of the Clockmakers' Company it was as a "Great Clockmaker."

An interesting point mentioned by Symonds in his book on Thomas Tompion was that when his father, also Thomas, died in 1665, he left the tools of his workshop, not to Thomas, as might be expected, but to his younger brother, James, which would seem to imply that Thomas had already by then given up his work in the smithy, presumably in favor of clockmaking. In July 1671, he paid a search fee to the Clockmakers' Company and was made a Free Brother on the 4th of September of that year, a significant time roughly corresponding with the invention of the anchor escapement and the introduction of the Royal or seconds beating pendulum.

Figure 471. *Thomas Tompion, 1639-1713.*

In 1674 Tompion moved to Water Lane at the sign of "The Dial and Three Crowns" where he was to continue his ever-expanding business for the rest of his life. Tompion was in his day a very highly respected figure, associating with leading scientists, for whom he sometimes made instruments, and members of the Court. The poet Matthew Prior (1666-1721) wrote, "Remember that Tompion was a Farrier and began his great knowledge in the Equation of Time by regulating the wheels of a common jack to roast meat."[1]

The rise to fame of Tompion can only be described as meteoric. In the same year that he moved into Water Lane, he met Dr. Robert Hooke who commissioned a quadrant from him for the Royal Society. Hooke was the leading physicist and Mathematician of the day; a man of great intellect and undoubtedly an inventive genius. It was his association and friendship which was to be the starting point for Tompion's rise to fame. It is also from his *Diaries 2, 3* that we learn much about Tompion's work.

In his preface to *The Artificial Clockmaker* William Derham states, "In the History of Modern Inventions, I have had (among some others) the assistance chiefly of the ingenious Dr. Hooke and Mr. Tompion. The former being the Author of some, and well acquainted with others, of the Mechanical Inventions of that fertile Reign of King Charles the II and the latter actually concerned in all, or most of the late inventions in Clock-work, by means of his famed skill in that, and other Mechanick operations." This gives some idea of Tompion's dominant position in English horology in the last quarter of the 17th century.

In 1675 Hooke asked for Tompion's help in proving that he had invented the spiral balance spring for watches prior to Huygens. The result of this was that Tompion was the first English maker to apply the spiral balance spring to a watch, making these more accurate than any produced by his competitors. Indeed, such was the demand for his watches throughout Europe that during his lifetime he, or to be more accurate, he and his employees, produced around 5,500. Charges for these are quoted as silver cased £11, gold cased £23, gold repeater £70, a very large sum of money in those days particularly when one bears in mind that in 1675 Tompion bought his new premises on the corner of Water Lane and Fleet Street for £80.

In 1676, only two years after Tompion had moved into Water Lane, he was commissioned by Sir Jonas Moore to provide two clocks for the new observatory being built at Greenwich which Flamstead was to run. The purpose of this observatory was to assist in navigation at sea by improving current knowledge of the fixed stars and the passage of the moon among them.

The concept of the clocks was, for their time, breathtaking. The dials were designed to appear in the room with the 13-foot, two-seconds-beating pendulums suspended above them. Moreover, they were of year duration.

Although Tompion produced a standard series of bracket and longcase clocks, he was always willing to undertake special commissions, however complex, some of which were way beyond the abilities of most of his contemporaries. One of the most remarkable of his achievements was a small and very beautiful spring or table clock which he made for William III which had a duration of one year. This is now known as "The Mostyn Tompion."

Besides clocks of long duration or with complex striking, he made one with an angle measure, another incorporating an astrolabe Figure 105 B, and several equation clocks, i.e. showing the difference between solar and mean solar (our) time (Figure 125).

In 1685 Tompion started numbering his clocks and watches, a practice which he continued until his death and which George Graham carried on after him. A list of the numbered clocks and watches is contained in *Clocks and their Values* by Donald de Carle, first published by N.A.G. Press in 1968.

In 1696 George Graham joined Tompion at "The Dial and Three Crowns" and ten years later married his niece Elizabeth. Edward Banger also worked with Tompion, and between 1701-1708 the products were signed in their joint names, but in 1708 Banger left Tompion for reasons unknown and from 1710 to 1713, when Tompion died, the clocks and watches were signed "Tompion & Graham."

Tompion was buried in Westminster Abbey where his gravestone may still be seen. He left the business and most of his estate to George Graham and his wife.

For a fascinating and superbly researched and documented account of this maker, the reader can do no better than refer to R.W. Symonds *Thomas Tompion, His Life and Work* -B.T. Batsford Ltd., London and New York.

---

[1] *Essays and Dialogues of the Dead*. A dialogue between Mr. John Lock and Seigneur de Montargne by Matthew Prior (Cambridge University Press).

[2] *The Diary of Robert Hooke*. Edited by A. W. Robinson and W. Adams between the years 1672-1680.

[3] *Early Science in Oxford* and *The Life and Work of Robert Hooke*, by R. T. Gunther.

312 Makers

Figure 472. *Year-Duration Spring Clock by Tompion.* This clock is regarded by many as Tompion's masterpiece. Not only was it the first striking spring driven clock to be produced of year duration, but artistically it is also of the highest order with superbly executed silver and gilt mounts. The photo makes the clock look larger but it is only 28 inches high to the top of the figure and the dial is just 5 inches square.

The clock was made for William III and went on his death to Henry Sydney, Earl of Romney. It was passed on by descent to Lord Mostyn and was purchased by The British Museum in 1982. Photo courtesy British Museum.

Makers 313

Figure 473 A, B, and C. *Year-Duration Equation Clock by Thomas Tompion.* This clock was made by Tompion for William II, his best patron, towards the end of the 17th century. The beautifully executed timepiece movement, i.e. without strike, has an inverted train with the pallets of the anchor escapement mounted upside down and the pendulum swinging in front of the weight of 60 pounds.

It is fitted with a 24-hour dial with a fixed chapter ring inscribed *Equal Time*, with I-XII twice for the hours, and delicate Arabic numerals marked 0-60 twice for the minutes.

Outside this is a further ring marked in minutes on which is engraved *Apparent Time* (solar time) which moves back and forth throughout the year thus enabling not only solar and mean solar (our time) to be read off against the minute hand, but also the "Equation of Time" to be calculated.

There is an arch to the dial, probably the first occasion on which this feature was used, which contains two apertures, the upper one showing the signs of the Zodiac and the suns position in the ecliptic, and the lower one with a year calendar. Mounted behind this may be seen the Equation cam. The winding square is situated below 12 o'clock and is protected by the shutter of the bolt and shutter maintaining power which is activated by a lever on the right side of the dial. To reduce friction, the minute hand is counterbalanced and the seconds hand omitted.

The finely figured walnut case, only seven feet three inches high excluding the finial, has a beautifully shaped, brass-bound lenticle to the trunk door which has spandrels set into the four corners. Despite the fact that the clock does not strike, there are frets to the top and sides of the case.

Reproduced by gracious permission of Her Majesty The Queen.

Figure 474. *George Graham 1673-1751.*

Figure 475. *An oak longcase regulator by George Graham, No. 767, circa 1735-1740.* Photo courtesy R.A. Lee Fine Arts.

## GEORGE GRAHAM

George Graham was born in Cumberland in 1673 and at the age of 15 came to London where on the 2nd July, 1688, he was apprenticed to Henry Aske. He gained his freedom on the 30th September, 1695 and immediately went to work for Thomas Tompion. Their relationship prospered and soon became one of mutual respect and friendship which was strengthened by Graham's marriage in 1704 to Tompion's niece Elizabeth Tompion.

From 1711 to 1713 when Tompion died, Graham was Tompion's partner and it is likely that for several years before this, due to Tompion's failing health, Graham had taken an increasing responsibility for the business. It is now almost impossible to know which clocks Graham may have been responsible for prior to 1711 as they all were signed by Tompion, except for the few which were signed by Tompion and Banger. This is the problem with so many early makers, a large number of whom worked for other clockmakers and whose pieces were then signed by them. In this respect it is interesting to note that in 1695 Tompion had 12 clockmakers in his household.

One of the most important inventions which Graham has been credited with is the "dead beat" escapement circa 1715. However, the report of the Research Laboratory for Archaeology and the History of Art of Oxford University (reproduced in *Precision Pendulum Clocks*, published by Derek Roberts Antiques, page 53) suggests that this escapement was invented at least by 1709 when Graham was still working for Tompion. This could mean either that Tompion invented the dead beat escapement or that Graham did but at an earlier date than originally thought.

Tompion, a bachelor, died at the age of 75 on the 20th November, 1713. In his will he referred to Graham as "my loving nephew" and he left him and his wife Elizabeth "All the rest and residue of my Goods and Chattels and Personal Estate whatsoever which shall remain after my funeral charge, debts and above mentioned Legacy, shall be paid and satisfied." On the 28th November Graham published the following announcement:

> George Graham, nephew of the Late Mr. Tho. Tompion, watchmaker, who lived with him upwards of 17 years, and managed his trade for several years last past; and whose name was joined with Mr. Tompion for some time before his death, and to whom he hath left all his stock and work, finished and unfinished; continues to carry on the said trade, at the sign of the "Dial and Three Crowns," at the corner of Water Lane and Fleet Street, London, where all persons may be accommodated as formerly.

Graham continued Tompion's business much as it had been before, concentrating on watches, which he sold all over Europe. But Graham also made around 170 clocks, most of them in the style which had already evolved when he was working with Tompion. However, no longer were many of the superb astronomical and long-duration clocks produced which had put Tompion in such high esteem.

Graham acquired a wide knowledge of astronomy, delivering papers on this subject to the Royal Society, of which he was a member. He also made many astronomical instruments including a quadrant of eight-feet radius for the Greenwich Observatory, a transit instrument, a Zenith Sector, and geodectric instruments which the French scientist de Mauperturs took to Sweden to discover the shape of the Earth.

The elimination of errors in timekeeping because of changes in the effective length of the pendulum due to variation in the temperature was one of Graham's goals and to this end he investigated the relative expansion of several different metals. It was this research which eventually gave rise to the mercury compensated pendulum circa 1726 and this, coupled with the dead beat escapement, resulted in a fine series of regulators, probably made for him by Shelton, the performance of which was little improved in the next 100 years.

Probably, one of Graham's most significant achievements was his nearly perfect cylinder escapement, circa 1725, which was a substantial improvement on the verge and made his watches much sought after. In 1751 George Graham died and was buried "On Mr. Tompion in the middle isle" of Westminster Abbey.

# DANIEL QUARE

Quare's life and work was in many ways very similar to that of Tompion. Both came from a Quaker background although Quare was the stricter of the two and he maintained his beliefs far more assiduously throughout his life. Although it seems to have done his career little harm, he was unable to be appointed Royal Clockmaker as he could not take the Oath of Allegiance.

He was born in Somerset circa 1647 and, like Tompion, he came to London and joined the Clockmakers' Company in 1671 as a "Great Clockmaker." In 1676 he married and throughout his life took a keen interest in the Clockmakers' Company where he was much respected, becoming its master in 1708.

One of his greatest achievements was the invention of the repeating watch, although this was surrounded by some acrimony and confusion as Edward Barlow (Booth), who invented rack striking for clocks which enabled repeat work to be provided, claimed this invention also. Only the adjudication of King James II decided the matter in Quare's favor.

Thomas Tompion was also very interested in providing repeat work for clocks and watches and was one of the first to do so. An extract from William Derham's *The Artificial Clockmaker* (1696) throws some light on the subject.

> Mr. Quare (a very ingenious Watch-maker in London) has some years before been thinking of the like Invention: but not bringing it to perfection, he laid by the thoughts of it, until the talk of Mr. Barlow's Patent revived his former thoughts; which he then brought to effect. This being known among the Watch-makers, they all pressed him to endeavour to hinder Mr. Barlow's Patent. And accordingly, applications were made at Court, and a Watch of each Invention, produced before the King and Council. The King, upon tryal of each of them, was pleased to give the preference to Mr. Quare's: of which, notice was given soon after in the *Gazette*.

Besides making a considerable number of watches he also made a series of very fine, portable, free-standing, pillar barometers to a design which he patented and which are now much sought after. Examples of these are in the Royal Collection. His bracket clocks are always of fine quality and of particular interest is that some of those signed by Quare are undoubtedly from Tompion's workshop which indicates that the two men must have worked closely together.

Quare made some fine longcase clocks of year duration, one of which is at Hampton Court Palace and another in the British Museum (Figure 476 A, B, and C). This latter clock and also several others by Quare incorporated equation work which it is thought to have been designed and made for him by Williamson. Undoubtedly one of Quare's strengths was his fine reputation across the whole of the Continent where he sold many fine pieces, in particular to some of the leading families in Spain, Portugal and Italy. For instance, there is a longcase clock of month duration and with equation work in the Palazzo Pitti in Florence which is thought to have been ordered by the Grand Duke of Florence.

Some indication of his stature in society was revealed when his daughter Ann married in 1705. The guests included the Envoys of Venice, Florence, Hanover, Portugal, Sweden, Prussia and Denmark. When his son and daughter married in 1712 the guests included the Earl of Orrery, the Duke of Argyll, the Venetian Ambassador, and the Count of Briancion. It was only for religious reasons that the Prince and Princess of Wales failed to attend the wedding of his daughter Elizabeth in 1715, although the Princess did go to the reception.

Quare died in Croydon in 1724 and was buried in the Quakers' Cemetery at Burnhill Fields, Finsbury, aged 77. Prior to this, in 1718, he had taken into partnership Stephen Horseman and from then on clocks and watches were signed in their joint names. The numbering system introduced by Quare was carried on by Horseman after his death and continued until the firm ceased trading following Horseman's bankruptcy in 1733.

Figure 476 A, B, and C. *Year-duration Equation clock by Daniel Quare*. A fine walnut clock with brass-bound doors and gilt mounts to the base and the top of the hood. At least two other clocks by Quare are known in which this equation work is set into the trunk door. Photo courtesy British Museum.

Figure 477. *Christopher Pinchbeck.*

## CHRISTOPHER PINCHBECK

Christopher Pinchbeck is known mainly for his zinc and copper alloy called "Pinchbeck gold" which he discovered while seeking better formula for brass with which, presumably, to improve his clock movements. This was so similar to gold in appearance and resistance to corrosion that he and his son seem to have had a prodigious business in the manufacture of trinkets and what we would term today "costume jewelry." It would appear, however, that he had in his day a fine international reputation as a clockmaker and certainly there are still in existence many clocks which bear this out such as that illustrated in Figure 478 A and B.

Christopher Pinchbeck was born in 1670 and, somewhat surprisingly, never became a member of the Clockmakers' Company. He started business in St. George's Court, subsequently known as St. John's Lane, and in 1727 moved to the sign of the "Astronomico-Musical Clock" in Fleet Street, a fact which he announced in Applebee's *Weekly Journal* on the 8th of July. This is quoted in full as it gives an insight not just into Christopher Pinchbeck and his products, but also the times in which he lived, and the way business was conducted:

> Notice is hereby given to Noblemen, Gentlemen, and Others, that Chr. Pinchbeck, Inventor and Maker of the famous Astronomico-Musical Clocks, is removed from St. Georges Court, St. Johns Lane, to the sign of the Astronomico-Musical Clock in Fleet Street near the Leg Tavern. He maketh and selleth Watches of all sorts and Clocks, as well for the exact Indication of Time only, as Astronomical, for showing the various Motions and Phenomena of planets and fixed stars, solving at sight several astronomical problems, besides all this a variety of Musical Performances, and that to the greatest Nicety of Time and Tune with the usual graces; together with a wonderful imitation of several songs and Voices of an Aviary of Birds so natural that any who saw not the Instrument would be persuaded that it were in Reality what it only represents.

Figure 478 A and B. *Four-sided Astronomical Clock by Christopher Pinchbeck.* This clock was supplied to George III by Christopher Pinchbeck Junior in 1768 and now resides in the music room in Buckingham Palace. The four dials show:

1. Mean and solar time with small orrery showing the movements of Mercury, Venus, Earth, Mars, Jupiter and Saturn. It also has a thermometer.

2. The tides at 43 ports.

3. A rotating globe moon and subsidiaries for strike/silent and slow/fast.

4. The position of the sun in the Zodiac and days of the month with a planisphere for the position of certain stars.

He makes Musical Automata or Instruments of themselves to play exceeding well on the Flute, Flaggelet or Organ, Setts of Country dances, Minuets, Jiggs and the Opera Tunes, or the most perfect imitation of the Aviary of Birds above mentioned, fit for the Diversion of those in places where a Musician is not at Hand. He makes also Organs performing of themselves Psalm Tunes with two, three or more Voluntaries, very Convenient for Churches in remote Country Places where Organists cannot be had, or have sufficient Encouragement. And finally he mends Watches and Clocks in such sort that they will perform to an Exactness which possibly thro' a defect in finishing or other Accidents they formerly could not.

He is recorded as having made a fine musical clock for King Louis XIV of France at a cost of £1,500, an enormous sum for those days, and exhibited complex clocks at various fairs.

Christopher Pinchbeck had five children and one of these, Edward, continued the business after Christopher's death in 1732. Another son, named Christopher like his father, started up his own business in Cockspur Street. He supplied a complex astronomical clock to George III in 1768, and in 1781, at the age of 71, joined the Clockmakers' Company. He died in 1783.[2]

Footnotes

---

[2] Shenton, R. *Christopher Pinchbeck and his Family*, Brant Wright Associates, 1976 and Roberts, D. and Lister, R. *A Month Going Longcase Clock* by Christopher Pinchbeck, *Antique Collecting,* March, 1986, Vol. 20, No. 10, pages 40, 41, 42.

## JOHN ELLICOTT, FRS

There are two John Ellicotts, father and son, who worked as clockmakers in the 18th century and although we are mainly concerned with the latter, who was the more eminent of the two, it may help a little if we discuss the father in the first instance.

*John Ellicott Senior,* like many fine clockmakers, hailed from the West Country, coming up from Bodmin in Cornwall to London around 1687. He started his apprenticeship to John Waters on the 5th September, 1696. He commenced business in Austin Friars and later moved to St. Swithins Alley, Royal Exchange. Undoubtedly, he was a very fine craftsman, one of his feats being to produce a watch only one-fifth of an inch thick between the plates and another to fit one with a center-sweep seconds hand. He continued to work in St. Swithins Alley until he died in 1733.

In 1706 his son, *John, FRS*, was born and in 1728 is recorded as working in his father's premises in Sweetings Alley, also known as St. Swithins, which is where the statue of Rowland Hill now stands. His home was in Allhallows, north of London Wall. Somewhat surprisingly, the son seems to have had little contact with the Clockmakers' Company, presumably learning his craft from his father. However, not only was he the possessor of a fine pair of hands, but he was also academically brilliant, indeed there can be few other 18th-century clockmakers who were his equal in that respect. In 1738 he was elected to the Royal Society, a singular honor at that time and indeed even today. His sponsors were Sir Hans Sloane, president of the society and Royal Physician, the antiquary Martin Ffolkes, John Senex the eminent globemaker, and John Hadley, astronomer. He also became a friend of another eminent astronomer James Ferguson and in due course built his own observatory in Hackney. He was on the Council of the Royal Society for three years and during his time as a member delivered several papers on horology.

Figure 479. *John Ellicott.*

### The influence of two pendulums upon each other

One of the most interesting of Ellicott's papers was that concerning the influence of two pendulums upon each other. He describes how two pendulums, each with a 23-pound bob, were set going only two feet apart and within two hours one, termed number 1, always stopped. He felt that this must be due to the influence of one pendulum on the other and thus he watched them closely after they were set swinging. He found that as the arc of swing on number 2 increased, that of number 1 diminished until finally it stopped. At that point the swing of number 2 had risen from its normal arc of three to five degrees.

He now stopped the pendulum of number 2 and set number 1 going with as wide an arc of swing as possible. Within a short while the pendulum of number 2 began to move until after 17 minutes and 40 seconds it described an arc of 2 degrees 10 seconds, at which point the clock started going. It continued to increased its arc until, after 45 minutes, it reached five degrees and number 1 came to rest.

The account of this experiment, somewhat abbreviated here, has been of great practical significance to clockmakers, particularly those concerned with precision timekeeping, and emphasizes how important it is to secure clocks as rigidly as possible.

### The variability of the length of the pendulum with latitude

Another useful piece of research carried out by Ellicott was tabulating the effective length of seconds-beating pendulums at different latitudes; for instance in London it is 39.1393 but, because the magnetic attraction at the Pole is stronger, thus making the pendulum beat faster, its length has to be increased there to allow for this.

### Compensated Pendulum

Possibly the work for which Ellicott is best remembered is his compensated pendulum. In this he tried to eliminate the basic disadvantages of Harrison's gridiron which were:

a) That the rods tended to seize up where they passed through the blocks, and
b) The only way of varying the degree of compensation being achieved was to take the pendulum to bits and re-pin it and possibly, at the same time, vary the length of the rods.

Ellicott overcame this problem by providing a variable system of compensation. Basically, a flat bar of brass is fixed to the iron pendulum rod by screws passing through slots in the brass and allowing a free-sliding movement of the two relative to each other. The pendulum rod is extended beyond the brass and widened. To this are fitted two pivoted levers on the outside of which rest two screws which support the bob. The brass rod presses on the inner ends of the levers and thus, on expansion, depresses them and raises the bob. The relative lengths required of the two ends of the lever will depend on the difference in coefficients of the two metals, but is readily adjusted by the screws provided.

Figure 462 C shows one of Ellicott's beautifully constructed pendulums, which were very expensive to make, and in an article by Evans and Evans[3] a further form of compensation devised by Ellicott is described. Probably one of the most famous clocks to incorporate Ellicott's pendulum was Cumming's amazing clock/barometer which was made for the Royal Household and is still in their possession.

Ellicott, like Cumming, was a Royal Clockmaker, although there is now only one example of his work in their collection. He made many fine and complex clocks including a month-duration equation clock which was in the Iden Collection. This shows seconds in the arch, has a 24-hour dial with the minute hand going round every two hours, and there is an effigy of the sun which traverses the heavens every 24 hours. Shutters rise and fall throughout the year so that the effigy always rises and sets at the correct time. Another fine example of his work is shown in Figure 480.

In 1760 John Ellicott FRS took into partnership his son *Edward*, who subsequently became master of the Clockmakers' Company. Initially they just signed their products *Ellicott,* but from 1769 to 1788 this was changed to *John Ellicott and Son.*

Edward Ellicott died in 1791 and the business was continued by his son *Edward.* The name of the business was changed to *Edward Ellicott and Sons,* and from 1811 until 1830 they traded under the name of *Ellicott and Taylor,* after

Figure 480. Astronomical Bracket Clock by John Ellicott. A dark green and gilt lacquer quarter chiming astronomical clock with rotating globe moon, signed on the dial and backplate Ellicott, London.

The arched brass dial has wheatear engraving around its upper border and a raised chapter ring and spandrels. Within the arch is contained the Astronomical disc.

It reads from within out:

1. The days of the month.
2. The months of the year in Spanish.
3. The days of each Zodiac.
4. The months of the Zodiac.

5. and 6. These lines indicate the position of the sun in the ecliptic and the times of sunrise throughout the year.

7. The suns declination North or South of the Equator and the times of the Equinox.
8. Below this is an arched brass strip carrying Ellicott's signature and inside this is indicated the age of the moon.

which it was known as *Ellicott and Smith.*

In 1838 Sweetings Alley and the Royal Exchange were destroyed and from that time until 1842 business was continued in Lombard Street.

John Ellicott started his own numbering system, the lowest number known to exist being a watch, number 123, made in 1728. It would appear that the original numbering was continued on throughout the various changes of ownership of the business. A rough guide to their dates is as follows:

123—1728
400—1730
1800—1740
3250—1750
4770—1760
6435—1770
7620—1780
8450—1790
8760—1800
9074—1810

Prior to 1760 it is likely that most of Ellicott's clocks and watches were made by their firm, but after this time an increasingly large number were provided by outside contractors. Thwaites and Reid supplied many of the clock movements.

John Ellicott FRS will be remembered as a clockmaker who throughout his life continued to try to improve the standards of clockmaking in an era when all too many merely slavishly repeated the designs originated by their forefathers.

[3] Evans and Evans N. *A Remarkable Clock*, by John Ellicott. *Antiquarian Horology,* 7th September, 1973, pages 402-405.

## EARDLEY NORTON

Eardley Norton (1760-1794) was undoubtedly one of the most talented clockmakers of the second half of the 18th century. He made a large number of watches, many of which were for export and also some fine longcase (see Figure 229) and bracket clocks. Many of the latter were quarter chiming or musical, and sometimes of considerable complexity. One of the finest was a beautiful gilt metal musical and quarter-chiming clock only 12 inches high (Figure 481). It was even fitted with Grande Sonnerie repeat at will. A considerable number of his finer clocks were for export and possibly for this reason were fitted with automata.

Undoubtedly, the finest clock Eardley Norton produced, and for which he is justly famous, is the astronomical clock he made for George III for which he was paid £1,042, an enormous sum of money in those days. This spring clock, which is still in the Royal Collection in Buckingham Palace, has four white enamelled dials. These were described by the *Gentleman's Magazine* for June 1765 as follows:

> The first and principal face shews the true apparent time, with the rising and setting of the sun in several parts of the world by a moving horizon;
> The second front shews the motion of the planets in their orbits, according to the system of Copernicus;
> The third, the age and the different phases of the moon, with the time of the tide at 32 different sea ports;
> The fourth, by a curious retrograde motion in a spiral, shews every day of the month and year, with the number and days of the week in proper emblems.
> The calculations and numbers for the wheels for the solar system were given by Dr. Bevis, and the designs for the dial plates, with the numbers and calculations and mode of performing, etc., by Mr. Ferguson.

Figure 481. *A small George IIIrd fire-gilt, musical, quarter striking and Grande Sonnerie repeating clock by Eardley Norton.*

## THE KNIBB FAMILY

Several members of the Knibb Family are recorded as clockmakers and there can be some confusion identifying them, particularly as they worked in both London and Oxford.

*Samuel Knibb* was born in 1625, the third son of John Knibb, Yeoman of Claydon and Warborough. He started work in Newport Pagnell circa 1655 and moved to London in 1662, joining the Clockmakers' Company by redemption in the following year. He is believed to have died in 1670. Only five clocks by Samuel Knibb are known to have survived and each of these is relatively complex. The explanation for this is thought to be that he worked in very close association with Ahaseurus Fromanteel and that most of the clocks he made were for Fromanteel.

*Joseph Knibb (Senior)*, the fifth son of Thomas Knibb of Claydon, was born in 1640 and was apprenticed to his cousin Samuel in 1655. Seven years later he moved to Oxford, where he initially met considerable opposition as he was not a Freeman of the city. However, within three years he seems to have made his peace with the other clockmakers in Oxford and, on paying his fine of £6-13-4 and "a leather bucket," was granted the freedom of the City. While in Oxford he was concerned with the development of the seconds beating pendulum together with his "crossbeat" and the anchor escapement. Examples of the latter were fitted to the turret clock of St. Mary the Virgin circa 1670 and also to the Wadham College Clock. In 1670 Joseph Knibb moved to London and was granted the freedom of the Clockmakers' Company. It is believed that at that time he took over Samuel's workshop.

Unlike Tompion and Quare, Joseph Knibb's output of watches seems to have been relatively small and it has been postulated that even those watches that are signed by him or other members of the Knibb Family may have been bought in. Joseph Knibb is best known for the beautiful longcase clocks he produced which reflect the strong artistic flare he undoubtedly possessed. The movements have a delicacy not usually seen with other makers and the dials are exquisitely laid out with finely fretted out and chased hands, delicate chapter rings, frequently skeletonized and with each minute numbered, and beautifully executed cherub spandrels. The cases are nearly always small and of excellent proportions and unlike most other makers, they did not get larger as the century progressed nor did the designs of the dial get heavier.

Joseph Knibb, like Tompion, produced a standard series of longcase and bracket clocks. He also made some clocks of long duration and specialized in unusual forms of strike work. He produced both longcase and bracket clocks with Grande Sonnerie striking and also clocks with Dutch Striking, i.e. the half hour would be struck in full on a high pitched bell, followed by the hour on a larger bell. Other methods of striking which he employed were:

a) Passing strike in which one blow is activated from the going train at each hour,
b) Hour and quarter hour striking using a two-train movement,
c) Roman notation in which two different bells are used of varying tones. The smaller one indicates "one" and the larger one, "five." Two blows on the large bell indicate "ten," and
d) Double six hour striking, whereby the clock strikes "one" to "six," and then from seven to 12 o'clock repeats this process (on "seven" it strikes "one" and so on). A single half-hour strike was also sometimes incorporated in this.

Besides the foregoing, the Knibbs also produced bracket clocks with pull-quarter repeat work, usually with cords on either side of the case. However, on the earlier clocks, before the invention of the rack strike, an external count wheel was used which was usually numbered so that it could be seen that the hours were being struck correctly. Joseph Knibb did not produce any of the complex clocks for which Tompion rightly became famous; however, he did make several night clocks and also some wall and lantern clocks.

A. Lee, in his excellent book on the Knibb family to which all who seek further knowledge on this maker are strongly recommended, suggests that approximately 200 of his clocks still exist, possibly indicating that in all around 400 were made.

A letter written in 1675 from Sir Richard Legh to his wife in Cheshire and quoted by R.A. Lee in his book, gives some idea of the standing of Knibb at that time and the cost of his longcase clocks:

I went to the famous Pendulum maker Knibb, and have agreed for one, he having none ready but one dull stager which was £19; for £5 more I have agreed for one finer than my Father's, and it is to be better finish'd with carved capitalls gold, and gold pedestalls with figures of boys and cherubimes in brass gilt. I wold have had itt Olive Wood (the case I mean) but gold does not agree with that colour, so took their advice to have it black Ebony which suits your Cabinett better than Walnut tree wood, of which they are mostly made. Lett me have thy advice herein by the next.

In the London *Gazette* of the 22-26th April, 1697, the following advertisement appeared:

At the Clock Dyal in Suffolk-street near Charing cross, on Monday the 26th instant, will begin the sale of a good parcel of very good pendulum clocks made by Joseph Knibb; some do go a year, some a month, some a week, and some 30 hours; some are table clocks, some repeat themselves, some by pulling repeat the hours and quarters: they are made and to be sold by Joseph Knibb at his House at the Dyal in Suffolk-street aforementioned, where the sale will continue until Whitsuntide, unless all be sold sooner. There are also some watches to be then and there sold, a good Pennyworths.

Shortly after this Joseph Knibb retired to Hanslope in Buckinghamshire and died there on the 4th May, 1712. A few clocks from this period exist marked *Joseph Knibb of Hanslope* or *Joseph Knibb at Hanslop*.

*John Knibb* was born in 1650 and from 1664 worked with his brother Joseph in Oxford, continuing the business after he moved to London. Shortly after this he received the freedom of the City of Oxford and in 1700 became its mayor. He died in 1722.

It is clear that John and Joseph Knibb worked closely together, for instance John sometimes used the same cabinetmaker as Joseph and it seems that at least some movements were made in London for John and possibly occasionally vice versa. That John had a substantial business is indicated by the fact that he is recorded as having taken on ten apprentices. Most of his clocks are very similar to those produced by Joseph at the same time.[1]

Footnotes

## JOSEPH AND THOMAS WINDMILLS

Joseph Windmills and his son Thomas were two of the most prolific clockmakers of the last quarter of the 17th and the first quarter of the 18th centuries.

*Joseph Windmills,* like Tompion and Quare, became Free of the Clockmakers' Company as a "A Great Clockmaker" in 1671 and also rose to become its master in 1702. He worked first at St. Martin le Grand and in 1687 moved to Mark Lane End, Tower Street (London). During his life he is believed to have collaborated with both Tompion and Quare with whom he was on the Court of The Clockmakers' Company. However, although he seems to have produced a similar range to Tompion, including watches, bracket, longcase and lantern clocks, he did not make so many complex and important pieces but catered for a much wider spectrum of the public with a relatively large number of simpler and less expensive pieces. There are three musical longcase clocks recorded as having been made by Windmills and a fine month-duration quarter-chiming longcase clock which was made for Cusworth Hall in 1714 (see Figure 130). In the Victoria and Albert Museum are two impressive clocks by him, one a lacquer longcase clock and the other a giant table clock decorated with mirror glass.

*Thomas Windmills* was born in 1672 and became apprenticed to his father on the 17th January, 1686 and gained his freedom of the Clockmakers' Company on the 20th January, 1695. In 1714 his father took him into partnership after which the clocks were signed either *J & T Windmills, London* or just *Windmills*. In 1718 he became master of the Clockmakers' Company and following the death of his father on the 4th March, 1723, took over running the family business. During his life time he had various partnerships: Thomas Bennett circa 1720-1730, William Elkin circa 1725, and Joseph Wightman circa 1725-1730. Thomas Windmills died on the 23rd May 1737 leaving some £18,000 excluding any real and other personal estate, a clear indication of just how successful the Windmills family had been.

[1] Ronald A. Lee, *The Knibb Family, Clockmakers* The Manor House Press, Byfleet, Surrey.

Figure 482. *Examples of the work of the three different members of the Knibb Family:*

Left, John; Center, Joseph; Right, Samuel. Photo courtesy R.A Lee Fine Arts, London

Until recently very little had been written about the Windmills Family, but during the last two years two excellently researched articles by J.A. Neale have appeared in *Antiquarian Horology* (June 1987 and winter 1988) and it is from these that much of the foregoing information has been obtained. It is to be hoped that in time these articles may be expanded into a book.

## CHARLES CLAY

Charles Clay was born in Flockton in the West Riding of Yorkshire, although exactly when is not known. However, a fine longcase clock decorated in "all over" arabesque marquetry which may be dated circa 1710 appears in *Britten's Old Clocks and Watches and Their Makers 7th Edition* and in 1716 he petitioned the King for a patent on the repeating mechanism for a watch. This was opposed by the Clockmakers' Company of which he was not a member and also by Quare, who had fought a similar battle 30 years previously, and it was Clay, in the end, who lost. In 1720 Clay set up business near St. Mary le Strand Church (London). By 1723 he had become Clockmaker to His Majesty's Board of Works and in 1731 he is recorded as having made a clock for the Gatehouse of St. James's Palace.

Although Clay made some relatively conventional clocks and watches, he is known today for the great individuality of many of the pieces he produced, a typical example being the fine longcase clock seen in Figure 220 with such an interesting and beautifully balanced dial layout. His greatest works are undoubtedly the musical and in particular the organ clocks which he produced of which by far the most spectacular was that known as "The Temple of the Four Great Monarchies" which Augustus, Prince of Wales, acquired circa 1743. It is illustrated in W.H. Pyne's *History of the Royal Residences* (1819) when it was on display in the Cupola Room in Kensington Palace and is also extensively described and illustrated in *Royal Clocks* by Cedric Jagger. Although Clay undoubtedly made this clock it was left to George Pyke to finish it in 1743, three years after Clay died.

To construct the clock Clay enlisted the help of the finest craftsmen, artists and musicians of the day, the music for instance being composed specially for the clock by Geminiani, Handel and Corelli and then properly adapted to the machine by Geminiani. Amigoni is recorded as having painted the dial while a group of figures representing the Liberal Arts are described as "Made of silver in Alto Relievo, by Mr. Rysbrack." This masterpiece, having been put in storage for many years and sadly neglected, is now in a sorry state, the original movement, dial, and base all having been lost and replaced with inferior substitutes.

For those who would wish to learn more about the clockmakers discussed in this chapter the following books, from which much of the information was taken, are recommended.

(i) Atkins S.E., and Overall W.H.: *Some Accounts of Worshipful Company of Clockmakers of The City of London*. Privately printed in 1881.
(ii) Baillie G.H., Clutton C. and Ilbert C.: *Britten's Old Clocks and Watches and Their Makers, 7th Edition* E. & F.N. Spon Ltd, 1956.
(iii) Dawson P, Drover C.B. and Parkes D.W.: *Early English Clocks*, Antique Collectors Club, 1982.
(iv) Jagger C.: *Royal Clocks* and *The British Monarchy and its Timekeepers 1300-1900*, Robert Hale, London, 1983.
(v) Loomes B: *The Early Clockmakers of Great Britain*, N.A.G. Press Ltd., London, 1981.
(vi) Symonds R W: *Thomas Tompion, His Life and Work*, B.T. Batsford Ltd, London.
(vii) *The Company of Clockmakers—Register of Apprentices, 1631-1931*. Privately printed for the company, 1931.

For more general references of recorded clockmakers the reader is referred to:
(viii) Baillie G H: *Watchmakers and Clockmakers of the World.*, Vol. I, N.A.G. Press Ipswich. Last reprinted in 1985.
(ix) Loomes B: *Watchmakers and Clockmakers of the World.*, Vol. II, N.A.G. Press, Ipswich, 1976.

(These two volumes contain some 70,000 names. However, this list is far from complete and it must be remembered that there are still many clockmakers, probably tens of thousands, who for one reason or another have not so far been recorded. Should you know of any of these, it is suggested that you communicate them to Mr. B. Loomes c/o N.A.G. Press.)

326 *Hands and Spandrels*

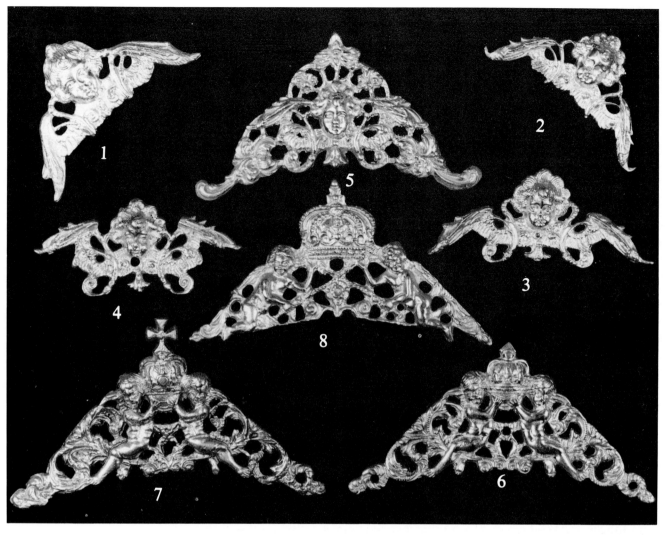

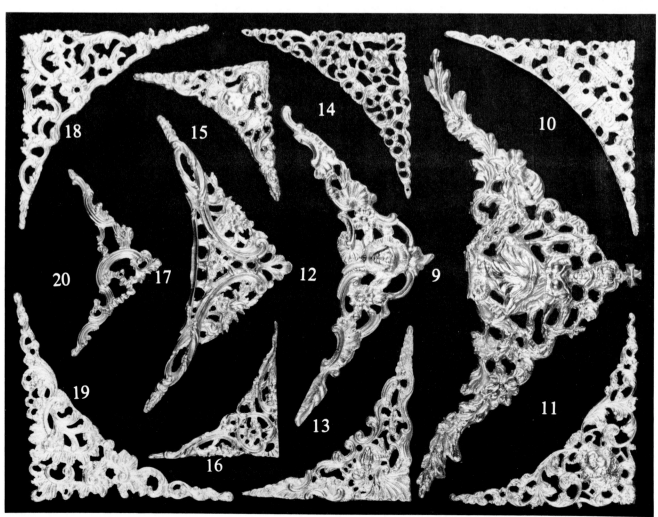

# Chapter 18.
# Hands and Spandrels
## SPANDRELS

Spandrels were used on longcase clocks from their very inception and indeed were in common use on other clocks a long time prior to this. On a few of the early longcase clocks the corners were just engraved (Figure 33 A and B) but this practice, while most pleasing, had largely ceased by 1675.

The cherub spandrel was used on all the early clocks, the first design being that seen in No. 1 with Nos. 2,3,4 following. These early designs have never been improved upon. An attractive feature of the early spandrels was that they were relatively small with plenty of space about them so that they were displayed to the maximum advantage (Figures 34 A, 35, and so on), whereas later spandrels were much larger and occupied virtually the whole corner.

What is often known as "the late cherub spandrel," No. 5, a larger and more elaborate affair, came into use circa 1680 and continued until around 1710. The spandrel in which twin cherubs are seen upholding the crown was introduced circa 1700, possibly as a result of the accession of William and Mary to the throne. The earliest form is that seen in No. 6 with those in Nos. 7 and 8 being introduced maybe ten years later.

No. 9 is a beautiful spandrel, reputed to represent Queen Anne, which Edward Cockey designed for his celebrated astrononomical clocks, one of which was actually made for Queen Anne. No. 10 is the urn spandrel, one of the most popular designs ever produced, which came into use circa 1710 and lasted until the 1740s. Many different forms of mask spandrel, both male and female, were introduced from 1710-15 onwards of which just two are shown here Nos. 11 and 12.

As the century progressed the design of the spandrel became far more open, flowers, foliage and seashells often being represented and by 1760 what might be termed the rococo spandrel had become popular (No. 18). During the 18th century various designs seem to have originated, such as Nos. 15 and 16 which were usually only used on country clocks, one of the most attractive being that of a girl in four different poses and clothing representing the seasons.

With the introduction of the arch to the dial the dolphin appeared (Nos. 21 and 22) flanking either the name disc or the strike/ silent ring and its use continued throughout the whole century. Various other designs were introduced (Nos. 25-28) often with floral influence, from 1750 onwards. An amusing one (No. 24) depicts the moon, another design which was largely confined to country clocks. Spandrels, like many components were largely bought in from specialist suppliers and thus it is almost impossible to be able to attribute a clock, from the design of the spandrels, to a particular maker; although Tompion does seem to have introduced a few individual designs of his own.

The quality of spandrels varies enormously and this is related to four factors: (1) the quality of the original pattern, (2) the quality of the mould from which it was taken, (3) how many times that mould was used (the longer the run of the spandrels taken from the mould, the more their quality deteriorates), and (4) how carefully the spandrel was finished up after casting. This is a time-consuming process and thus expensive; it is surprising how on even the better London clocks the spandrels are left relatively unfinished with "flash" protruding from their edges.

Frequently, it will be seen that even the quality of different spandrels on the same dial varies appreciably. This does not usually indicate that one or more have been changed but merely that they were cast when the mould had started to deteriorate.

Although the type of spandrel is a good general guide to the age of a clock, assuming they are original to the dial, it, like so many other features of a clock, will give only an approximation to the correct date and should be used in conjunction with many other features. This is because many different designs persisted for quite a long time, up to 30 to 40 years, and patterns at one time popular in London, may reappear in the country some 20 to 30 years later.

Opposite page:
Figure 483. *Single and twin cherub spandrels, 1670-1715.*
Figure 484. *18th century corner spandrels.*

## 328 Hands and Spandrels

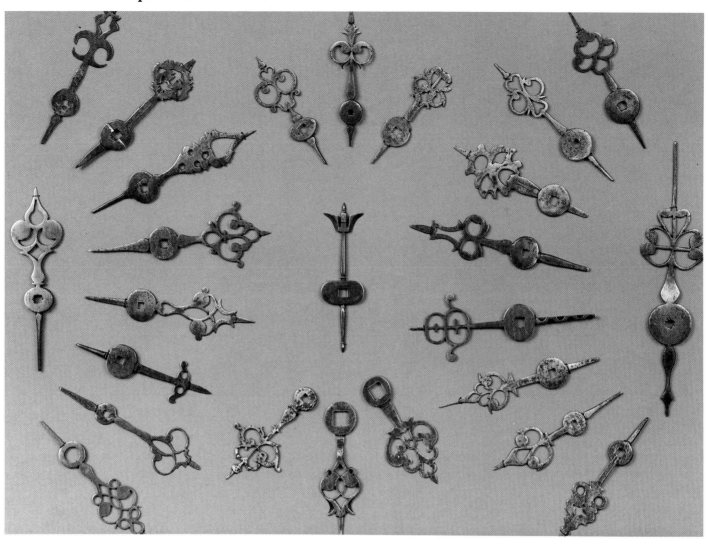

Opposite page:
Figure 485. *Early lantern clock hands, circa 1600-1670.* Note the resemblance of many of them to weapons such as lances, spears and arrows.

Figure 486. *A selection of hands, mostly dating from 1680-1710.* Some would have come from lantern clocks.

Figures 487 and 488. *Late 17th century clock hands.* Note the beautiful way in which they have been chased and faceted.

# 330 Hands and Spandrels

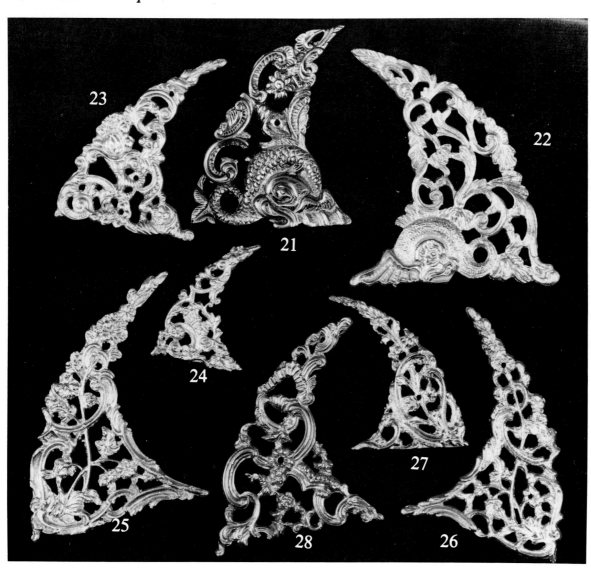

Figure 489. *Arch spandrels.*

*Opposite page:*
*Figure 490.* A miscellaneous collection of clock hands. Shown on the bottom row is the evolution of the minute hand from the simple S base and the long pointer through to the far more ornamental hands of the 1770s where the decoration was carried right up to the tip.

Figure 491. *Longcase and lantern hands, 1700-1720.*

## CLOCK HANDS

An extensive collection of clock hands dating from the beginning of the 17th century through circa 1860-1870 are illustrated here. By showing a large number, readers are able to appreciate how they were evolved and restorers are able to see what to replace on part or all of a clock hand.

The designs of the early clock hands (Figure 485) evolved from weapons such as spears, lances and arrows, and it is likely that in some instances they were both made by the same craftsmen.

The earliest hands shown probably come from lantern clocks and predate the longcase clock by up to 50 years. Because the lantern clock has only one hand, considerable force has to be applied when turning it. For this reason they are robustly constructed and usually have a tail to simplify holding them. Following the evolution of the longcase clock with two hands, the designs changed rapidly. The hour hand became far more delicate and detailed in its construction. However, the minute hand was kept as a simple pointer with an S-design at the base and it was only towards the end of the 17th century that it started to become more decorative. Gradually the ornamentation spread up the hand until by circa 1760 the entire hand was often involved. To match this, the hour hand had gradually become more open in its design.

With the advent of the "all-over" engraved and silvered brass dial and the white dial, far simpler blued steel hands appeared such as those seen in Figure 498 and these were probably in use from 1790-1840. It was not to be long after the white dial appeared before brass hands started to be used. These could never be used with a brass dial because there would be no contrast. At first they were quite delicate in their design and followed that of steel hands. They were also pierced in a similar way but this was somewhat easier because brass is a much softer metal than steel. However, to compensate for the lack of strength they had to be kept a little thicker.

## Hands and Spandrels 331

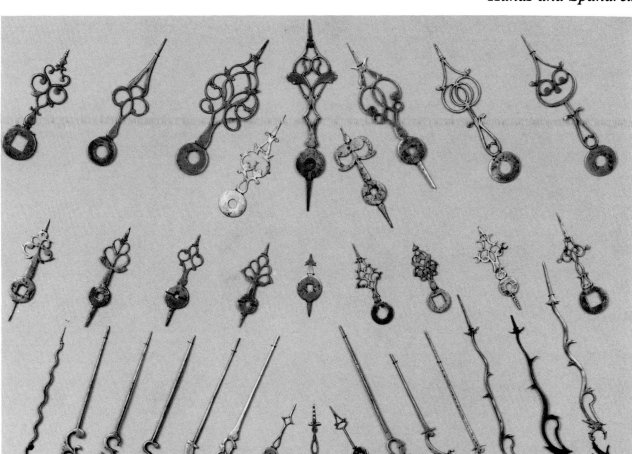

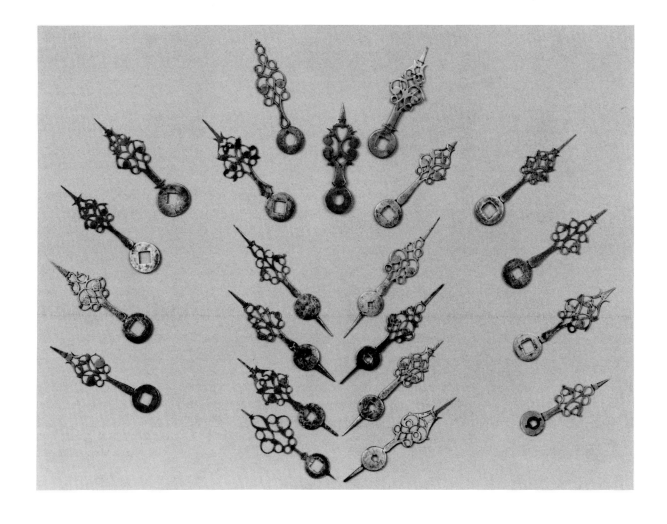

332  Hands and Spandrels

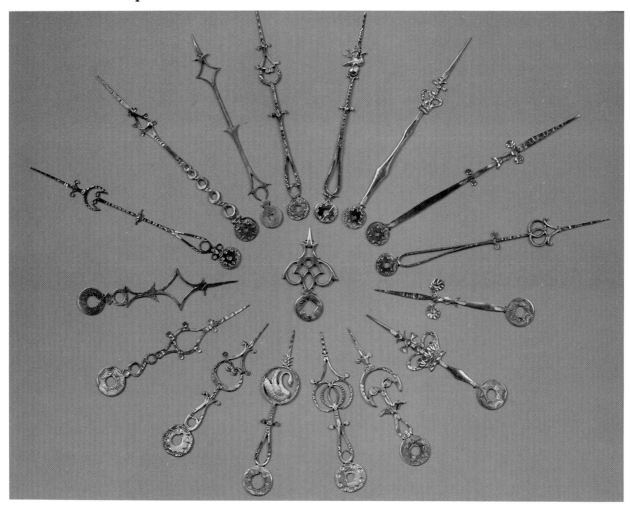

Figure 492.

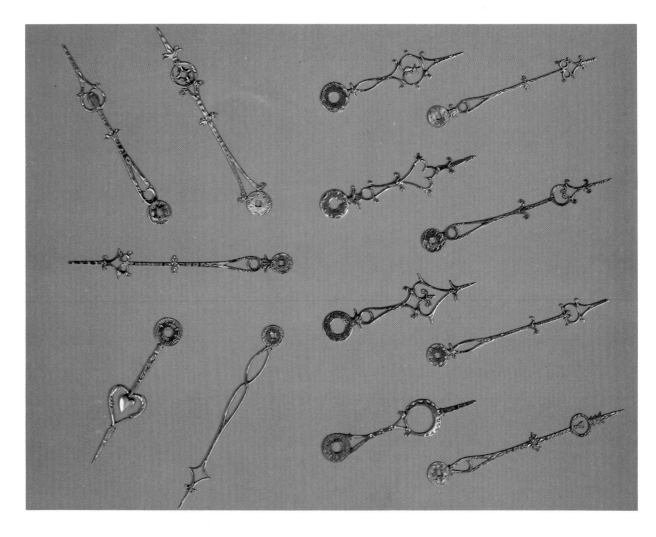

Figure 493.

By around 1830 the laborious process of piercing the brass hands had given way to stamping them out on a fly press, a much quicker and cheaper operation. Sometimes to make them more decorative they had castings soldered to them such as the swan seen in Figure 494.

Probably, some of the finest hands were made in the period 1680-1705. The amount of time taken in chasing and faceting hands such as those in Figures 487 and 488 must have been enormous. Although some clockmakers might have been capable of this work, it is likely that in the vast majority of cases it would have been carried out by a specialist hand manufacturer.

Figure 494.

Figures 492, 493, and 494. *Decorative gilded brass hands circa 1795-1870.* The early examples were similar to the steel hands of the same period but the later ones were far more crudely made, usually being stamped out. Ornamental castings were sometimes applied. In Figure 497 may be seen a hand with a swan mounted on it.

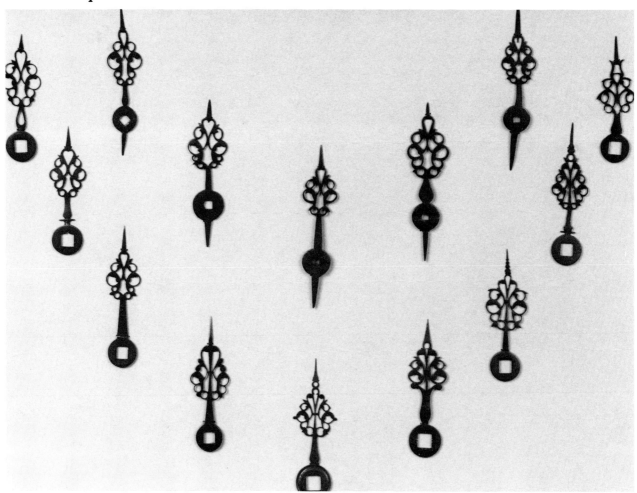

Figure 495. *Longcase and lantern hands, 1710-1730.*

Figure 496. The somewhat more open designs of the hands which appeared as the century progressed.

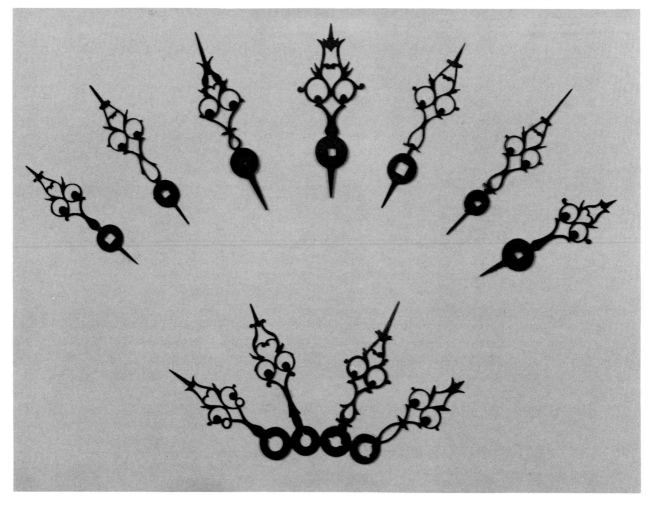

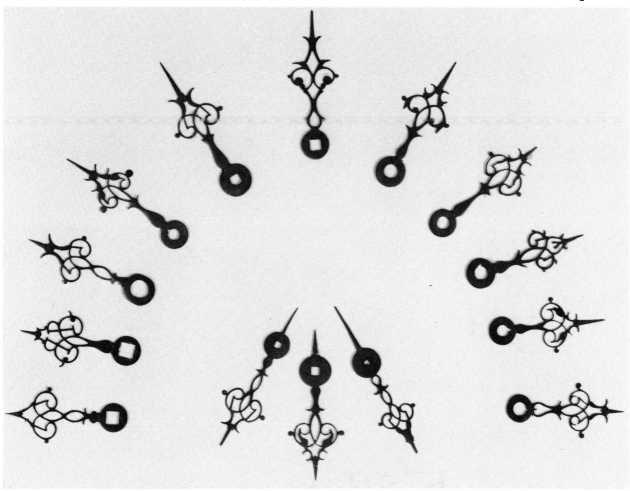

Figure 497. Circa 1760-70.

Figure 498. Much simpler steel hands had started to be used by 1780 and continued in favor until circa 1840.

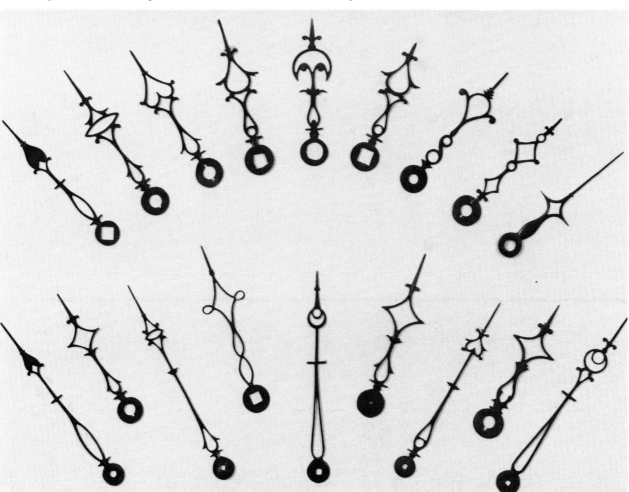

336 *Care*

Figures 499 A and B, 500 A and B, and 501 A and B. Three country brass faced thirty hour clock dials before and after restoration including resilvering of the chapter ring (Note how much more detail can be seen). In the case of that in Figure 500a new spandrel had to be cast up using one of the existing ones as a pattern.

Figure 502 A, B. A flat brass thirty hour dial by Sam Pearce of Honiton, before and after silvering. Some of the darker flecks are due to corrosion of the brass which has occurred during the last 200 years.

Figure 503 A and B. A painted or white dial before and after cleaning and restoration, including the reinstatement of the makers name.

# Chapter 19.
# Care of a Longcase Clock
## —including setting up and dial restoration

Whenever a longcase clock was produced, be it in the 17th, 18th, or 19th centuries, it was designed and made to last for several hundred years, provided reasonable care was taken of it. We find this concept difficult to comprehend in this throwaway age where five years is considered quite a long life for many mechanical and electrical goods. Longcase clocks are rugged and well engineered. However, an understanding of the basic care and maintenance they require is desirable if we are to pass them on to successive generations in at least as good a condition as when we received them. This is really a very small price to pay for the pleasure they have given us during our lifetimes.

### Clock Dial Restoration
### Brass Dials

When a brass dial was made by a clockmaker, the dial plate and spandrels almost invariably were polished and then protected with a coat of clear lacquer to prevent them from tarnishing. The only exception to this is the use of fire gilt on some of the finest early clocks. This is the deposition of a layer of gold on the dial by making an amalgam of gold and mercury, applying this to the dial, and then driving off the mercury by heating it up to a high temperature over a charcoal fire. Because of the obvious danger to the worker's health because of the mercury fumes, this process has been banned for many years. After the dial and spandrels were gilded, the high spots were burnished to give depth and lustre to the finish. Fire gilding as opposed to electroplating will last for many hundreds of years and should on no account ever be removed from a dial.

The chapter, seconds, and date rings of dials and the name plaque (where present) were almost invariably silvered in the first instance to make them easier to read. This was achieved by the chemical deposition of silver onto the surface of the brass followed by polishing with cream of tartar and then protecting it with a coat of clear lacquer or varnish. This process should not be confused with electroplating, which must never be used on old clock dials.

The question "Won't I reduce the value and originality of a clock if I restore the dial?" is often asked. The answer to this is that the dial will almost certainly already have been restored several times during its lifetime, so the finish you are looking at is not original anyway. It is the originality of the dial itself which counts.

Brass and silver, when they do tarnish, often do so unevenly, possibly because they have been handled or the protective lacquer has broken down in places. They gradually turn a brown/black color which can only be regarded as unsightly. Nothing is to be gained by retaining this any more than you would leave a pair of brass candlesticks unpolished. Whereas wood mellows with age and takes on warmth, metals such as brass and silver just oxidize and discolor and thus need regular attention. To give some idea of the results, we show examples of dials before and after they have been restored (Figures 499-503).

Once restored a dial must not be polished otherwise the protective lacquer will be breached and the metal will start to tarnish. On no account should a dial ever be cleaned or restored unless it is stripped down first. To attempt to do so will inevitably leave chemicals underneath the chapter ring, spandrels, etc which will in time cause corrosion.

### Painted Dials

The term painted dial is somewhat misleading as it gives the impression that the dials were painted in a similar way to that which we would do today. However, such is not the case and far more time and trouble was expended on the meticulous preparation of the background in particular, heat being used to produce a hard finish. Layer upon layer of paint was used, each one being rubbed down when hard and a further coat applied until gradually a perfect serface was achieved.

The description "enamelled" rather than "painted" dial is probably the more suitable term as it gives a better idea of the technique used and the finish obtained but it should not be confused with the hard enamels employed for clock dials on the Continent. These were fired, usually onto copper, at much higher temperatures.

An extract from "A visit to a Clerkenwell Clock Factory" which appeared in *The Illustrated London News*, 20th September 1851 gives some insight into the process:

> "The dial faces are coated with what is technically termed white-flake, a superior kind of white-lead, which is ground down with the finest description of Copal Varnish, and then put into a stove, similar to that used by Japanners, until it becomes sufficiently hard to receive a polish, which is effected chiefly with pumice stone, by which means a most perfect surface is produced, ready to receive figures, which are painted with lamp black, varnish and turpentine."

The restoration of a painted dial is very like that of a painting and a similarly cautious approach should be employed. So far as possible the original background, which usually is in reasonable condition, should always be retained, touching in where necessary and similar remarks apply to the corner decorations (spandrels). However, the numerals and the name which are not as durable, will often have become worn. If this is not too bad then they can just be sympathetically restored but if extensive then they may have to be completely redone. Fortunately this is relatively easy to do because of the way in which the dial was originally painted.

The remains of the existing numerals can be removed fairly easily, just their shadows remaining to act as a guide when repainting.

Quite frequently when restoring a dial, a makers name can be detected on what was thought to be an anonymous clock. The best way of doing this is to vary the angle of the dial relative to the light source until the shadows of the original name can be read. Once one has picked up some of the letters, reference to guide books listing clockmakers such as Baillies & Loomes should give you a clue to any which are difficult to decipher.

### The Movement

The movement of the longcase clock is designed so that steel pinions mesh with brass wheels. This combination requires no lubrication; in fact, if oil is applied, considerable damage will ensue over a period of time as dust and dirt will stick to the oil and be forced between the wheels and pinions, greatly increasing their wear. Oil or thin grease is only required where there is a sliding friction between two parts, and then only sparingly. Examples of this are the faces of the pallets which engage the 'scape wheel and possibly the block on the pendulum which passes through the fork on the crutch. These can safely be attended to by the owner every few years. A thin grease is also often used on other components which slide against another, such as parts of the strikework, but this can only be carried out by the clockmaker when the clock is dismantled. It is difficult to guess how long a clock may be left before cleaning as it varies so much with temperature, humidity, and the dust and dirt in the atmosphere. However, probably a reasonable regime is to oil it every three to five years and strip and clean it every ten years. Although the clock will continue to go after this, it has by then become dirty and is grinding this dirt into its wheels and pinions.

The only other part of a longcase clock which requires attention from time to time is the sinks around the pivot holes. Just use a little oil there. However, as this necessitates the removal of the dial to reach those behind the front plate, the work is usually best left to a clockmaker.

### The Case

Never subject a clock case to a dry atmosphere. In cold climates where central heating is used in the winter, conditions that are too dry can occur. As soon as the relative humidity (RH) drops below 55 percent, the wood starts to shrink, the doors twist, veneers crack and great damage is done to the case, much of it, sadly, irreversible. In the last 20 years more damage to antique furniture in general has probably been caused in this manner than has occurred in the previous 300 years due to cold, damp, draughty houses and general neglect. The pity is all the greater in that it can so easily be avoided by the use of humidifiers, the cost of which is far less than the cost of restoring fine antiques. Indeed, it seems to be a pattern of advancing civilization that we damage so much of our environment and heritage, not willfully or by neglect, but by thoughtlessness. So far as the preservation and even the improvement of a clock case is concerned, the best approach is to apply a beeswax polish a few times a year and spend the rest of your time just polishing and burnishing it. If too much wax is used, dust and dirt will stick to it and in time obscure the figuring and color of the wood.

A practice which has been advocated now for several hundred years is to gently wash the case once a year with a weak solution of vinegar water. This lifts the surface wax and with it the dirt which has accumulated without going through the deeper layers of wax and reaching the wood. After this has been done the case is carefully dried, rewaxed, and polished.

Besides applying wax to the outside of the case, the inside can also be treated and left unpolished. The advantage of this is that if the wood starts to dry out, then the wax will be drawn into it and reduce, to some extent, the risk of further damage to the case.

**Dismantling a Longcase Clock**

A longcase clock should always be dismantled before it is moved. First stick a small piece of paper on the backboard immediately behind the tip of the pendulum. Then stop the pendulum and mark the position of its tip on the piece of paper. When the clock is reassembled, if it has been put together correctly and nothing (particularly the crutch) has been bent or disturbed, then, provided the tip of the pendulum corresponds with the mark on the paper, the clock should go. The pendulum acts as a plumb line and determines that the case is upright.

Having noted the position of the pendulum, the next step is to remove the hood. This may be locked in place by a turn buckle or bolt fixed to the trunk immediately below the front of the hood. This is reached by opening the trunk door and feeling up inside. Sometimes a wooden bar is used which will usually be apparent when the door is opened.

The hood may now be slid forward and removed. The only exception is on a few very early clocks made prior to circa 1690 which still have lift-up hoods and are held by a catch in their raised position for winding. Although they are usually locked down by a spoon latch, this is automatically released when the trunk door is opened.

To save the lines supporting the weights from getting tangled, particularly inside the movement, these are best let down. By far the easiest way of doing this is to keep the clock running until it stops with the weights at the bottom of the case. Should this be impractical, it is best to remove the pendulum first. This is achieved by lifting it up about one-half inch, if possible by the small brass or steel block *above* the steel strip, then moving it back so that the steel strip comes out of the slot in which it is engaged. Great care is needed at this stage not to damage it. Finally, the pendulum may gently be lowered, allowing the suspension to pass down through and out of the crutch. It can now be removed from the case and carefully fastened to a length of wood to prevent it from being damaged. Particular attention should be paid to the steel suspension; a useful ploy being to fix two small pieces of wood on either side of it.

Before unhooking the weights it is wise to get a friend to hold the movement steady in the case. The movements are front-heavy due to the dial, and as soon as the weights are removed, tend to fall forward out of the case, which does neither the movement nor the owner any good if it lands on his head. Nearly all clock dealers and restorers have suffered at least once in this way.

Having taken off the weights, the lines can be let down. This is done by releasing the clicks engaging the barrels, either by pushing the other end of what is usually a pivoted lever or, if this is not possible, by lifting the active end of the lever out of engagement with the dog-toothed ratchet wheel. Sometimes this is simplified by slightly winding the wheel with the key so as to lift the end of the lever up to the tip of a tooth where it can be more easily engaged.

As soon as the lever is disengaged, the lines can be gently pulled off the barrel. It is not normally advisable to let the lines down with the weights still on as they may crash down to the bottom of the case.

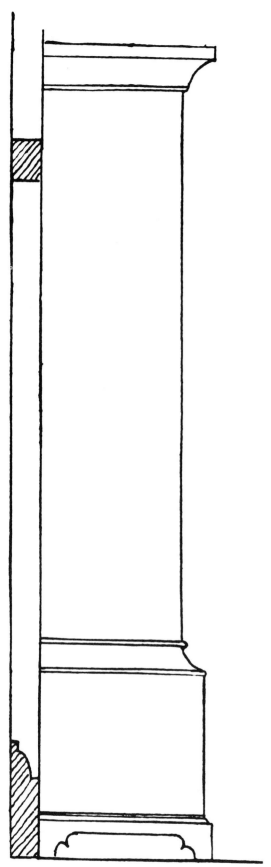

Should the lines get tangled, untie the free ends, remove the pulley, and uncoil the line from the barrel turn by turn. The movement, complete with its seat board, may now be removed from the case and put safely in a box, care being taken to see that the crutch cannot be damaged. Occasionally the seatboard will be held in place with screws which will obviously have to be removed first. Where the seatboard is nailed down, particularly if this is with the original clout nails, then it should not be disturbed for fear of damage. In these instances it is usually best to untie the lines where they pass through the seatboard, pull them out, remove the pulleys, release the two bolts or hooks holding the movement down, and then remove it.

**Setting Up Your Clock**

It is nearly always desirable to fix your clock to the wall. If this is not done, apart from the risk of its being knocked over, when the weights are roughly half way down (nearly level with the pendulum bob), they will pick up its motion and rock from side to side and result in the pendulum stopping.

On a solid floor it is sometimes possible to avoid fixing the clock to the wall by just chocking up the front feet approximately three-eights of an inch off the floor. The weight of the clock will then be thrown back against the wall and make it relatively stable.

A longcase is usually best situated flat against the wall as opposed to across a corner. It can be fixed there by screwing a piece of timber approximately eight-by-two-by-one inches horizontally to the wall and then fixing the clock to this. (See Figure 504.)

Should your skirting board be thicker than one inch, the piece of timber will have to be increased in thickness correspondingly. The case may be conveniently fixed to the piece of wood, either behind the hood, or more simply, behind the trunk door. Frequently there are holes already in the back of the case. In such instances the batten can be fixed at the same height.

When a clock has to be set up across a corner, a bracket such as that shown in Figure 505 should be used.

Figure 504.

Adopt the following procedure after getting the clock home.

1. Decide where you want to place the clock and then fit a convenient batten.

2. Place the case in front of this, get it upright, and fix it to the piece of wood.

3. Place the movement in the case, preferably with the lines which hold the weights fully extended. Get a second person to hold the movement while one of the weights is put on to steady it.

4. Put the hood on. Then, by sliding your hand up inside the case, position the movement so that the dial is correctly positioned inside the hood with the latter fully seated.

5. Remove the hood, taking care not to disturb the movement, and put on the second weight (if applicable). The pendulum may now be fitted by putting it inside the case through the trunk door and then feeding the strip of spring steel behind the movement through the steel loop or fork at the back of the movement (known as the crutch). Then hook it onto the brass bracket above by sliding the steel through the slot and letting it down until the small brass block at the top of the spring rests on the bracket (Figure 506).

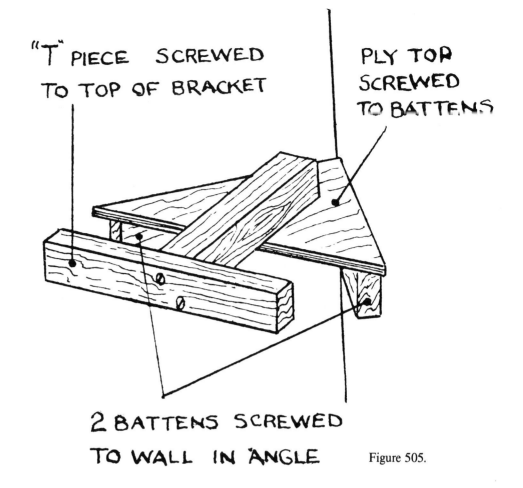

Figure 505.

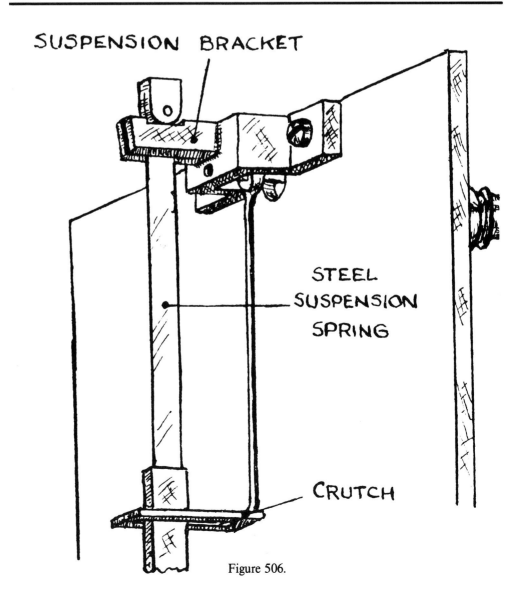

Figure 506.

6. Carefully wind up the weights, making certain that the lines are not twisted and are winding evenly onto the barrels.

7. *Do not* wind the weights up as far as possible. Always leave a gap of about two inches between the underneath of the clock and the pulleys.

8. Mark on the backboard of the clock the position where the tip of the pendulum hangs, then move the pendulum first to one side and note or mark the position when the clock "ticks" or the second hand just goes on one second and then repeat the procedure on the other side.

The "ticks" should occur evenly on either side of the original mark where the pendulum hangs. If it is too far to the left, then the whole clock must be tilted a little to the right until the "beat" is even on both sides.

However, for obvious reasons such as appearance and the pendulum hitting the side of the case when it swings, this can only be done to a very limited extent. The best method is to bend the crutch at the back of clock. This is the way in which a clockmaker would normally put a clock "in beat," but it does require care—otherwise the crutch or the escapement may be damaged.

The crutch is bent in the opposite direction to the side on which the pendulum has to move farthest before the clock ticks. This is a delicate operation and the crutch will usually require several adjustments before the beat is even. (Figure 507). Unfortunately, the strength of crutches varies greatly, some bending easily and others only with appreciable force.

**Regulation—If the Clock Loses:**

1. Stop the pendulum.

2. Raise the bob a little by turning the regulator nut below it (see Figure 508).

3. Move the hands forward to the correct time and start the clock by swinging the pendulum.

4. For precise regulation of the clock, let the pendulum swing until the seconds hand is at top dead center (on 60 or 12). Move the minute hand until it is exactly on a minute division. Start the clock by holding the pendulum to one side and releasing it when the time signal is heard on the telephone, radio, or television.

**If the Clock Gains**

1. Stop the pendulum.

2. Lower the bob a little by turning the regulation nut below it.

3. Wait until the clock reads correct time and then restart it. Alternatively it can be left until it reads slow and then the minute and hour hands advanced to correct time before restarting the clock.

NEVER MOVE THE HANDS BACKWARDS.

NB: Regulators and all 30-hour clocks have a form of maintaining power which keeps them going whilst they are being wound, but the movement of nearly all others stops and loses this amount of time. Ideally, if it takes (for example) 20 seconds to wind the going side of a clock, then it should be adjusted to gain 20 seconds during the week to allow for this.

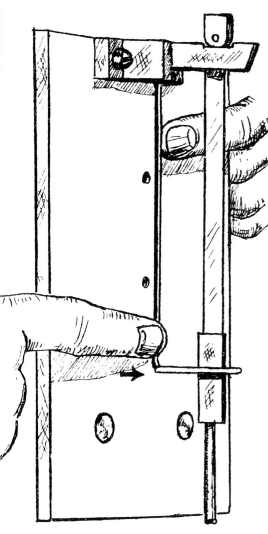

Figure 507.

## Striking

1. When moving the hands always let the clock strike each hour. Similarly, if it chimes at the quarter-hour, let it chime each quarter in full. It is essential with all clocks fitted with a count wheel (sometimes called a locking plate) strike to let them sound out each hour in full. Otherwise, the strike will go out of synchronization with the time and have to be corrected.

It is extremely important in clocks fitted with rack striking to let them strike 12 in full as this allows the clock to reset itself for a further series of 12-hour strikes.

2. When turning the hands always move them slowly for a few minutes before you come to the hour. When a "click" is heard pause for a few seconds to allow the clock to prepare itself to strike the hour and then gradually move the hand up to the hour. Allow the hour to be struck in full before moving the hands further.

### Rack Strike

With eight-day longcase clocks made in London after circa 1700 and in the country at a somewhat later date, a system of striking known as "Rack Strike" was employed in which the striking side of the clock was permanently synchronized to the going (timekeeping) side. Thus, even if the hands are turned round inadvertently, without allowing the clock to strike certain hours, on the next hour it will still strike correctly. However, it must not be taken beyond 12 o'clock without letting it strike, otherwise permanent damage may ensue.

Because it is essential that a clock with rack strike must always strike 12, both weights of the clock should be wound up together so that there is always sufficient power for it to strike.

It is not possible to have a clock with rack strike going with no strike because of potential damage to the striking mechanism. The only exception to this is if there is strike/silent regulation on the clock or the movement is modified.

### Count Wheel (or Locking Plate) Strike

"Count Wheel" is probably the more correct term for this form of strike. It is the earliest way of controlling the strike on longcase and indeed all clocks and was employed on most London clocks made before circa 1700, most country clocks prior to 1720, and virtually all 30-hour clocks.

With eight-day clocks it gives the advantage that if you do not want it to strike all you have to do is not wind the weight on the striking side (normally the one on the left as you look at the clock). No harm will come to the striking mechanism—as would be the case with rack strike. However, the strike will have to be resynchronized with the hands when the striking mechanism is put back into operation.

### Correction of the Count Wheel

To adjust the strike, the hooked bar (detent) which engages the locking plate is lifted and then released when the clock will strike the next hour (Figure 509). This process will have to be repeated until the strike is correct (for example, if the clock strikes two when the clock reads 5:00, then the bar will have to be lifted three times to set the strike correctly).

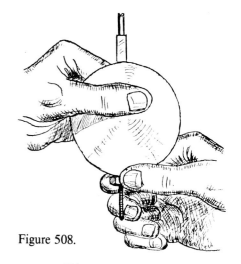

Figure 508.

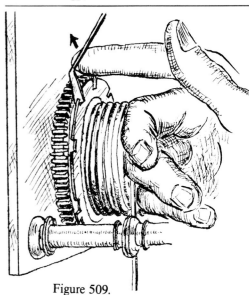

Figure 509.

### Position of the Count Wheel

On all 30-hour clocks and most eight-day clocks made before 1690, the Count Wheel is mounted on the back of the movement. (Figure 510). With eight-day clocks after 1690, it is usually mounted inside the movement either in front of or behind the great wheel on the striking side of the clock (the left side as you look at it).

### Winding: Thirty-Hour Clocks

The term "30-hour clock" does not necessarily mean that it will go for 30 hours, but merely more than 24. Some may go for 36 hours, whereas others will stop if you do not wind them within 26 hours. Occasionally, an early single-handed longcase clock will be encountered which requires winding twice a day, as did the first of the lantern clocks. The chain or rope which has to be pulled down to raise the weight is usually that on the right of the clock. However, to ease the strain on the movement, the left hand should be used to partially lift the weight, either by holding the weight itself or by raising the left-hand rope (Figure 511).

### Eight-Day Clocks

It is essential that the trunk door is opened before winding so that one can see that the gut lines are not twisted, that the lines are on the pulley, and that the weight rises smoothly and is not wound up too far. If it is then the pulley will hit the seatboard on which the movement rests, resulting in either 1) the pulley tilting and the gut line coming off; 2) the line breaking and the weight crashing to the bottom of the case; or 3) teeth being bent or broken on the great wheel.

On all eight-day longcase clocks with rack strike, which probably comprises 95 percent of all those made, both the striking (left side) when looking at the clock and going weight (right side) must always be fully wound, otherwise the striking mechanism may be damaged. Ideally, the striking weight should always be wound a little higher than the going weight.

Figure 510.

Figure 511.

## Calendar Work

If the clock has either a hand or an aperture with a revolving disc or ring behind it which indicates the day of the month, this will need to be set correctly when the clock is installed. It will also require advancing by one day on the four months of the year when there are only 30 days in the month and by three days every February (except on leap years, when it will be only two).

Some date indicators, usually the disc type, are driven from a part of the clock connected to the hour hand. These change every 12 hours, whereas others, usually the rings, are moved on by a pin on a separate wheel geared to turn it every 24 hours. The pin or lever which changes the date is usually only engaged for three-to-four hours a day so that only one number stays in the center of the aperture for practically the whole day. The date should be corrected when the mechanism is not engaged.

The clock should have been set up by the clockmaker so that the date starts changing at or a little before midnight. By 4 o'clock the mechanism is disengaged and from then until 8 or 9 o'clock it is safe to move it.

With a clock with a ring which moves only once every 24 hours, it may be necessary to turn the hand round 12 hours so as to get the date to change at 12 midnight instead of 12 noon. Once this is set correctly, the date can be altered at any time between 5 A.M. and 8 P.M.

To actually correct the date, all that needs to be done (if it is indicated by the hand) is to open the hood door and move the hand to the correct position. With a date aperture it is usually necessary to remove the hood to gain access to the date ring or disc behind the dial before it can be rotated.

Unfortunately, although all date mechanisms should be set up to commence changing at midnight and thus be free by 4 or 5 A.M., this is by no means always the case. Thus, if more than slight resistance is felt in moving the disc or ring it will mean that the disc is engaged and you must then delay changing the date until it is free. By looking behind the dial, you can often see whether or not the pin or lever driving the date mechanism is actually engaged.

## Moon Discs

When a clock has a moon disc (which is usually situated in the arch of the dial), this also will need to be set correctly. This is simply done by removing the hood and rotating the disc in a clockwise direction (never counterclockwise) until the correct age of the moon is shown. As with the date ring the moon is moved roughly once every 24 hours; (it goes round 29 ½ times a month) and is only engaged for a few hours of that period. It is important not to try and move the moon disc when resistance can be felt because the mechanism will actually be turning the disc. On no account should you try to rotate the moon disc without removing the hood (for example, by pushing it from the front), as either the disc is likely to become distorted or the pin on which it rotates will be bent and the mechanism will then fail to work.

## Summary

1. Make certain the clock is stable. Fix it to the wall if possible.

2. Always wind up both weights of the clock.

3. Never turn the hands backwards.

4. Always let it strike each hour.

5. To slow it down, lower the bob. To make it go faster, raise the bob.

6. Only correct the date between 4 and 8 o'clock.

7. Always open the trunk door before winding up the weights to insure that the lines are not twisted.

8. Stop winding weights well before the pulleys touch the underside of the clock.

# Appendix

### Maps and List of Place Names

To enable readers to identify where a particular clock was made, we are including an overall picture of England (showing the counties); six maps; and an index listing the principal villages, towns, and cities where the clockmakers worked.

We are extremely grateful to N.A.G. Press for allowing us to use the maps and index which have been extracted from C.H. Ballie's *Watchmakers and Clockmakers of the World* volume 1.

Figure 512. The counties of England as they were in the 17th, 18th, and 19th centuries and prior to their revision circa 1950. Reproduced by kind permission of the publishers of *This England*.

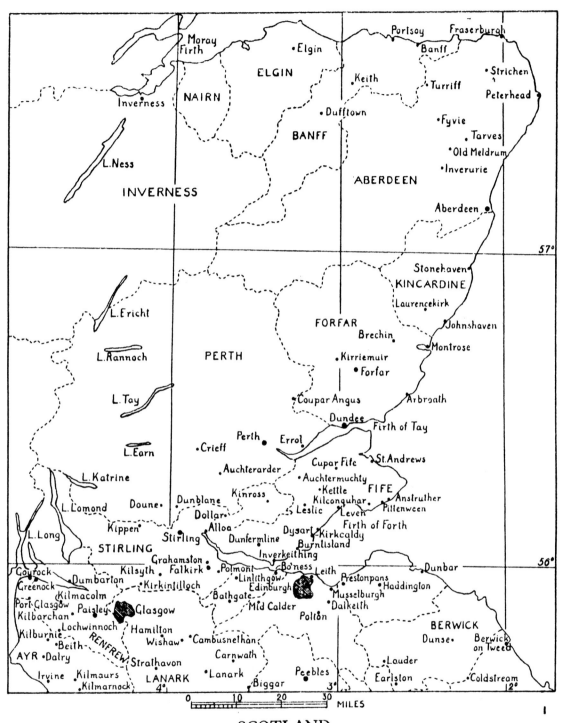

SCOTLAND

348  Appendix

**NORTH-WEST ENGLAND**

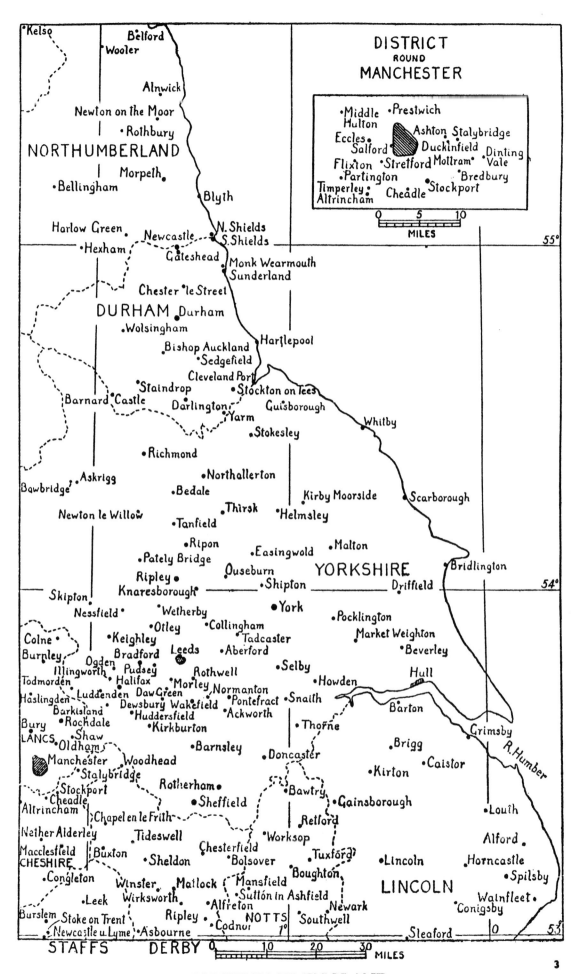

NORTH-EAST ENGLAND

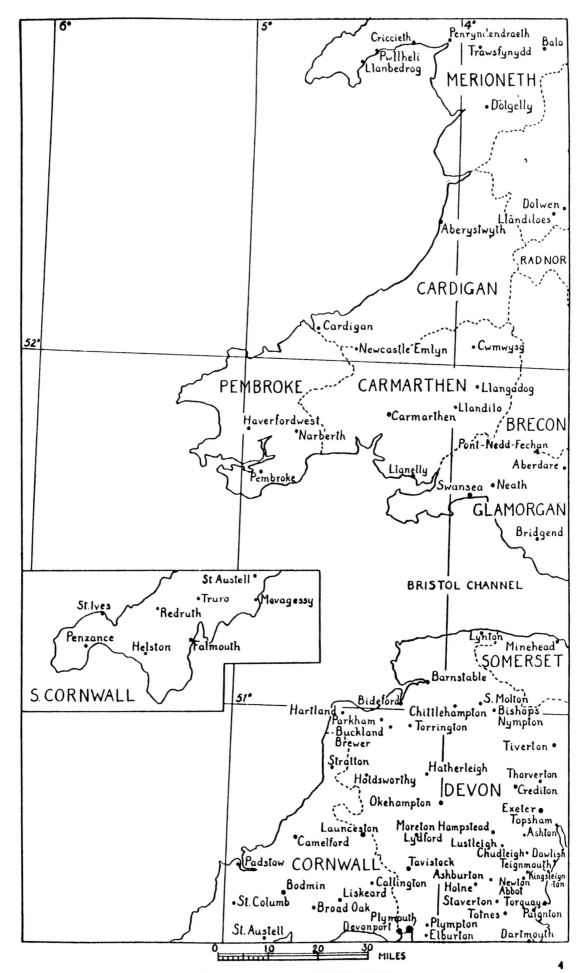

**SOUTH-WEST ENGLAND**

SOUTHERN ENGLAND

SOUTH-EAST ENGLAND

*Abbots Leigh* Vil. adj. Bristol   5
*Aberdare* T. Glam   4
*Aberdeen*   1
*Aberford* Vil. York's 8 m. E. of Leeds   3
*Abergavenny* T. 18 m. W. of Monmouth   5
*Aberystwyth* T. W. coast of Cardigan   4
*Abingdon* T. Berks, 6 m. SW. of Oxford   5
*Ackworth* t. Yorks, 13 m. SE. of Leeds   3
*Adderbury* Vil. 18 m. N. of Oxford   5
*Albrighton* t. 18 m. NW. of Birmingham   5
*Alcester* t. 17 m. S. of Birmingham   5
*Aldeburgh* t. E. coast of Suffolk   6
*Alderbury* t. 3 m. SE. of Salisbury   5
*Alderton* Vil. 11 m. E. of Ipswich   6
*Alford* t. 29 m. E. of Lincoln   3
*Alfreton* t. 12 m. N. of Derby   3
*Alloa* T. 6 m. E. of Stirling   1
*Almondsbury* t. Glos, 17 m. N. of Bristol   5
*Almwych, Amlwch* t. N. coast of Anglesey   2
*Alnwick* T. Northumberland near E. coast   3
*Alresford* T. 7 m. NNE. of Winchester   5
*Alston* t. 20 m. E. of Carlisle   2
*Althorp* Vil. 6 m. NW. of Northampton   5
*Alton* t. 15 m. ENE. of Winchester   5
*Altrincham* T. 8 m. S. of Manchester   3
*Alverton* Vil. 14 m. E. of Nottingham   6
*Amersham* T. Bucks, 23 m. WNW. of London   6
*Amesbury* Vil. 6 m. N. of Salisbury   5
*Ampthill* T. 8 m. S. of Bedford   6
*Andover* T. 12 m. NNW. of Winchester   5
*Annan* T. S. coast of Dumfriesshire   2
*Anstruther* T. E. coast of Fife   1
*Appleby* County town Westmorland   2
*Appleton* Vil. near Warrington, Lancs   2
*Arborfield* Vil. 13 m. SW. of Windsor   6
*Arbroath* T. on Forfarshire coast   1
*Arundel* T. 18 m. W. of Brighton   6
*Ashbourne* T. 13 m. NW. of Derby   3
*Ashbrook* Vil. E. Glos   5
*Ashburnham* t. 25 m. E. of Brighton   6
*Ashburton* T. 16 m. SW. of Exeter   4
*Ashby* Vil. Westmorland   2
*Ashby-de-la-Zouche* t. Leics, 12 m. S. of Derby   5
*Ashford* T. 12 m. SW. of Canterbury   6
*Ashton, Ashton-under-Lyne, and over-Lyne* T. 6 m. E. of Manchester   3 inset
*Ashton (Devon)* Vil. 5 m. S. of Exeter   4
*Ashwell Thorp* Vil. 10 m. SW. of Norwich   6
*Askrigg* Vil. NW. Yorks   3
*Aston* Vil. 7 m. W. of Chester   2
*Atherstone* t. 20 m. N. of Warwick   5
*Attleborough* t. 15 m. WSW. of Norwich   6
*Auchterarder* T. 12 m. SW. of Perth   1
*Auchtermuchty* T. in Fife 10 m. SE. of Perth   1
*Aughton* Vil. 10 m. N. of Liverpool   2
*Axbridge* t. 15 m. SW. of Bristol   5
*Axminster* T. 23 m. E. of Exeter   5
*Aylesbury* T. Bucks 36 m. NW. of London   6
*Aylsham* t. 12 m. N. of Norwich   6
*Ayton Banks* Vil. near Gateshead, Durham   3
*Ayr*   2
*Baddeley* Vil. 16 m. SE. of Chester   2

*Bagshot* t. Surrey, 10 m. S. of Windsor   6
*Bala* t. Merioneth   4
*Baldock* t. Herts, 32 m. N. of London   6
*Bampton* Vil. 12 m. W. of Oxford   5
*Banbury* T. 21 m. N. of Oxford   5
*Banff* T. 38 m. NNW. of Aberdeen   1
*Bangor* T. NW. coast of Wales   2
*Bardfield* Vil. N. Essex   6
*Barkisland* Vil. 18 m. NE. of Manchester   3
*Barnard Castle* T. 20 m. SW. of Durham   3
*Barnet* T. 10 m. N. of London   6
*Barnsley* T. 11 m. N. of Sheffield   3
*Barnstable* T. N. Devon   4
*Barrow* Coast T. 19 m. W. of Lancaster   2
*Barton-in-the-Beans* Vil. 11 m. W. of Leicester   5
*Barton-on-Humber* T. opposite Hull   3
*Barwell* t. 10 m. WSW. of Leicester   5
*Basingstoke* T. 17 m. NNE. of Winchester   5
*Bath* T. 11 m. WSW. of Bristol   5
*Bathgate* t. 17 m. W. of Edinburgh   1
*Battle* t. 27 m. ENE. of Brighton   6
*Bawtry* t. 18 m. E. of Sheffield   3
*Beaminster* t. 14 m. WNW. of Dorchester   5
*Beaulieu* t. 8 m. S. of Southampton   5
*Beaumaris* t. Anglesey   5
*Beccles* T. 16 m. SE. of Norwich   6
*Bedale* t. 29 m. NW. of York   3
*Bedford*   6
*Bedminster* T. adj. Bristol   5
*Beeston* T. near Nottingham   5
*Beith* T. 16 m. SW. of Glasgow   1
*Bedford* Vil. 14 m. SSE. of Berwick-on-Tweed   3
*Bellingham* t. 28 m. WNW. of Newcastle-on-T.   3
*Berkhamsted* t. Herts, 30 m. NW. of London   6
*Bermondsey* SE. London   6
*Berwick-on-Tweed* T. E. coast border of England and Scotland   1
*Berwick St. John* Vil. 14 m. WSW. of Salisbury   5
*Bethesda* t. N. Carnarvon   2
*Beverley* T. 8 m. NNW. of Hull   3
*Bewdley* t. 12 m. N. of Worcester   5
*Bexley* T. Kent, 14 m. ESE. of London   6
*Bicester* T. 11 m. N. of Oxford   5
*Bickenhall* Vil. 5 m. SE. of Taunton   5
*Bickerstaffe* t. 9 m. N. of Liverpool   2
*Bideford* T. N. Devon   4
*Biggar* T. 25 m. SSW. of Edinburgh   1 and 2
*Biggleswade* t. 9 m. E. of Bedford   6
*Billericay* Vil. 7 m. S. of Chelmsford   6
*Billingborough* t. S. Lincs. near Grantham   6
*Bilston* t. 9 m. NW. of Birmingham   5
*Birmingham*   5
*Bishop Auckland* t. 9 m. S. of Durham   3
*Bishop's Castle* t. 18 m. SSW. of Shrewsbury   5
*Bishop's Lydeard* t. near Taunton   5
*Bishop's Nympton* Vil. 22 m. NNW. of Exeter   4
*Bishop's Stortford* t. 11 m. NE. of Hertford   6
*Bishop's Waltham* t. 10 m. SE. of Winchester   5
*Blackburn* T. 21 m. NNW. of Manchester   2
*Blackheath* SE. Sub. of London   6
*Blandford* t. 17 m. NE. of Dorchester   5
*Bletchingley* Vil. 17 m. S. of London   6

*Blewbury* Vil. 12 m. S. of Oxford  5
*Blyth* Coast t. 12 m. N. of Newcastle-on-T.  3
*Bodmin* T. 16 m. WNW. of Plymouth  4
*Bognor* Coast t. W. Sussex  6
*Bold* Vil. adj. Farnworth, near Widnes  2
*Bolsover* t. 22 m. N. of Derby  3
*Bolton, Bolton-le-Moors* T. 11 m. NW. of Manchester  2
*Boothfold* Vil. E. Lancs.  2
*Bootle* T. adj. Liverpool  2
*Borrowstounness, Bo'ness* T. on Firth of Forth  1
*Boston* T. Lincs on The Wash  6
*Bosworth* t. 11 m. W. of Leicester  5
*Bottisdale, Botesdale* Vil. 21 m. NNW. of Ipswich  6
*Boughton* t. 18 m. NNE. of Nottingham  3
*Bourne* Vil. 14 m. NNW. of Peterborough  6
*Bowbrigge* prob. Bowbridge Vil. NW. York  3
*Brackley* t. 20 m. N. of Oxford  5
*Bradbury* adj. Stockport, Cheshire  3
*Bradford (Wilts)* t. 16 m. ESE. of Bristol  5
*Bradford (Yorks)* T. 8 m. W. of Leeds  3
*Bradwell* t. 9 m. ENE. of Buckingham  6
*Braintree* t. 11 m. N. of Chelmsford  6
*Brampton* t. 8 m. ENE. of Carlisle  2
*Brandon* t. 35 m. NW. of Ipswich  6
*Brechin* T. E. Forfar  1
*Brecknock*  5
*Bredbury* Vil. 6 m. SE. of Manchester  3 inset
*Brentford* Sub. W. London  6
*Brentwood* t. 20 m. ENE. of London  6
*Brewham* Vil. 23 m. SSE. of Bristol  5
*Bridge* Vil. near Canterbury  6
*Bridgend* T. S. Glam  4
*Bridgewater* T. 28 m. SW. of Bristol  5
*Bridgin* Prob. *Bridgen* Vil. near Bexley, Kent  6
*Bridgnorth* t. 22 m. W. of Birmingham  5
*Bridlington* Coast t. E. of York  3
*Bridport* t. 14 m. W. of Dorchester  5
*Brigg* t. 15 m. S. of Hull  3
*Brighthampton* Vil. 8 m. W. of Oxford  5
*Brighton* T. Sussex coast  6
*Bristol*  5
*Broad Oak* Prob. Vil. 15 m. WNW. of Plymouth  4
*Bromley (Kent)* T. 11 m. SE. of London  6
*Bromley (Midd.)* Sub. E. London  6
*Bromsgrove* t. 12 m. SW. of Birmingham  5
*Broseley* t. 12. m. SE. of Shrewsbury  5
*Broughton* Prob. Sub. W. Manchester  2
*Bruton* t. 24 m. SSE. of Bristol  5
*Buckingham*  5
*Buckland Brewer* Vil. 34 m. WNW. of Exeter  4
*Builth* t. N. Brecon  5
*Bungay* t. 13 m. SSE. of Norwich  6
*Bunny* Vil. 7 m. S. of Nottingham  5
*Buntingford* t. 10 m. N. of Hertford  6
*Burford* t. 17 m. W. of Oxford  5
*Burlington* Same as *Bridlington*
*Burnley* T. 21 m. N. of Manchester  3
*Burntisland* T. in Fife on Firth of Forth  1
*Burslem* T. Staffs, 30 m. S. of Manchester  3
*Burton* t. S. Westmorland  2
*Burton-on-Trent* T. 10 m. S. of Derby  5
*Burwash* Vil. 26 m. ENE. of Brighton  6
*Bury* T. 8 m. N. of Manchester  2 and 3
*Bury St. Edmunds* T. 22 m. NW. of Ipswich  6
*Buxton* T. Derby, 20 m. SE. of Manchester  3
*Caerphilly* T. 7 m. N. of Cardiff  5
*Caistor*. t. 16 m. S. of Hull  3
*Callington*. Vil. 12 m. NW. of Plymouth  4
*Calne*. t. 25 m. E. of Bristol  5
*Cambridge*  6
*Cambusnethan*. t. 15 m. ESE. of Glasgow  1
*Camelford*. t. N. Cornwall  4
*Camerton*. Vil. 11 m. SE. of Bristol  5
*Campbeltown*. t. in Kintyre, Scotland  1
*Cannock*. T. 10 m. SSE. of Stafford  5
*Canterbury*  6
*Cardiff*. T. Glam on Severn  5
*Cardigan*  4
*Carlisle*. County T. Cumberland  2
*Carmarthen*. County T. SW. Wales  4
*Carnarvon*. Coast T. NW. Wales  2
*Carnwath*. T. 22 m. SW. of Edinburgh  1
*Cartmel, Cartmel Fell*. t. 12 m. NW. of Lancaster  2
*Castle Donnington*. Vil. 8 m. S. of Derby  5
*Castle Douglas*. t. 13 m. SW. of Dumfries  2
*Caton*. Vil. near Lancaster  2
*Cerne, Cerne Abbas*. Vil. 7 m. NNW. of Dorchester  5
*Cerney*. Vil. 18 m. SE. of Gloucester  5
*Chalford*. Vil. 10 m. S. of Gloucester  5
*Challock*. Vil. 10 m. SW. of Canterbury  6
*Chapel-en-le-Frith*. t. 17 m. SE. of Manchester  3
*Chard*. T. 25 m. NE. of Exeter  5
*Chardstock*. Vil. 4 m. S. of Chard  5
*Charing*. Vil. 12 m. SW. of Canterbury  6
*Charlton*. Sub. SE. London  6
*Charminster*. Vil. near Dorchester  5
*Chatham*. T. 28 m. ESE. of London  6
*Chatteris*. t. 18 m. N. of Cambridge  6
*Cheadle*. t. 6 m. S. of Manchester  3
*Cheltenham*. T. 7 m. ENE. of Gloucester  5
*Chepstow*. t. 11 m. S. of Monmouth  5
*Cheriton*. Vil. 2 m. W. of Folkestone  6
*Chertsey*. t. 20 m. WSW. of London  6
*Cheshunt*. t. Herts, 14 m. N. of London  6
*Chester*  2
*Chesterfield*. T. 11 m. S. of Sheffield  3
*Chester-le-Street*. t. 6 m. N. of Durham  3
*Chewstoke*. Vil. 7 m. SSW. of Bristol  5
*Chichester*. T. 27 m. W. of Brighton  6
*Chigwell*. t. Essex, 12 m. NE. of London  6
*Childwell*. Vil. near Liverpool  2 inset
*Chippenham*. t. 20 m. E. of Bristol  5
*Chipping Norton*. t. 18 m. NW. of Oxford  5
*Chipping Ongar*. t. Essex, 20 m. NE. of London  6
*Chipstead*. Vil. 5 m. SW. of Croydon  6
*Chiswick*. Sub. W. London  6
*Chittlehampton*. Vil. 27 m. NW. of Exeter  4
*Chorley*. T. 20 m. NW. of Manchester  2
*Christchurch*. t. 20 m. SW. of Southampton  5
*Chudleigh*. t. 9 m. S. of Exeter  4
*Church Lawford*. Vil. near Rugby  5
*Cirencester*. T. 15 m. SE. of Gloucester  5
*Clapham*. Sub. S. London  6
*Clapton*. Sub. E. London  6

*Clare.* Vil. Suffolk, 24 m. W. of Ipswich   6
*Cleeve Prior.* Vil. 23 m. S. of Birmingham   5
*Cleobury.* Vil. 25 m. SSE. of Shrewsbury   5
*Cleveland,* Prob. *Cleveland Port.* Vil. on Tees mouth, Yorks   3
*Clifton (Cumb.)* Vil. near Penrith   2
*Clifton.* T. adj. Bristol   5
*Clitheroe.* t. 27 m. N. of Manchester   2
*Cockermouth.* t. 23 m. SW. of Carlisle   2
*Cockshutt.* Vil. 12 m. NW. of Shrewsbury   5
*Codnor.* Vil. 9 m. NNE. of Derby   3
*Coggeshall.* t. 13 m. NE. of Chelmsford   6
*Colchester.* T. NE. Essex   6
*Coldstream.* t. in Berwick   1
*Coleshill.* t. 8 m. E. of Birmingham   5
*Collingham.* Vil. 14 m. WSW. of York   3
*Colne.* T. 25 m. N. of Manchester   3
*Coltishall.* Vil. 8 m. N. of Norwich   6
*Colyford.* Vil. 20 m. E. of Exeter   5
*Compton Dando.* Vil. 6 m. SSE. of Bristol   5
*Congleton.* T. 22 m. S. of Manchester   3
*Coningsby.* t. 17 m. ESE. of Lincoln   3
*Cookham.* t. 7 m. NW. of Windsor   6
*Corscombe.* Vil. W. Dorset   5
*Corsley.* Vil. 22 m. WNW. of Salisbury   5
*Coupar-Angus.* t. 12 m. NE. of Perth   1
*Cowes.* t. Isle of Wight   5
*Cranborne.* Vil. NE. Dorset   5
*Cranbrook.* t. 12 m. S. of Maidstone   6
*Crawley.* Vil. 26 m. S. of London   6
*Crediton.* t. 7 m. NW. of Exeter   4
*Crewkerne.* t. S. Somerset   5
*Criccieth.* Vil. S. coast of Carnarvon   4
*Crieff.* t. 15 m. W. of Perth   1
*Cronton.* Vil. 6 m. E. of Liverpool   2 inset
*Croydon.* T. 13 m. S. of London   6
*Cubberley.* Vil. 8 m. E. of Gloucester   5
*Cuckfield.* t. 13 m. N. of Brighton   6
*Culchett.* Vil. 8 m. W. of Manchester   2 inset
*Cullompton.* t. 11 m. NNe. of Exeter   5
*Cupar-Angus* v. *Coupar-Angus*   1
*Cupar-Fife.* T. 10 m. S. of Dundee   1
*Curry-Mallet.* Vil. S. Somerset   5
*Cwmwsyg.* Vil. Carmarthen   4
*Dalkeith.* t. 6 m. SE. of Edinburgh   1
*Dalmellington.* t. 13 m. SE. of Ayr   2
*Dalry.* t. Ayr, 20 m. SW. of Glasgow   1
*Darlaston.* T. Staffs, 8 m. NW. of Birmingham   5
*Darlington.* T. 17 m. S. of Durham   3
*Dartford.* T. 14 m. E. of London   6
*Dartmouth.* t. on SE. Devon coast   4
*Datchet.* Vil. near Windsor   6
*Daventry.* T. 12 m. W. of Northampton   5
*Daw Green.* Vil. 6 m. S. of Leeds   3
*Dawlish.* t. 10 m. S. of Exeter   4
*Deal.* t. 8 m. NE. of Dover   6
*Dean.* Prob. *Deane.* Vil. SE. Lancs.   2
*Deddington.* Vil. 18 m. N. of Oxford   5
*Dedham.* Vil. 6 m. NE. of Colchester   6
*Denbigh.* County T. N. Wales   2
*Deptford.* Sub. SE. London   6
*Derby*   5
*Dereham.* t. 15 m. NW. of Norwich   6
*Devizes.* T. 26 m. E. of Bristol   5
*Devonport.* T. adj. Plymouth   4
*Dewsbury.* T. 9 m. S. of Leeds   3
*Didsbury.* Sub. S. Manchester
*Dinting Vale.* Vil. 10 m. E. of Manchester
*Diss.* t. 30 m. SW. of Norwich   63 inset
*Ditton.* t. 10 m. E. of Liverpool   2 inset
*Dolgelly.* T. Merioneth, W. Wales   4
*Dolwen.* Vil. near Llanidloes, Montgomery   4
*Dollar.* t. 11 m. ENE. of Stirling   1
*Doncaster.* T. 16 m. NE. of Sheffield   3
*Dorchester.* County t. Dorset   5
*Dorking.* t. Surrey, 26 m. SW. of London   6
*Douglas.* T. Isle of Man   2
*Doune.* Vil. Perth, 6 m. NNW. of Stirling   1
*Dover*   6
*Downham, Downham Market.* Vil. 39 m. W. of Norwich   6
*Draycott.* Prob. Vil. 6 m. E. of Derby   5
*Driffield.* t. 25 m. E. of York   3
*Droitwich.* t. 6 m. NE. of Worcester   5
*Dudley.* T. 8 m. W. of Birmingham   5
*Duffield.* Vil. 5 m. N. of Derby   5
*Dufftown.* Vil. 43 m. NW. of Aberdeen   1
*Dukinfield.* T. 8 m. E. of Manchester   3 inset
*Dumbarton.* T. on the Clyde   1
*Dumfries*   2
*Dunbar.* t. 27 m. E. of Edinburgh   1
*Dunblane.* Vil. 9 m. N. of Stirling   1
*Dundee*   1
*Dunfermline.* T. 14 m. WNW. of Edinburgh   1
*Dunmow.* Vil. 11 m. NW. of Chelmsford   6
*Dunse.* County T. of Berwick   1
*Dunstable.* T. 33 m. NNW. of London in Beds   6
*Durham*   3
*Dursley.* Vil. 15 m. SSW. of Gloucester   5
*Dutton.* Vil. 11 m. NE. of Chester   2
*Dysart.* t. Fife on Firth of Forth   1
*Eamont Bridge.* Vil. 18 m. SSE. of Carlisle   2
*Earlston.* Vil. 29 m. SE. of Edinburgh   1 and 2
*Easingwold.* Vil. 12 m. NNW. of York   3
*Eastbourne.* T. Sussex coast   6
*East Dereham* v. *Dereham*   6
*East Grinstead.* t. 26 m. S. of London   6
*East Pennard.* Vil. 19 m. S. of Bristol   5
*Eastry.* Vil. 9 m. N. of Dover   6
*East Sheen.* t. 8 m. W. of London
*Ecclefechan.* Vil. 12 m. E. of Dumfries   2
*Eccles.* T. adj. Manchester   2 and 3 inset
*Eccleshall.* t. 7 m. NW. of Stafford   5
*Eccleston.* t. 10 m. E. of Liverpool   2 inset
*Edinburgh*   1
*Edmonton.* Sub. N. London   6
*Egham.* T. 21 m. W. of London in Surrey   6
*Elburton.* Vil. near Plymouth   4
*Elgin.* T. 35 m. E. of Inverness   1
*Elham.* Vil 9 m. WNW. of Dover   6
*Ellesmere.* t. 20 m. NNW. of Shrewsbury   5
*Elmstead.* Vil. near Bromley, Kent   6
*Eltham.* Sub. SE. London   6
*Emsworth.* Vil. 8 m. NE. of Portsmouth   5
*Enfield.* t. 10 m. N. of London   6
*Epping.* t. 15 m. NNW. of London   6

*Epsom.* T. Surrey, 17 m. SSW. of London   6
*Erith.* T. Kent, 14 m. E. of London   6
*Errol.* Vil. 8 m. E. of Perth   1
*Essington.* t. 12 m. NNW. of Birmingham   5
*Etchells.* t. 8 m. S. of Manchester   3
*Eton.* t. near Windsor   6
*Ettrick.* Vil. 34 m. S. of Edinburgh   2
*Evesham.* t. 13 m. ESE. of Worcester   5
*Ewell.* t. 16 m. SSW. of London in Surrey   6
*Ewhurst.* Vil. Sussex, 32 m. SSW. of London   6
*Exeter*   4
*Eye.* t. 19 m. N. of Ipswich   6
*Fairford.* Vil. Glos, 25 m. WSW. of Oxford   5
*Fakenham.* t. 23 m. NW. of Norwich   6
*Falkirk.* T. 22 m. W. of Edinburgh   1
*Falmouth.* T. W. Cornwall   4 inset
*Fareham.* T. 6 m. NW. of Portsmouth   5
*Faringdon.* Vil. Berks, 15 m. WSW. of Oxford   5
*Farington.* Vil. near Preston   2
*Farnham.* T. Surrey, 38 m. SW. of London   6
*Farnsfield.* Vil. mid. Notts   5
*Farnworth, Farnworth-within-Widnes.* T. near Widnes   2
*Farringdon.* Vil. 15 m. E. of Winchester   5
*Faversham.* T. 8 m. WNW. of Canterbury   6
*Fazackerley.* Sub. N. Liverpool   2 inset
*Fenny Stratford.* t. Bucks, 17 m. SW. of Bedford   6
*Feversham.* Prob. *Faversham*   6
*Fifield.* Vil. 18 m. WNW. of Oxford   5
*Flixton.* t. 6 m. SW. of Manchester   3 inset
*Folkestone*   6
*Fordingbridge.* Vil. 9 m. S. of Salisbury   5
*Forfar*   1
*Framlingham.* Vil. 13 m. NNE. of Ipswich   6
*Frampton.* Vil. Lincs on The Wash   6
or 5 m. NW. of Dorchester   5
*Franche.* Vil. near Kidderminster   5
*Frant.* Vil. Sussex near Tunbridge Wells   6
*Fraserburg.* T. on coast 47 m. N. of Aberdeen   1
*Fritwell.* Vil. 13 m. N. of Oxford   5
*Frodsham.* t. 10 m. NE. of Chester   2
*Frome.* T. 19 m. SSE. of Bristol   5
*Fyvie.* Vil. 22 m. NNW. of Aberdeen   1
*Gainsborough.* T. 15 m. NW. of Lincoln   3
*Garnock.* Vil. near Glasgow   1
*Garstang.* Vil. 10 m. S. of Lancaster   2
*Garston.* Sub. SE. Liverpool   2 inset
*Gateshead.* T. adj. Newcastle-o.-T.   3
*Gee Cross.* Vil. E. Cheshire   3
*Girvan.* t. 17 m. SW. of Ayr   2
*Glasgow*   1
*Glastonbury.* t. 22 m. S. of Bristol   5
*Gloucester*   5
*Gloverstone.* Sub. of Chester   2
*Godalming.* T. Surrey, 34 m. SW. of London   6
*Godstone.* Vil. Surrey 26 m. S. of London   6
*Goldbourne.* Prob. *Golborne.* Vil. 6 m. N. of Warrington   2
*Gosport.* T. adj. Portsmouth   5
*Gourock.* t. on mouth of Clyde   1
*Grahamston.* t. near Falkirk   1
*Grantham.* T. 22 m. S. of Lincoln   6
*Gravesend.* T. Kent, 22 m. E. of London   6

*Great Boughton.* Sub. of Chester   2
*Great Haseley.* Vil. 7 m. ESE. of Oxford   5
*Great Marlow* v. *Marlow*   6
*Great Missenden.* t. Bucks, 25 m. WNW. of London   6
*Greenock.* T. on mouth of Clyde   1
*Grimsby.* T. on mouth of Humber   3
*Grove.* Vil. Berks, 12 m. SW. of Oxford   5
*Guildford.* T. 26 m. SW. of London in Surrey   6
*Guisborough.* t. 27 m. SE. of Durham   3
*Haddington.* t. 18 m. S. of Edinburgh   1
*Hadleigh.* t. 8 m. W. of Ipswich   6
*Hagburn.* Vil. Bucks, 11 m. S. of Oxford   5
*Hale.* t. 7 m. ESE. of Liverpool   2 inset
*Halesowen.* t. near Birmingham   5
*Halesworth.* t. 25 m. NE. of Ipswich   6
*Halewood.* t. 8 m. SE. of Liverpool   2 inset
*Halifax.* T. 22 m. NE. of Manchester   3
*Halkin.* Vil. 12 m. W. of Chester   2
*Hall Green.* Sub. of Birmingham   5
*Halstead.* Vil. 15 m. SE. of London in Kent   6
*Halton.* Vil. 10 m. NNE. of Chester   2
*Hamilton.* T. 11 m. SE. of Glasgow   1
*Hampton Court.* Sub. W. London
*Hampton Wick.* t. adj. Kingston, Surrey   6
*Hanslop.* Vil. 9 m. S. of Nottingham   5
*Harbledown.* Vil. adj. Canterbury   6
*Harleston.* Vil. 15 m. S. of Norwich   6
*Harlow Green.* Vil. 10 m. W. of Newcastle-o.-T.   3
*Harrington.* t. Cumberland on coast   2
*Harringworth.* Vil. 16 m. N. of Peterborough   6
*Harsfield.* Vil. 5 m. S. of Gloucester   5
*Hartland.* Vil. N. Devon   4
*Hartlepool.* T. 17 m. SE. of Durham   3
*Harwich.* Coast T. Essex   6
*Haslemere.* t. Surrey, 43 m. SW. of London   6
*Haslingden.* T. 14 m. N. of Manchester   2 and 3
*Hastings.* Coast t. E. Sussex   6
*Hatfield.* T. Herts, 18 m. N. of London   6
*Hatherleigh.* Vil. 23 m. WNW. of Exeter   4
*Hatton.* Vil. near Warwick   5
*Havant.* t. 7 m. NE. of Portsmouth   5
*Haverfordwest.* t. 9 m. N. of Pembroke   4
*Haverhill.* Vil. Suffolk, 16 m. ESE. of Cambridge   6
*Hawick.* T. 50 m. SSE. of Edinburgh   2
*Hawkshead.* Vil. 12 m. W. of Kendal   2
*Hay.* Vil. 13 m. NE. of Brecknock   5
*Helmdon.* Vil. 14 m. SW. of Northampton   5
*Helmsley.* Vil. 20 m. N. of York   3
*Helston.* t. SW. Cornwall   4 inset
*Hemel Hempsted.* t. Herts, 23 m. NW. of London   6
*Henley-in-Arden.* Vil. 8 m. W. of Warwick   5
*Henley-on-Thames.* t. 13 m. W. of Windsor   6
*Hereford*   5
*Hertford*   6
*Hexham.* t. 19 m. W. of Newcastle-on-T.   3
*Higham Ferrars.* Vil. 13 m. ENE. of Northampton   6
*Highworth.* Vil. Wilts, 21 m. WSW. of Oxford   5
*High Wycombe.* T. Bucks, 34 m. NW. of London   6
*Hinckley.* T. 14 m. SW. of Leicester   5
*Hindley.* T. near Wigan   2 inset
*Hindon.* Vil. 15 m. W. of Salisbury   5
*Hingham.* Vil. 11 m. W. of Norwich   6
*Hitchin.* T. 13 m. NW. of Hertford   6

*Hoddesdon.* t. near Hertford   6
*Hogsden.* Prob. *Hogston.*
Vil. 9 m. SE. of Buckingham   5
*Holbeach.* t. Lincs near The Wash   6
*Holbeck.* T. adj. Leeds   3
*Holdsworthy.* Vil. 35 m. W. of Exeter   4
*Holmes.* Prob. Sub. of Rotherham   3
*Holne.* Vil. 19 m. SW. of Exeter   4
*Holt.* Vil. 20 m. NNW. of Norwich   6
*Holyhead.* T. NW. Anglesey   2
*Holywell.* Vil. Flint, 14 m. WNW. of Chester   2
*Honiton.* t. 16 m. ENE. of Exeter   5
*Hopton.* Vil. near Yarmouth   6
*Horncastle.* t. 17 m. E. of Lincoln   3
*Horsham.* T. Sussex, 32 m. SSW. of London   6
*Hoveton.* Vil. 4 m. NE. of Norwich   6
*Howden.* Vil. 17 m. SSE. of York   3
*Hoxton.* Sub. NE. London   6
*Huddersfield.* T. 14 m. SW. of Leeds   3
*Hungerford.* Vil. Berks, 9 m. E. of Marlborough   5
*Huntley.* Vil. 7 m. W. of Gloucester   5
*Hurley.* Vil. 10 m. WNW. of Windsor   6
*Hursley.* Vil. 5 m. SW. of Winchester   5
*Hurst.* Vil. SE. Lancs or E. Berks or SE. Kent   6
*Huxley.* Vil. 7 m. SE. of Chester   2
*Hythe.* t. S. Kent coast   6
*Ilkeston.* T. 7 m. ENE. of Derby   5
*Illingworth.* t. 5 m. WSW. of Leeds   3
*Ilminster.* Vil. 10 m. SE. of Taunton   5
*Ince.* T. Lancs, 7 m. NE. of Chester   2
*Ingatestone.* Vil. Essex, 23 m. NE. of London   6
*Inverkeithing.* t. Fife on Firth of Forth   1
*Inverness*   1
*Inverurie.* t. 14 m. NW. of Aberdeen   1
*Irvine.* t. 10 m. N. of Ayr   1 and 2
*Isleworth.* Sub. W. London   6
*Ixworth.* Vil. 21 m. NW. of Ipswich   6
*Jedburgh.* County t. of Roxburgh   2
*Johnshaven.* Coast vil. 26 m. SSW. of Aberdeen   1
*Kegworth.* t. NW. Leics   5
*Keighley.* 15 m. W. of Leeds   3
*Keith.* t. Banff, 42 m. NW. of Aberdeen   1
*Kelso.* t. 37 m. SE. of Edinburgh   2
*Kendal.* County T. of Westmorland   2
*Keswick.* t. 21 m. SSW. of Carlisle   2
*Kettering.* T. 13 m. NE. of Northampton   6
*Kettle.* Vil. Fife, 15 m. S. of Dundee   1
*Keynsham.* t. 5 m. SE. of Bristol   5
*Kidderminster.* T. Worcs, 16 m. WSW. of Birmingham   5
*Kilbarchan.* t. 12 m. W. of Glasgow   1
*Kilburnie, Kilbirnie.* t. 20 m. SW. of Glasgow   1
*Kilconquhar.* Vil. Fife, 18 m. S. of Dundee   1
*Kilmacolm.* t. 14 m. W. of Glasgow   1
*Kilmarnock.* T. 11 m. NNE. of Ayr   1 and 2
*Kilmaurs.* t . 13 m. NNE. of Ayr   1 and 2
*Kilsyth.* T. 11 m. NE. of Glasgow   1
*Kimbolton.* Vil. 9 m. W. of Huntingdon   6
*Kingsbridge.* Vil. 17 m. SE. of Plymouth   4
*Kings Lynn.* T. 37 m. W. of Norwich   6
*Kingsteignton.* Vil. 12 m. S. of Exeter   4
*Kingston (Kent).* Vil. 5 m. SE. of Canterbury   6
*Kingston (Surrey).* T. 12 m. SW. of London   6
*Kingswood.* Sub. E. Bristol   5
*Kington.* Vil. NW. Herefordshire   5
*Kinnoul.* Sub. of Perth   1
*Kinross.* Vil. 13 m. S. of Perth   1
*Kippen.* Vil. 9 m. W. of Stirling   1
*Kirby Kendal v. Kendal*   3
*Kirby Moorside.* Vil. 22 m. N. of York   3
*Kirkburton.* t. 15 m. S. of Leeds   3
*Kirkby Lonsdale.* Vil. S. Westmorland   1 and 2
*Kirkby Stephen.* Vil. E. Westmorland   1 and 2
*Kirkcaldy.* T. Fife on Firth of Forth   1
*Kirkconnel.* Vil. 24 m. E. of Ayr   2
*Kirkcudbright.* t. S. coast of Scotland   2
*Kirkdale.* t. adj. Liverpool   2
*Kirkham.* t. 25 m. N. of Liverpool   2
*Kirkintilloch.* T. 7 m. NE. of Glasgow   1
*Kirriemuir.* t. 14 m. N. of Dundee   1
*Kirton.* Vil. 17 m. N. of Lincoln   3
*Knaresborough.* t. 1 5m. W. of York   3
*Knowle.* Vil. 9 m. SE. of Birmingham   5
*Knowsley.* Vil. 7 m. NE. of Liverpool   2 inset
*Knutsford.* t. Cheshire, 13 m. SSW. of Manchester   2
*Lamberhurst.* Vil. 12 m. SSW. of Maidstone   6
*Lambourn.* Vil. 10 m. NE. of Marlborough   5
*Langholm.* Vil. 25 m. ENE. of Dumfries   2
*Langport.* Vil. 12 m. E. of Taunton   5
*Lauder.* Vil. 23 m. SE. of Edinburgh   1
*Launceston.* t. NE. Cornwall   4
*Laurencekirk.* Vil. 25 m. SSW. of Aberdeen   1
*Lavenham.* Vil. 16 m. W. of Ipswich   6
*Lavington.* Vil. 17 m. N. of Salisbury   5
*Leamington Spa.* T. near Warwick   5
*Leatherhead.* t. 18 m. SW. of London in Surrey   6
*Ledbury.* Vil. 13 m. E. of Hereford   5
*Leeds*   3
*Leek.* T. Staffs, 27 m. SSE. of Manchester   3
*Leicester*   5
*Leigh.* T. 21 m. NE. of Liverpool   2 inset
*Leighton, Leighton Buzzard.*
   t. Beds, 37 m. NW. of London   6
*Leith.* T. adj. Edinburgh   1
*Lenham.* Vil 8 m. E. of Maidstone   6
*Leominster.* t. 12 m. N. of Hereford   5
*Leslie.* t. Fife, 18 m. N. of Edinburgh   1
*Leven.* t. Fife on Firth of Forth   1
*Lewes.* T. 7 m. ENE. of Brighton   6
*Lewisham.* Sub. SE. London   6
*Leyton.* Sub. NE. London
*Lichfield.* t. 14 m. SE. of Stafford   5
*Limpsfield.* t. Surrey, 18 m. S. of London   6
*Linlithgow.* t. 5 m. W. of Edinburgh   1
*Liskeard.* t. 15 m. W. of Plymouth   4
*Liverpool*   2
*Llanbedrog.* Vil. Carnarvonshire   5
*Llancarfan.* Vil. 8 m. WSW. of Cardiff   5
*Llandilo.* Vil. 14 m. E. of Carmarthen   4
*LLanelly.* T. 10 m. WNW. of Swansea   4
*LLanfair-Caereinion.* Vil. 9 m. NW. of Montgomery   5
*Llanfrothen.* Vil. NW. Merionethshire   4
*Llanfyllin.* Vil. N Montgomeryshire   5
*Llangadog.* Vil. E. Carmarthenshire   4
*Llangollen.* Vil. Denbighshire   5

*Llanhilleth.* Vil. 16 m. N. of Cardiff   5
*Llanidloes.* Vil. S. Montgomeryshire   4
*Llanerch-Coedlan.* Vil. NE. Denbighshire   2
*Llanerch-y-Medd.* Vil. Anglesey   2
*Llanrwst.* Vil. 16 m. W. of Denbigh   2
*Lochwinnoch.* t. 14 m. WSW. of Glasgow   1
*Long Buckby.* Vil. 10 m. NW. of Northampton   5
*Long Melford.* Vil. 18 m. W. of Ipswich   6
*Longtown.* t. 9 m. N. of Carlisle   2
*Loughborough.* T. 10 m. NW. of Leicester   5
*Louth.* t. 24 m. ENE. of Lincoln   3
*Lowestoft.* T. Norfolk coast   6
*Luddendon* t. 20 m. NE. of Manchester   2
*Ludlow.* t. Salop, 35 m. W. of Birmingham   5
*Lussley.* Prob. *Lustleigh.* Vil 10 m. SW. of Exeter   4
*Luton.* T. 28 m. NNW. of London in Beds   6
*Lutterworth.* Vil. 6 m. N. of Rugby   5
*Lydford.* Vil. 19 m. N. of Plymouth   4
*Lymington.* t. 12 m. SW. of Southampton   5
*Lynn v. Kings Lynn*   6
*Lynton.* Vil. N. coast of Devon   4
*Macclesfield.* T. 17 m. SSE. of Manchester   3
*Madeley Wood.* t. 13 m. ESE. of Shrewsbury   5
*Maidenhead.* T. Berks 24 m. W. of London   6
*Maidstone.* T. Kent, 41 m. SE. of London   6
*Malden.* T. Surrey, 10 m. SW. of London   6
*Maldon.* t. Essex, 9 m. E. of Chelmsford   6
*Malling.* t. near Maidstone   6
*Malmesbury.* t. N. Wilts   5
*Malton, New Malton.* t. 17 m. NE. of York   3
*Manchester*   2 and 3
*Manningtree.* Vil. 8 m. NE. of Colchester   6
*Manfield.* T. 14 m. N. of Nottingham   3
*March.* T. Cambs, 14 m. E. of Peterborough   6
*Margate.* Coast T. NE. Kent   6
*Market Drayton.* t. 18 m. NE. of Shrewsbury   5
*Market Harborough.* t. 16 m. SE. of Leicester   5
*Market Weighton.* Vil. 17 m. ESE. of York   3
*Marlborough.* t. 24 m. N. of Salisbury   5
*Marlow.* t. Bucks, 32 m. W. of London   6
*Maryport.* T. Cumberland coast   2
*Matlock.* t. 13 m. N. of Derby   3
*Maybole.* t. 8 m. SW. of Ayr   3
*Melford v. Long Melford*   6
*Melksham.* t. W. Wilts   5
*Melrose.* t. 30 m. SE. of Edinburgh   2
*Melton Mowbray.* T. 14 m. NE. of Leicester   5
*Membury.* Vil. 22 m. ENE. of Exeter   5
*Merthyr Tydvil.* T. 24 m. NNW. of Cardiff   5
*Mevagessy.* Vil. S. Cornwall   4 inset
*Mid Calder.* Vil. 11 m. W. of Edinburgh   1
*Middle Hulton.* Vil. 7 m. NW. of Manchester   3 inset
*Middleton.* Sub. of Manchester   3
*Middleton Cheney.* Vil. 11 m. N. of Oxford   5
*Middlewich.* t. 18 m. E. of Chester   2
*Midhurst.* Vil. 10 m. N. of Chichester   6
*Milnthorpe.* Vil. 13 m. N. of Lancaster   2
*Milton.* Part of Gravesend   6
*Minchinhampton.* t. 12 m. S. of Gloucester
*Minehead.* t. 33 m. N. of Exeter   4
*Missenden v. Great Missenden*   6
*Mitcham.* T. Surrey, 9 m. S. of London   6
*Modbury.* Vil. 12 m. E. of Plymouth   4
*Moffat.* Vil. 19 m. NNE. of Dumfries   2
*Mold.* t. Flint, 10 m. W. of Chester   2

*Monk Wearmouth.* t. adj. Sunderland   3
*Monmouth*   5
*Montgomery*   5
*Montrose.* T. Forfarshire coast   1
*Moreton.* Vil. 7 m. E. of Dorchester or in Staffs   5 or in Cheshire on NW. coast   2
*Moreton Hampstead.* Vil. 10 m. WSW. of Exeter   4
*Moreton-in-Marsh.* Vil. 25 m. NW. of Oxford   5
*Morley.* T. 5 m. SW. of Leeds   3
*Morpeth.* t. 15 m. N. of Newcastle-o.-T.   3
*Mortimer.* Vil. 7 m. S. of Reading   5
*Mottram.* t. 12 m. E. of Manchester   3 inset
*Musselburgh.* T. 5 m. E. of Edinburgh   1
*Nailsworth.* Vil. 14 m. S. of Gloucester   5
*Nantwich.* t. 17 m. ESE. of Chester   2
*Narberth.* Vil. 11 m. NNE. of Pembroke   4
*Neath.* T. 8 m. NE. of Swansea   4
*Needham Market.* Vil. 16 m. NW. of Ipswich   6
*Nessfield.* Vil. 15 m. NW. of Leeds   3
*Nether Alderley.* Vil. 14 m. S. of Manchester   3
*Newark.* T. 17 m. NE. of Nottingham   3
*Newbury.* T. Berks, 25 m. S. of Oxford   5
*Newcastle-on-Tyne*   3
*Newcastle Emlyn.* Vil. 26 m. NW. of Carmarthen   4
*Newcastle-under-Lyme.* T. 14 m. NNW. of Stafford   3 and 5
*Newington.* Sub. S. London
*Newmarket.* t. 15 m. E. of Cambridge   6
*Newnham.* Vil. 11 m. SW. of Gloucester   5
*Newport (Mon).* T. 10 m. NE. of Cardiff   5
*Newport (Salop).* t. 17 m. ENE. of Shrewsbury   5
*Newport Pagnell.* t. Bucks, 12 m. SE. of Northampton   6
*Newton Abbot.* T. 20 m. S. of Exeter   4
*Newton Bushel.* Part of Newton Abbot   4
*Newton-le-Willows.* t. 5 m. N. of Warrington   2 inset
*Newton-on-the-Moor.* Vil. E. Northumberland   3
*Newton Stewart.* Vil. Wigtonshire   2
*Newtown.* T. 7 m. SW. of Montgomery   5 or one of many small villages
*Normanton.* Vil. 9 m. SE. of Leeds   3
*Northallerton.* t. 29 m. NNW. of York   3
*Northampton*   5 and 6
*Northiam.* Vil. 29 m. S. of Maidstone   6
*North Leith v. Leith*   1
*North Shields.* T. on mouth of the Tyne   3
*North Walsham.* t. 16 m. N. of Norwich   6
*Northwich.* t. 16 m. E. of Chester   2
*Northwood.* Vil. NW. Salop   5
*Norton.* Many villages of this name
*Nottingham*   5
*Norwich*   6
*Nuneaton.* T. 17 m. E. of Birmingham   5
*Nunland.* Perhaps Vil. in N. Berwickshire   1
*Oakhampton, Okehampton.* t. 25 m. W. of Exeter   4
*Oakingham.* Perhaps *Wokingham.* t. near Reading   6
*Odiham.* t. 20 m. NE. of Winchester   5 and 6
*Ogden.* Vil. 4 m. N. of Halifax   3
*Old Cumnock.* t. 15 m. E. of Ayr   2
*Oldham.* T. 7 m. NE. of Manchester   3
*Old Meldrum.* Vil. 15 m. NNW. of Aberdeen   1
*Old Swinford.* Vil. adj. Stourbridge   5
*Olney.* Vil. Bucks, 11 m. SE. of Northampton   6
*Ormskirk.* T. 11 m. N. of Liverpool   2
*Oswestry.* t. 17 m. NW. of Shrewsbury   5

*Otley.* T. 9 m. NW. of Leeds   3
*Ottery St. Mary.* Vil. 12 m. NE. of Exeter   5
*Oundle.* t. 13 m. SW. of Peterborough   6
*Ouselbridge.* Prob. *Ouseburn.* Vil. 11 m. WNW. of York   3
*Oversley.* Vil. 24 m. S. of Birmingham   5
*Overton.* Vil. 13 m. N. of Winchester   5
*Oxbridge.* Vil. W. Devon   4
*Padstow.* Vil. W. coast of Cornwall   4
*Paignton.* T. 20 m. S. of Exeter   4
*Painswick.* Vil. 6 m. SE. of Gloucester   5
*Pangbourne.* Vil. 5 m. NW. of Reading   5
*Parham.* Vil. 13 m. NE. of Chichester   6
*Parkham.* Vil. 30 m. WNW. of Exeter   4
*Partington.* Vil. Cheshire, 10 m. WSW. of Manchester   33 inset
*Pateley Bridge.* t. 22 m. NNW. of Leeds   3
*Peasley Cross.* Vil. adj. St. Helens   2
*Peasemarsh.* Vil. E. Sussex   6
*Peckham.* Sub. S. London   6
*Peebles.* T. 22 m. S. of Edinburgh   1 and 2
*Pembridge.* Vil. 12 m. NW. of Hereford   5
*Pembroke*   4
*Pengelly.* Vil. W. Cornwall   4
*Penrith.* t. 17 m. SSE. of Carlisle   2
*Penrhyndeudraeth.* Vil. NW. Merionethshire   4
*Pentonville.* Part of N. London   6
*Penzance.* T. extreme W. of Cornwall   4 inset
*Pershore.* 7 m. SE. of Worcester   5
*Perth*   1
*Peterborough*   6
*Peterhead.* Coast T. 32 m. NNE. of Aberdeen   1
*Petersfield.* t. 16 m. NNE. of Portsmouth   5
*Petersham.* adj. Richmond, Surrey   6
*Petworth.* Vil. 12 m. NNE. of Chichester   6
*Picehurst.* Perhaps *Ticehurst*
*Pinkneys.* Vil. near Maidenhead   6
*Pittenween.* Vil. Fife on Firth of Forth   1
*Plymouth*   4
*Plympton.* t. 4 m. E. of Plymouth   4
*Pocklington.* t. 12 m. E. of York   3
*Polmont.* t. 19 m. W. of Edinburgh   1
*Polton.* Vil. 7 m. S. of Edinburgh   1
*Pontefract.* T. 21 m. SSW. of York   3
*Pont Nedd Fechan.* Vil. Brecon near Aberdare   4
*Pontypool.* T. 15 m. SW. of Monmouth   5
*Poole.* T. SE. Dorset coast   5
*Poplar.* Part of E. London   6
*Port Glasgow.* T. 20 m. W. of Glasgow   1
*Portsea.* Part of Portsmouth   5
*Portsmouth*   5
*Portsoy.* Vil. N. coast of Banffshire   1
*Potton.* Vil. 11 m. E. of Bedford   6
*Poulton.* Vil. 17 m. SSW. of Lancaster   2
*Prescot.* T. 7 m. E. of Liverpool   2
*Preston* T. 27 m. NW. of Manchester   2
*Prestonpans,* t. 8 m. E. of Edinburgh   1
*Prestwich.* T. 4 m. N. of Manchester   3 inset
*Puddletown.* Vil. 5 m. NE. of Dorchester   5
*Pudsey.* T. 5 m. W. of Leeds   3
*Pwllheli.* Vil. 20 m. SW. of Carnarvon   4
*Quernmmore.* Vil. 3 m. SE. of Lancaster   2
*Quinton.* Vil. near Stratford-on-Avon   5
*Rainhill.* Vil. 9 m. E. of Liverpool   2 inset
*Ramsey.* t. Isle of Man   2
*Ramsgate.* T. 15 m. N. of Dover   6
*Reading.* T. Berks   5
*Redbourn.* Vil. 14 m. W. of Hereford   6
*Redmile.* Vil 24 m. NNE. of Leicester   5 and 6
*Redpath.* Vil. near Melrose   2
*Redruth.* T. W. Cornwall   4 inset
*Reigate.* T. Surrey, 24 m. S. of London   6
*Retford.* T. Notts, 21 m. E. of Sheffield   3
*Richmond (Surrey).* T. 10 m. WSW. of London   6
*Richmond (Yorks).* T. 38 m. NW. of York   3
*Ringwood.* t. 16 m. WSW. of Southampton   5
*Ripley.* Vil. Derbyshire or Yorks   3 or vil. Surrey
*Ripon.* t. 21 m. NW. of York   3
*Roby-within-Huyton.* Vil. 5 m. E. of Liverpool   2 inset
*Rochdale.* T. 10 m. N. of Manchester   3
*Rochester.* T. Kents, 27 m. E. of London   6
*Romford.* T. Essex, 13 m. E. of London   6
*Romney, New Romney.* t. S. coast of Kent   6
*Romsey.* t. 7 m. NW. of Southampton   5
*Ross.* t. 12 m. SE. of Hereford   5
*Rothbury.* Vil. 25 m. NNW. of Newcastle-o.-T.   3
*Rotherham.* T. 6 m. NE. of Sheffield   3
*Rotherhithe.* Sub. E. London   6
*Rotherwas.* Vil. near Hereford   5
*Rothweill.* T. 4 m. SE. of Leeds   3 or t. 12 m. N. of Northampton   5 and 6
*Royston.* t. 12 m. SSW. of Cambridge in Herts   6
*Ruabon.* t. Denbigh, 15 m. SSW. of Chester   5
*Rugby.* T. Warwick   5
*Rugeley.* t. 8 m. SE. of Stafford   5
*Rushbury.* Vil. 13 m. S. of Shrewsbury   5
*Ruthin.* Vil. 8 m. SE. of Denbigh   2
*Rye.* t. E. Sussex coast   6
*Saffron Walden.* t. 24 m. NNW. of Chelmsford   6
*St. Albans.* T. Herts, 21 m. N. of London   6
*St. Andrews.* t. E. coast of Fifeshire   1
*St. Austell.* t. 28 m. W. of Plymouth   4
*St. Columb.* t. 35 m. W. of Plymouth   4
*St.Helens.* T. 12 m. ENE. of Liverpool   2
*St. Helier.* Vil. Surrey, 13 m. SW. of London   6
*St. Ives.* Coast t. extreme W. Cornwall   4 inset
*St. Ives (Hunts).* Vil. near Huntingdon   6
*St. Neots.* t. Hunts, 16 m. W. of Cambridge   6
*St. Ninians.* Sub. of Stirling   1
*Salford.* T. adj. Manchester   3 inset
*Saltcoats.* T. 13 m. N. of Ayr   2
*Sandbach.* t. 23 m. E. of Chester   2
*Sandwich.* t. 10 m. N. of Dover   6
*Sanquhar.* t. 35 m. S. of Glasgow   2
*Saxmundham.* Vil. 18 m. NE. of Ipswich   6
*Scarborough.* T. on Yorks coast   3
*Seagrave.* Vil. 6 m. N. of Leicester   5
*Seaton.* Vil. near Axminster   5
*Sedberg.* t. Yorks, 9 m. E. of Kendal   2
*Sedgefield.* t. 9 m. SSE. of Durham   3
*Selby.* T. 14 m. S. of York   3
*Selkirk.* t. 30 m. SSE. of Edinburgh   2
*Selling.* Vil. 7 m. W. of Canterbury   6
*Settle.* Vil. 21 m. E. of Lancaster   2
*Sevenoaks.* T. Kent, 21 m. SE. of London   6
*Shaftesbury.* t. 18 m. W. of Salisbury   5

*Shaw.* t. 9 m. NE. of Manchester   3
*Sheerness.* T. on mouth of Thames   6
*Sheffield*   3
*Shefford.* Vil. 10 m. SE. of Bedford   6
*Sheldon.* Vil. Derbyshire   3
*Shepton Mallet.* t. 18 m. S. of Bristol   5
*Sherborne.* t. 18 m. N. of Dorchester   5
*Shifnal.* t. 17 m. ESE. of Shrewsbury   5
*Shipton.* Vil. 5 m. NW. of York   3
or 11 m. E. of Gloucester   5
*Shipston-on-Stour.* Vil. Worcs, 14 m. S. of Warwick   5
*Shoreditch.* Part of NE. London   6
*Shrewsbury.* County T. of Salop   5
*Sidbury.* Vil. 14 m. E. of Exeter   5
*Sidmouth.* t. on coast 14 m. ESE. of Exeter   5
*Silchester.* Vil. 28 m. S. of Oxford   5
*Sittingbourne.* T. kent, 38 m. E. of London   6
*Skipton, Skipton-cum-Craven.* T. 26 m. NW. of Leeds   3
*Skircoat.* Part of Halifax   3
*Sleaford.* t. 17 m. S. of Lincoln   3 and 6
*Snaith.* Vil. 18 m. S. of York   3
*Solihull.* T. 6m. SE. of Birmingham   5
*Sombourne.* Vil. 7 m. W. of Winchester   5
*Somersham.* Vil. 8 m. ENE. of Huntingdon   6
*Somerton.* Vil. 17 m. E. of Taunton   5
*Sorn.* t. 13 m. E. of Ayr   2
*Sotherton.* Vil. 26 m. NE. of Ipswich   6
*Southam.* Vil. 9 m. E. of Warwick   5
*Southampton*   5
*South Kilworth.* Vil. 9 m. NE. of Rugby   5
*South Mimms.* t. 12 m. N. of London   6
*Southminster.* Vil. 16 m. S. of Colchester   6
*South Molton.* Vil. 25 m. NW. of Exeter   4
*South Moulsham.* Part of Chelmsford
*South Shields.* T. on mouth of Tyne   3
*Southwark.* Part of S. London   6
*Southwell.* Vil 16 m. NE. of Nottingham   3
*Southwold.* Coast Vil. 29 m. NE. of Ipswich   6
*Spalding.* T. 35 m. SSE. of Lincoln   6
*Speke.* t. adj. Garston near Liverpool   2
*Spilsby.* Vil. 26 m. E. of Lincoln   3
*Stadhampton.* Vil. 7 m. SE. of Oxford   5
*Stafford*   5
*Staindrop.* Vil. 17 m. SSW. of Durham   3
*Staines.* t. 19 m. SW. of London   6
*Stalbridge.* Vil. 17 m. N. of Dorchester   5
*Stalham.* Vil. 14 m. NE. of Norwich   6
*Stalybridge.* T. Cheshire, 17 m. E. of Manchester   3
*Stamford.* t. Lincs, 11 m. WNW. of Peterborough   6
*Stanford.* Vil. Kent, Norfolk, or Northants   5
*Stanmore.* Vil. 13 m. NW. of London   6
*Staverton.* Vil. 11 m. W. of Northampton
or 5 m. NE. of Gloucester   5
or 10 m. W. of Torquay   4
*Steyning.* Vil. 9 m. WNW. of Brighton   6
*Stickland.* Vil. 15 m. NE. of Dorchester   5
*Stilton.* Vil. 12 m. NNW. of Huntingdon   6
*Stirling*   1
*Stockport.* T. 6 m. SE. of Manchester   3
*Stockton, Stockton-on-Tees.* T. 17 m. SE. of Durham   3
*Stoke.* Prob. *Stoke-on-Trent.* T. 16 m. N. of Stafford   3
*Stoke Newington.* Part of London   6

*Stokesley.* Vil. 35 m. N. of York   3
*Stone.* t. 7 m. N. of Stafford   5
*Stonehaven.* Coast T. 13 m. S. of Aberdeen   1
*Stony Stratford.* Vil. 13 m. S. of Northampton   6
*Stourbridge.* T. worcs, 11 m. W. of Birmingham   5
*Stourton Caundle.* Vil. 16 m. N. of Dorchester   5
*Stouting.* Vil. 11 m. S. of Canterbury   6
*Stow-on-the-Wold.* Vil. Glos, 22 m. NW. of Oxford   5
*Stowmarket.* t. 11 m. NW. of Ipswich   6
*Stranaer.* t. SW. coast of Scotland   2
*Stratford (Essex).* Sub. E. London   6
*Stratford-on-Avon.* T. 7 m. SW. of Warwick   5
*Strathavon.* t. 15 m. SE. of Glasgow   1
*Stratton.* t. N. Cornwall   4
*Stretford.* T. adj. Manchester   3 inset
*Strichen.* Vil. NE. Aberdeenshire   1
*Stroud.* t. 12 m. S. of Gloucester   5
*Sudbury.* t. 18 m. W. of Ipswich   6
*Sunderland.* T. on coast of Durham   3
*Sutton Coldfield.* T. 7 m. N. of Birmingham   5
*Sutton-in-Ashfield.* T. 13 m. NNW. of Nottingham   3
*Sutton (Lancs).* T. adj. St. Helens   2
*Sutton Maddock.* Vil. 16 m. ESE. of Shrewsbury   5
*Swaffham.* Vil. 25 m. W. of Norwich   6
*Swansea.* T. Glamorgan on Bristol Channel   4
*Swindon.* T. NE. Wilts   5
*Tadcaster.* t. 9 m. SW. of York   3
*Tamworth.* t. Staffs, 13 m. NE. of Birmingham   5
*Tanfield.* t. 28 m. N. of Leeds   3
*Tarporley.* t. 10 m. ESE. of Chester   2
*Tarves.* Vil. 16 m. N. of Aberdeen   1
*Tavistock.* t. 12 m. N. of Plymouth   4
*Teignmouth.* T. 12 m. S. of Exeter   4
*Temple Cloud.* Vil. 10 m. S. of Bristol   5
*Tenbury.* Vil. 17 m. NW. of Worcester   5
*Tenterden.* t. 14 m. SSE. of Maidstone   6
*Tetbury.* Vil. 17 m. S. of Gloucester   5
*Twekesbury.* t. 10 m. N. of Gloucester   5
*Thakeham.* Vil. 14 m. WNW. of Brighton   6
*Thame.* t. 12 m. E. of Oxford   5
*Thaxted.* Vil. 16 m. NNW. of Chelmsford   6
*Thetford.* t. 26 m. SW. of Norwich   6
*Thirsk.* t. 22 m. NNW. of York   3
*Thomaston v. Plymouth Hollow*   4
*Thornbury.* t. 12 m. N. of Bristol   5
*Thorne.* T. 13 m. S. of York   3
*Thorverton.* Vil. 6 m. N. of Exeter   4
*Thrapston.* Vil. 21 m. NE. of Northampton   6
*Three Mile Cross.* Vll. 3 m. S. of Reading   5
*Ticehurst.* Vil. 28 m. NE. of Brighton   6
*Tideswell.* t. 14 m. SW. of Sheffield   3
*Timperley.* t. Cheshire, 7 m. SW. of Manchester   3 inset
*Tiverton.* t. 13 m. N. of Exeter   4
*Todmorden.* T. Yorks, 17 m. NE. of Manchester   3
*Tonbridge.* T. 12 m. WSW. of Maidstone   6
*Tong.* Vil. 19 m. E. of Shrewsbury   5
*Topsham.* t. 5 m. SE. of Exeter   4
*Torquay.* T. E. Devon coast   4
*Torrington.* t. 30 m. NW. of Exeter   4
*Totnes.* t. 22 m. SSW. of Exeter   4
*Towcester.* Vil. 8 m. SW. of Northampton   5
*Town Malling.* Vil. 4 m. W. of Maidstone   6
*Toxteth Park.* Part of Liverpool   2
*Trafford.* Vil. near Chester

*Trawsfynydd.* Vil. W. Merionethshire   4
*Tredrea.* Vil. near St. Ives, Cornwall   4
*Tring.* t. Herts, 31 m. NW. of London   6
*Trowbridge.* T. WIlts, 18 m. ESE. of Bristol   5
*Truro.* T. SW. Cornwall   4 inset
*Tullialan.* Vil. Fifeshire   1
*Tunbridge Wells.* T. 10 m. SW. of Maidstone   6
*Turdoff.* VII. Dumfriesshire   1
*Turriff.* t. 30 m. NNW. of Aberdeen   1
*Tutbury.* Vil 18 m. ENE. of Stafford   5
*Tuxford.* Vil. 22 m. NNE. of Nottingham   3
*Twickenham.* T. 11 m. W. of London   6
*Uckfield.* t . 14 m. NE. of Brighton   6
*Ulverstone.* t. 15 m. NW. of Lancaster   2
*Upminster.* t. Essex, 16 m. E. of London   6
*Upper Broughton.* Vil. 10 m. SSE. of Nottingham   5
*Upper Holland.* Vil. in Lancs   2 inset
*Uppingham.* Vil. 18 m. E. of Leicester   6
*Upton.* Many small villages of this name
*Uttoxeter.* t. 13 m. NE. of Stafford   5
*Uxbridge.* T. 17 m. W. of London   6
*Wainfleet.* Vil. 33 m. E. of Lincoln   3
*Wakefield.* T. 9 m. S. of Leeds   3
*Walford.* Vil. 13 m. SSe. of Hereford   5
*Wallingford.* Vil. Berks, 11 m. SE. of Oxford   5
*Walsall.* T. Staffs, 8 m. NNW. of Birmingham   5
*Walsham* v. *North Walsham*
*Walsingham.* Vil. 25 m. NE. of Norwich   6
*Waltham Abbey.* t. Herts, 13 m. N. of London   6
*Walthamstow.* Part of E. London   6
*Walton.* Vil. 20 m. WSW. of Norwich   6
*Walton-on-Thames.* T. Surrey, 17 m. SW. of London   6
*Walton-on-Trent.* Vil. 14 m. SSW. of Derby   5
*Wandsworth.* Part of SW. London   6
*Wangford.* Vil. 25 m. NE. of Ipswich   6
*Wantage.* t. Berks, 13 m. SSW. of Oxford   5
*Ware.* T. near Hertford   6
*Wareham.* Vil. 15 m. E. of Dorchester   5
*Warfield.* Vil. Berks, 25 m. W. of London   6
*Warminster.* t. 18 m. NW of Salisbury   5
*Warrington.* T. 15 m. E. of Liverpool   2
*Warwick*   5
*Watford.* T. Herts, 16 m. Nw. of London   6
*Watlington.* Vil. 15 m. SE. of Oxford   5
*Wattisfield.* Vil. 20 m. NNW. of Ipswich   6
*Watton.* Vil. 20 m. W. of Norwich   6
*Wavertree.* T. adj. Liverpool   2
*Weaverham.* t. 13 m. ENE. of Chester   2
*Wellingborough.* T. 10 m. ENE. of Northampton   6
*Wellington (Salop).* t. 10 m. E. of Shrewsbury   5
*Wellington (Somerset).* t. 7 m. SW. of Taunton   5
*Wells (Norfolk).* t. on N. coast of Norfolk   6
*Wells (Somerset)* . t. 17 m. S. of Bristol   5
*Welshpool.* t. Montgomeryshire, 17 m. WSW. of Shrewsbury   5
*Wem.* Vil. 12 m. N. of Shrewsbury   5
*Wenlock.* Vil. 12 m. SE. of Shrewsbury   5
*Weobley.* Vil. 10 m. NW. of Hereford   5
*West Bromwich.* T. 3 m. NW. of Birmingham   5
*Westbury.* t. 24 m. NW. of Salisbury   5
*West Cowes.* t. Isle of Wight   5
*West Derby.* Part of Liverpool   2 inset
*Westerham.* t. Kent, 25 m. S. of London   6

*West Tarring.* Vil. 8 m. E. of Brighton   6
*Wetherby.* Vil. 11 m. NE. of Leeds   3
*Weymouth.* T. on Dorset coast   5
*Whalley.* Vil. 23 m. N. of Manchester   2
*Whetstone.* Vil. near Leicester or Barnet   6
*Whiston.* Vil. adj. Prescot   2
*Whitby.* T. NE. Yorks coast   3
*Whitchurch (Hants).* Vil. 12 m. N. of Winchester   5
*Whitchurch (Salop).* t. 18 m. N. of Shrewsbury   5
*Whitehaven.* T. Cumberland coast   2
*Wickham Market.* Vil. 11 m. NE. of Ipswich   6
*Widcombe.* t. 5 m. S. of Bristol   5
*Widnes.* T. 12 m. SE. of Liverpool   2
*Wigan.* T. 17 m. NE. of Liverpool   2
*Wigton.* t. 10 m. WSW. of Carlisle   2
*Wigtown.* Vil. SW. coast of Scotland   2
*Willenhall Field.* T. 10 m. NW. of Birmingham   5
*Williton.* VII. 13 m. NW. of Taunton   5
*Wimbourne.* t. 20 m. SSW. of Salibury   5
*Wincanton.* Vil. 28 m. S. of Bristol   5
*Winchester*   5
*Winden.* One of several small villages. Prob. near Carlsruhe
*Windsor.* T. Berks, 20 m. W. of London   6
*Winster.* Vil. 19 m. NNW. of Derby   3
*Winwick.* t. 14 m. E. of Liverpool   2 inset
*Wirksworth.* t. 12 m. NNW. of Derby   3
*Wisbech.* T. 31 m. N. of Cambridge   6
*Wishaw.* T. 15 m. ESE. of Glasgow   1
*Wissett.* Vil. 20 m. SSE. of Norwich   6
*Wiston.* Vil. 11 m. WNW. of Brighton   6
*Witham.* t. 9 m. NE. of Chelmsford   6
*Witney.* t. 10 m. WNW. of Oxford   5
*Wiveliscombe.* Vil. 8 m. W. of Taunton   5
*Woburn.* Vil. 12 m. SSW. of Bedford   6
*Wokingham.* t. Berks, 32 m. W. of London   6
*Wolsingham.* Vil. 12 m. W. of Durham   3
*Woodborrow.* Prob. *Woodborough.* Vil. Notts or 18 m. N. of Salisbury   5
*Woodbridge.* t. 8 m. ENE. of Ipswich   6
*Woodford.* T. Essex, 9 m. NE. of London   6
*Woodhead.* Vil. 18 m. E. of Manchester   3
*Woodstock.*Vil. 8 m. NW. of Oxford   5
*Wooler.* Vil. N. Northumberland   3
*Woolton.* t. 5 m. SE. of Liverpool   2 inset
*Woolwich.* Sub. E. London   6
*Wootton Bassett.* Vil. 11 m. NW. of Marlborough   5
*Worcester.*   5
*Workington.* T. Cumberland coast   2
*Worksop.* T. 15 m. ESE. of Sheffield   3
*Worsted.* Vil. 12 m. NNE. of Norwich   6
*Wotton-under-Edge.* Vil. 20 m. SSW. of Gloucester   5
*Wrentham.* Vil. 17 m. SSW. of Norwich   6
*Wrexham.* T. Denbigh, 12 m. SSW. of Chester   2
*Wye.* Vil. 17 m. W. of Dover   6
*Wymondham.* t. 10 m. SW. of Norwich   6
*Yalding.* Vil. near Maidstone   6
*Yarm.* Vil. Yorks, 21 m. SSW. of Durham   3
*Yarmouth.* T. on coast 19 m. E. of Norwich   6
*Yeovil.* T. 20 m. E. of Taunton   5
*Yoel.* Vil. in Surrey   6
*York*   3
*Yoxford.* Vil. 20 m. NE. of Ipswich   6
*Ystrad.* t. 10 m. N. of Cardiff   5

# Glossary of Terms

*Anchor Escapement* The recoil escapement that has been in common use in longcase clocks since it was invented by Dr Hooke in 1670, and in bracket and wall clocks since circa 1800. See Figure 20.

*Arbor* This is an axle in a clock on which the wheels and pinions are mounted.

*Astronomical Clock*
  (a) One which indicates the movement of some of the heavenly bodies relative to each other or
  (b) One which indicates the Astronomical day, which is 23 hours 56 minutes and 4 seconds long (as measured by mean solar or our time.)
  (c) In the U.S.A. the term astronomical or "astro" clock is often used to refer to a longcase regulator (precision pendulum clock) when the dial is laid out with a centersweep minute hand and large seconds and hour rings usually, respectively below 12 o'clock and above 6 o'clock,

*Balance (wheel)* The oscillating wheel used on its own in early clocks and in conjunction with a hairspring after 1675 on watches and travelling clocks to regulate the timekeeping.

*Balance Cock* The bracket that carries the top pivot of the balance wheel staff.

*Balance Spring* The coiled spring, either flat or helical, which is used to control the speed of oscillation of the balance wheel.

*Barometric Error* The variations in the timekeeping of a clock caused by rising or falling barometric pressure which alters the density of the air through which the pendulum has to pass and thus its speed of oscillation. This is normally only of significance with Precision Pendulum Clocks (Regulators).

*Barrel* The circular container which houses the mainspring on spring-driven clocks or the drum onto which the line is wound on weight-driven clocks.

*Barrel Arbor* The axle or arbor for the barrel onto which the spring is fixed at its center.

*Bezel* The grooved ring of a watch or clock which holds the glass protecting the dial.

*Bluing* The process of turning steel blue by heating it to a high temperature and then usually wiping it with oil whilst still hot to protect it. It is used for most steel clock hands and also for other components such as screws, partly to make them more attractive but also to protect the steel.

*Bolt and Shutter* A form of maintaining power to keep the clock going while it is being wound and the driving power is removed. Refer to Figure 454.

*Center Seconds* The provision of a long seconds hand sweeping out from the center of the dial.

*Chapter Ring* The ring or dial on which the minutes and hours are marked out.

*Chiming Clock* One which sounds out the quarter hours on a set of four or more bells or gongs.

*Clepsydra* A clock which measures the passage of time by means of the flow of water.

*Count Wheel* A method of controlling the strike used on virtually all 30-hour and many longcase clocks up to circa 1720. It employs a disc with notches at its periphery at varying distances from each other which determines the number of hours which will be struck. Unlike rack striking, it is not synchronised with the hands. Hence, if for any reason the clock misses a strike, the number of hours struck will stay out until they are corrected. See Figures 22, 18 A and B, and 26.

*Crown Wheel* The 'scape wheel in a verge escapement.

*Crutch* This connects the pallet arbor, to which it is fixed, to the pendulum, the top of which usually passes through a loop in it.

*Dead Beat Escapement* One in which the 'scape wheel does not recoil. Generally used on regulators and other precision clocks.

*Equation of Time* The difference between solar (the sun's time) and mean time (our time), which may be as much as 16 minutes. Four times a year—25th Dec,

15th April, 14th June, and 31st August—the two coincide.

*Equation Tables* These tables, which show the variations in the equation of time throughout the year, are or were used to check the clock's time by using a sundial to find solar time and then adding the equation of time onto or substracting it from this. Refer to Figure 124.

*Fly* A fan, usually two or three bladed, which is used to regulate the speed at which a clock strikes or chimes. It is also occasionally employed in other circumstances, such as the rewinding of a remontoire and in gravity escapements.

*Fusee* A grooved, tapering cone onto which the line or chain from the spring barrel runs. It is designed to even out the force exerted by the spring on the movement as it runs down, and is used in nearly all English spring-driven clocks.

*Gong* A coiled or sometimes a straight rod of steel wire on which the hours or quarters may be struck. They usually sound a much-deeper note than a bell and are struck more slowly.

*Gravity Escapement* An escapement in which a constant force is imparted to the pendulum by lifting up a weight a set distance and then releasing it onto the pendulum to give it an impulse. The most famous example is the double three-legged gravity escapement designed and used by Lord Grimthorpe in the Westminster Palace clock known as Big Ben.

*Great Wheel* The largest wheel in a clock, which in longcase clocks is the one fixed to the barrel onto which the line supporting the weight is coiled.

*Gimbals* These are used on marine chronometers and barometers to suspend them in such a way that regardless of the ship's angle they will always remain upright.

*Invar* A metal, invented in 1897 by Dr Guillaume, which has an extremely low coefficient of expansion or contraction. Used in pendulums and balances in various forms to minimise any changes in timekeeping due to temperature changes.

*Isochronous* This means moving in equal times and may be used, for instance, in relation to the beat of a pendulum.

*Jewelling* Jewels are sometimes used in precision clocks and more generally watches because of their extreme hardness and ability to take a high polish. Their primary purposes are to reduce friction and wear. In regulators they are used for the pallet faces acting on the 'scape wheel, and occasionally the bearing surfaces (pivot holes) for the arbors (particularly those of the 'scape wheel and pallets). They are also employed as "end stones" to control the end float of the arbor and stop their shoulders butting up on the plates.

*Lever Escapement* The escapement most generally used on clocks and watches. It was invented by Thomas Mudge circa 1757 and employs two levers to interconnect the pallets and the balance largely superseded in the 1720s by Harrison's "going ratchet" (Figure 455).

*Matting* On longcase clocks this usually refers to the roughening of the dial center either by spiked rollers or punches.

*Mean Time* This is the time in general use. It is more correctly known as "mean solar time" as it is the average of all the solar days which are constantly varying in length throughout the year.

*Motion Work* The wheels, usually situated immediately beneath the dial on the frontplate of the movement, which gear down 12:1 so that the hour hand moves at the correct speed relative to the minute hand.

*Pillar* In longcase clocks this refers to the brass rods, usually attractively turned, which hold the two plates of the movement rigidly together. On country eight-day clocks there are usually four and on London clocks five, but this number is sometimes exceeded. On early clocks these pillars employed pivoted latches to lock the two plates together, but after circa 1700 pins were usually employed.

*Pinion* The small, toothed wheels, usually of steel, which mesh with the larger brass wheels in a clock.

*Pivot* The finely polished and reduced end of an arbor which rotates in a hole in the plates of a clock.

*Rack Striking* Rack striking was invented circa 1675 and means that the striking always stays synchronised relative to the hands. This is made it possible to provide repeat work on clocks. The basic principle consists of a segment with teeth on it which is dropped back an amount which varies with the number of hours to be struck, just before each hour. At the hour a pallet known as a "gathering pallet" advances the teeth one at a time, each one corresponding with one strike.

*Rating Nut* The nut immediately below the pendulum bob which is rotated to raise or lower it and thus regulate the timekeeping of the clock.

*Recoil Escapement* See *Anchor*.

*Regulator* A master clock used by Astronomers to measure the passage of the stars or by clockmakers for rating other clocks or chronometers. It usually incorporates a compensated pendulum to prevent any fluctuations in timekeeping with temperature change and maintaining power so that it keeps going whilst it is being wound. The regulators made in the 20th century were often contained in sealed containers at a constant pressure to prevent any variations in pressure affecting their timekeeping, and were also kept at a constant temperature.

*Remontoire* A device used to wind up a small weight or auxiliary spring at regular intervals, usually between 30 seconds and five minutes. This provides either (a) a constant force on the escapement and thus the pendulum, thereby improving timekeeping; or (b) only one mainspring may be provided and the power taken from this via the remontoire to drive a second train. This may be for the strike or time.

*Seat Board* The wooden board on which the movement of a longcase clock sits.

*Sidereal Day* See *Astronomical Clocks*.

*Silvering* In longcase clocks this refers to the chemical deposition of silver on components of the dial such as the chapter ring, which is then protected with lacquer. It should not be confused with electroplating.

*Snail* An eccentric cam which regulates the number of hammer blows struck in rack striking. See Figure 23.

*Solar Day* This is the time recorded between two successive passages of the sun over the meridian. The length of the solar day varies throughout the year by up to 30 minutes because of the uneven passage of the earth around the sun.

*Suspension Spring* The strip of steel used to support the pendulum.

*Timepiece* A clock which doesn't strike.

*Train* The series of wheels and pinions which connect the springbarrel or the barrel on which the lines of a longcase clock are wound to the escapement.

*Verge Escapement* The earliest escapement used on a mechanical clock, in which the axis of the pallets is at right angles to that of the 'scape wheel. See Figure 19.

Glossary 365

# The Components of a Clock

Figure 519. The Case.

Figure 520. The Dial.

## 366 Glossary

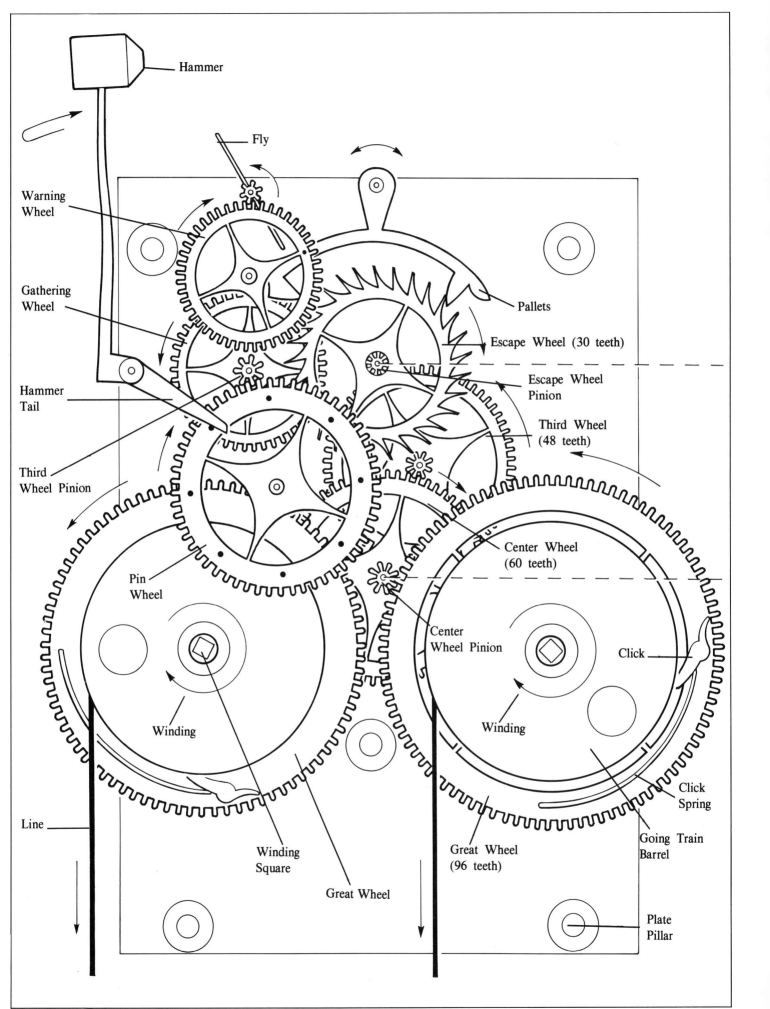

Figure 521. Eight-day movement.     Strike Work Omitted

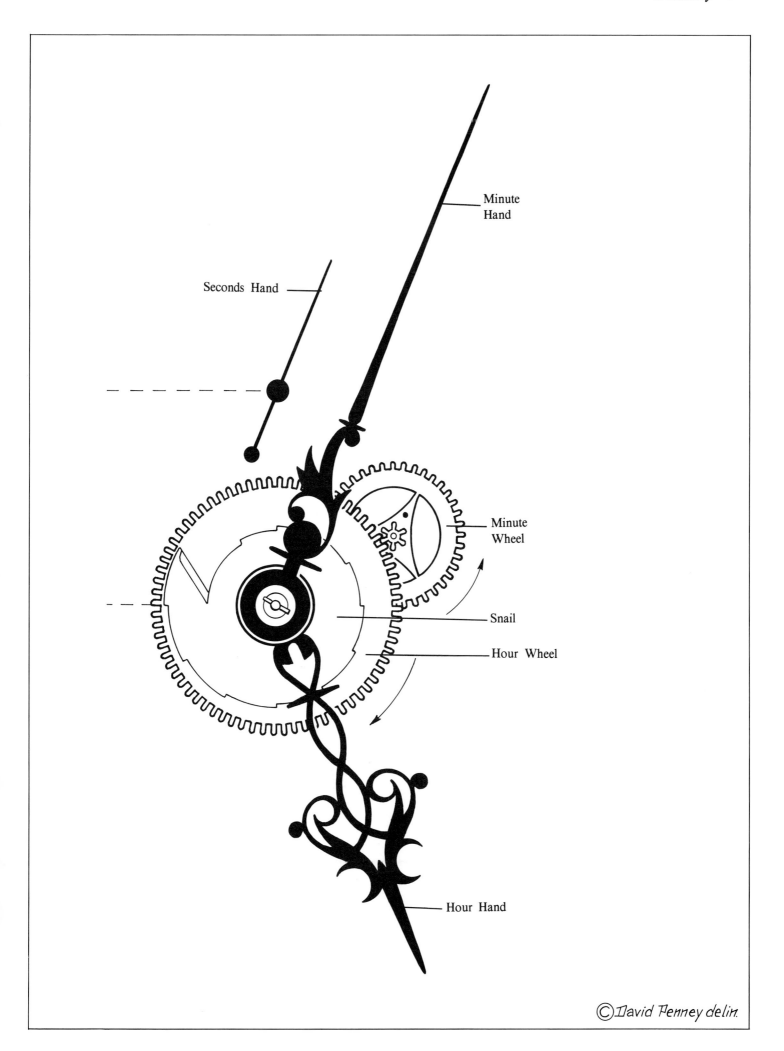

368 *Glossary*

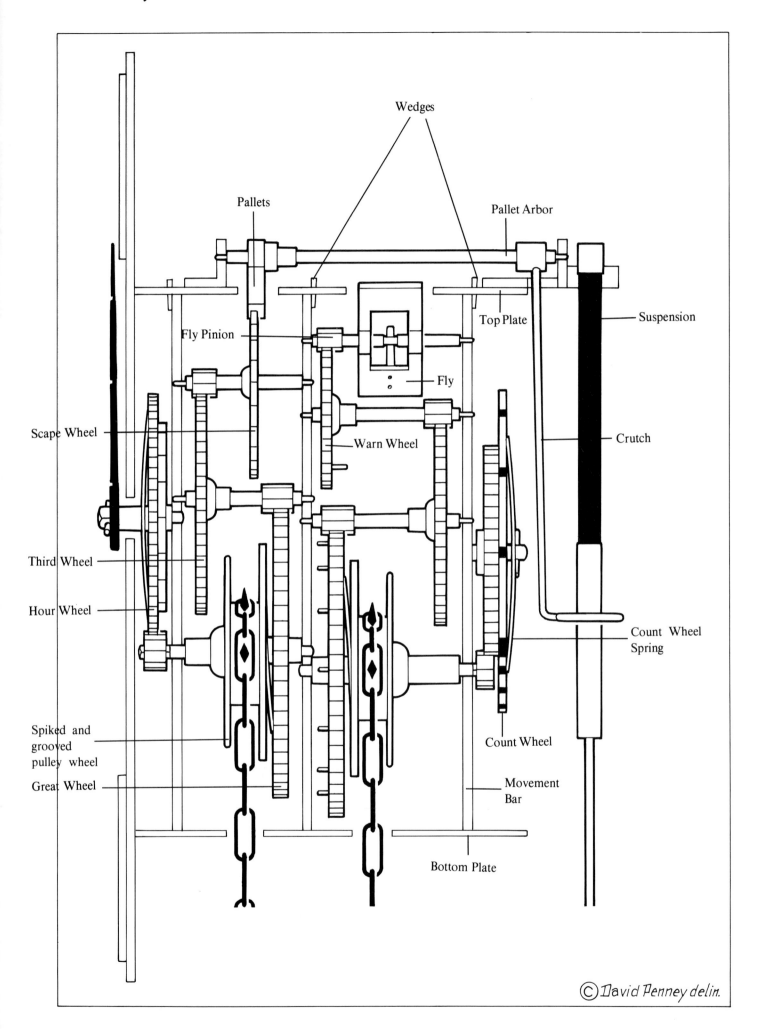

Figure 522. Thirty-hour birdcage movement.

# Makers Index

Agar - York, 191
Alexander & Son, William - Glasgow, 268
Allaway, John - London, 57
Allen, Charles - Bewdley, 254
Allen, Richard - Woolwich, 210
Andrews, John - 4, 58
Anthony, William - Truro, 184
Ashton, Thomas - Macclesfield, 182
Asselin, Stephen - London, 54
Audouin - London, 172
Auld, 264
Avenall, George - Farnham, 138
Avenall, Phillip - Farnham, 193

Bagnell - Talk, 185
Baker, Henry - Malling, 142
Baker, John - London, 150
Baker, J. - Hull, 205
Banger, Edward, 110, 311
Barber, Joseph - London, 122
Barchan - Tonbridge, 250
Barker - Wigan, 79
Barron, John - Aberdeen, 231
Barron, John - Hermitage, 133
Barwise - London, 159
Bates, Jos. - London, 62
Bayfield, Richard - Great Dalby, 198
Behoe, John - London, 124
Bennett, Thomas, 323
Benniworth - St. Albans, 176
Berry, James - Pontefract, 198
Biggs, Richard - Romsey, 215
Billings, John - London, 157
Bird, Edward - 51
Blackie, George - Musselburgh, 236
Blanchard, William - Hull, 275
Blaylock - Carlisle, 267
Boby, Obadiah - Battle, 244
Boffi, P. & A. - Hastings, 239
Booth, George - Manchester, 192
Bowtell, Samuel - London, 60
Bowen - London, 154
Bower, John - Kirriemuir, 231
Bradbar - Leyburn, 240
Brace, Joshua - Chepstow, 141
Branson - Hull, 275
Bromley - Horsham, 218
Brock - London, 252
Brock, James, 294, 295
Bruce, R. - Wigtown, 196
Bryson, Robert & Alexander, 264
Bryson, William - Kilmarnock, 212
Budgen, Thos. - Croydon, 79, 89, 147
Buchan, William - Stonehaven, 262
Buckney, Thomas, 292
Buffet, John - Colchester, 140
Burgess, John - Gossport, 142
Burgi, Jost, 283, 285
Burputt, Richard - London, 121
Burroughs, Edward - Farnham, 208
Burton & Pattison - Halifax, 272

Carter, William, 155
Cawdle, W. - Torquay, 251
Cawston, Russell, 225
Chapman, John - Loughborough, 234
Christian, John - Aylesham, 190
Christie, James - Perth, 233
Clark, James - Bulcombe, 206

Clark, Thomas - Warrington, 127
Clay, Charles - London, 164, **325**
Clement, Francis - 52
Clement, William - 10, 46, 52, **309**
Clower, James, 65, 132
Coates, John - Coventry, 211
Cockey, Edward - Warminster, 79, 81, 82, 83, 84, 86
Cohen, Emmanuel - Redruth, 234
Colley, Richard - London, 121
Condliff - Liverpool, 302
Cooper, Charles - Whitchurch, 190
Cornwell, Daniel - Billericay, 194
Coster, Soloman - 10, 283
Coulon, Charles - London, 146
Cowell, John - London, 149
Cranbrook, Stephen - Dover, 188
Crolee, John - London, 124
Crump - Canterbury, 150

Daking, Richard - Halstead, 212
Daring, Robert - Edinburgh, 207
Davey - Norwich, 230
Deane, William - London, 11
Delander, Daniel - London, 104, 113, 134
Denison, 282
Desiemer, George - Portsmouth, 227
de Stokes, Laurentius, 79
Dicker, Thomas - Silchester, 141
Dondi, Giovanni - 79
Donisthorpe - Birmingham, 216, 248
Du Chesne, Claudius - London, 36
Dunlop, Conyers - London, 148
Durward, J. - Edinburgh, 265

East, Edward - London, 4, 23, 28, 46, 70, **308**
Ebsworth, John, 5, 29
Elkin, William, 323
Elkins, Thomas - Barslem
Ellicott, John - London, 74, 117, 118, 165, 298, **319**, 320
Elliott - London, 256, 259, 296
Elliott & Son - Ashford, 228
Ellory, Thomas - St. Austle, 277
Ely, James - London, 163
Etherington, George - London, 16, 54
Evans, H. - Llangadock, 251
Everell, John - London, 62

Fannell, Joseph - Poole, 254
Farley, Richard - London, 160, 173
Favey, Robert - London, 159
Ferguson, Alexander - Edinburgh, 180, 264
Finney - Liverpool, 97
Fish, Henry - London, 132
Fletcher, Wm. - Gainsborough, 230
Flint, William - Charing, 244
Foster, Thos. - London, 63
Fowlds, Allan - Kilmarnock, 200
Frodsham, Charles - London, 282, 283, 285
Fromanteel, Abraham, 306
Fromanteel, Ahasuerus - London, 10, 23, 24, 25, 80, 165, **304**, 306, 307
Fromanteel, Daniel, 306
Fromanteel, John, 306
Fromanteel & Clark, 306
Fry, John - Kilmersden, 139
Fryer - Pockington, 250

Gandy - 90

Gandy, James I, 90
Gandy, James III, 90
Gatten, Edward - London, 174
Gilbert - Hythe, 237
Gillet & Johnson - Croydon, 256
Gillows - Lancs., 270
Gladding & Co - Brighton, 161
Glover, William - Worcester, 209
Goddard, Ben - London, 136
Goode, Charles - London, 32, 137
Gordon, John - Aberdeen, 204
Gould, Christopher, 49
Graham, George, 13, 22, 111, 112, 291, 296, 311, **314**
Gratton, Charles - London, 59
Gravel & Tolkein - London, 157
Gray, Benjamin, 38
Green - Liverpool, 271
Green, Jn. - London, 220
Gregg, Francis, 74
Gregory, Robert - London, 53
Guilliam - Falmouth, 277

Hall, Thomas, 142
Hall, Robert - Chichester, 170
Hardwick, Richard - Ashkirk, 203
Hardy, William, 264
Harley, Samuel - Salop, 189, 243
Harriman, Edward, 90
Harriman, John, 90
Harris, Step - Tonbridge, 242
Harrison, John, 22, 292
Haswell - London, 163
Hawkins, John - Southampton, 133
Hay, William - Airdrie
Hedge, Nathanel - Colchester, 186
Herbert, Cornelius - London, 135
Heslop, Jn. - Newcastle, 235
Hilderson, John, 23, 26
Hoadley, William - Rotherfield, 245
Holt, Richard - Newark, 238
Hopkins, John - Deptford, 126
Horseman, Stephen, 106, 140, 315
Hosmer - Tonbridge, 249
Hovil, John - London, 151
Howell, John - London, 155
Hubbard, John - London, 155
Hubert, David - London, 121
Hunter, Wm. - Dunfermline, 232
Hunt, John - London, 56
Hutchinson - Retford, 237
Hutton, James - Edinburgh, 206

Irish, James - Lewes, 244
Isaac, James - Carmarthen, 233

Jackson, J. - Boston, 224
Jacob, Moses - Redrith, 278
Jenkins, Henry - London, 156
Johnson, Benjamin - London, 55
Jones, John - Aberystwyth, 236
Jordan - Great Missenden, 232

Kay, John - Liverpool, 207, 272
Kellett, S. - Bradbury, 240
Kemp, N., 249
Kinnear, Chas. - Dunbar, 238
Knibb Family, **322**, 324
Knibb, Johannes, 47

# Makers Index

Knibb, John - Hanslope, 323
Knibb, Joseph - London, 26, 29, 30, 38, 47, 48, 49, 65, 67, 69, 71, **322**
Knibb, Samuel, **322**
Knight - Chichester, 208

Lambert, George - Blandford, 190
Lambert, L. - Haxey, 228
Lawson, Christopher - Edinburgh, 266
Lawson, William - Keighley, 245
Le Bond, Robert - London, 122
Lecomber - Liverpool, 302
Le Plastrier - Deal, 162, 218
Lee, Thomas - Pottern, 233
Lee, William - Leicester, 166
Lines, Samuel - Luton, 153
Lister - Halifax, 79
Lister, Thomas Jr. - Halifax, 96
Lister, Thomas - Luddenden, 248
Lloyd, Philip - Bristol, 94, 95
Long, John - London, 150
Longridge - Burwash, 249
Louch - St. Albans, 208
Lovelace, William - London, 123
Lumsden, George - Pittenweem, 268

Mallet, Peter - London, 61
Marshall - Dublin, 39
Markwick/Markham, 139
Mason, William - London, 149
Massingham - Fakenham, 187
Mathews - Leighton, 194
Mayhew, Wm. - Woodbridge, 195
McGregor, Thomas, 268
Meere, Stephen - Hampnell, 206
Mirgan - London, 256
Mitchell & Sons - Gorbals, 214
Monk I.- Hsing, 79
Monkhouse, John - London, 148
Moore, Thos. - Ipswich, 241
Morgan, John - Monmouth, 123
Morgan, Walter - Hereford, 138
Morrison, Theodore - London, 152
Moses, John - Bishops Auckland, 303
Motley, Richard - London, 56
Muirhead, James - Glasgow, 267

Nash, William - Bridge, 186
Neale, George - London, 140
Nevitt, Thomas - Bristol, 280
Newman, James - Lewes, 247
Nickisson - Newcastle, 231
Nicoli, James I - Edinburgh, 197, 264
Noon, A. - Ashby, 199, 229
Norton, Eardley, 14, 165, 175, **321**

Ogden, John - Darlington, 202
Oliver, John - Manchester, 186
Osborne & Wilson - Birmingham, 14, 219
Oxley, John - Norwich, 128

Pagnard, Pierre - Jersey, 232
Panchaud, Abel - London, 144
Pannell, Hugh - Northallerton, 209
Pare, Thomas - London, 32, 33
Parker, Robert - Derby, 212
Parr, B. - Grantham, 222, 223
Payne, William - Ludlow, 221, 273
Peachey, Newman, 156
Peyton, Richard - Gloucester, 178
Piggin, John - Norwich, 237
Pike, James - Newtown Abbott, 191

Pinchbeck, Christopher, 88, **316**, 317
Pitt, William - London, 161
Poignard, Pierre - Jersey, 232
Pollard, James - Plymouth, 217
Porthouse, William - Penrith, 125
Pridgen - Hull, 193
Pringle - Dalkeith, 263

Quare, Daniel, 37, 65, 80, 106, 140, **315**

Radeloff, 285
Rayment - Stamford, 177
Rayman, William - Dublin, 269
Reed, Richard - Chelmsford, 151
Reid, 264
Reid & Auld, 301
Reid, J.D. - Airdrie, 267
Reinman, Paulus - Normbergae, 8
Reith, James - London, 67
Rice, B. - Neath, 251
Rich, Andrew - Bridgewater, 232
Riefler, S., 291, 294
Ritchie, James, 264
Robb, William - Montrose, 99
Roberts, Peter - Newport, 209
Robertson, William - Dunbar, 227
Robinson, Francis - Northampton, 38
Robson Jr. - North Shields, 230
Rogers, Robert - Bristol, 203
Roskell, Robert - Liverpool, 160
Russell, John - Falkirk, 198

Salmon - London, 163
Sampson - Wrexham, 193
Sapley, John - Edinburgh, 119
Scaffe, William, 74
Scobell - Ottery, 233
Scott, James - Leith, 213
Scholfield, Major, 75
Sellers, William - London, 62, 120
Sellon, Humphrey - London, 141
Sewell, Greg Sr. - London, 145
Shapley - Stockport, 188
Sharly, Thomas - Leighton, 246
Sharpe, William - London, 55
Shelton, John - London, 296, 297
Sherwood, John - Hythe, 235
Simpson, Hector - London, 154
Smiths - Clerkenwell, 256
Smith, Edward - Newark, 200
Smith, E. - Richmond, 153
Smith, Gabriel -Barthomley, 35
Smith, John - Pittenweem, 199
Smith, Samuel - London, 181
Smith, William - London, 157
Smyth, James - Saxmundham, 196
Snelling, John - Alton, 187
Spark, William - Aberdeen, 262
Speakman, Thomas - London, 59
Spendlove, John - Thetford, 214
Standring, Jeremiah - Bolton, 201
Stewart, Thomas - Auchterarder, 224
Stevenson, Charles - Congleton, 246
Stokes, W., 68
Strattern, James - Hampton, 217
Sutton, Thomas - Maidstone, 194
Symonds, John - Reepham, 226

Taylor, John - London, 168
Terry - Manchester, 257
Thackall, W. - Banbury, 225
Thackwell - Ross, 247

Thwaites & Reid, 300, 321
Thomas, R. - Cornwall
Thompson, 92
Thompson, John - London, 137
Thompson, Joseph, 92
Thompson, Plaskett, 92
Thompson, William, 92
Thorn, James - Colchester, 184
Thornton, Yates - Romford, 210
Tobias, Evan - Llandilo, 225
Tompion, Thomas, 12, 20, 29, 36, 39, 65, 72, 73, 78, 80, 87, 107, 110, 111, 112, 113, 286, 287, **310, 312, 313**
Topping, John, 74
Townley, 111, 286
Trail, William - London, 156
Trattle, Joseph - Newport, 195
Tregent, Anthony / Enameller, 220
Tregent, James, 220
Trigg, Thomas - London, 33
Tucker, Charles - London, 118
Twigg, Thomas - Shilton, 228, 273

Upjohn, James - Brentford, 221
Usher, J. / Casemaker, 223

Valin, Nicholas, 165
Vokins, George - Newtown, 215
Vulliamy, Justin, 38, 153, 158

Waller, Thomas - Preston, 128
Walsh, Henry, 284
Walsh - Newbury, 80, 283
Walsham, William, 79
Warner, Thomas - London, 118
Warren, James - Canterbury, 224
Watson, 79
Webster, T. - London, 217
Webster, William, 87
Welch, Robert - Dalkeith, 221
Wells, H. - Framlingham, 276
Weston, Samuel - Stratford, 135
Whitelaw - Edinburgh, 266
Whitern, J. - Abingdon, 189
Whittaker, William, 211
Wightman, James - London, 50
Wightman, Joseph, 323
Wilcox, Edward, 183
Williams, Thomas - Chewstoke, 203, 280
Williamson, 80, 288
Williamson, Joseph, 108
Wills, William - Truro, 227, 231
Wilmshurst - Burwash, 237
Windmills, 115
Windmills, Joseph, **323**
Windmills, Thomas, **323**
Wise, John - London, 5, 26, 60, 61
Wright, Edmund - London, 65, 66

Youl, George - Edinburgh, 223

# General Index

Aberdeen, 264
Alarm, 15, 124, 241
Anti-Friction Rollers, 294, 295, 301
Aperture, Moon, 66
Apparent Time, 132, 286
Arbor Barrel, 358
Astrolabe Clock, 78
Astrological Circle, 95
Astronomical
    Clocks, **79**, 83, 86, 87, 88, **90**, 95, 96, 97, 317, 320, 358
    Dial, 99, 102
    Longcase, 84
Automata, 113, 181, 216, 237
    Musical, 319
    Rocking Ship, **210**

Balance
    Cock, 358
    Spring, 358
    Wheel, 10, 358
Barometers, Pillar, 315
Barlow, Edward, 315
Barometric
    Compensation, 293
    Error, 358
    Pressure, 292
Barrel, 358
Bartrum, C.O., 294
Bezel, 358
Big Ben, 166
Blacksmiths, 304
Bluing, 358
Bone-Stained, 45
Bristol, 94, 203, 279
British Museum, London, 6, 83, 86
Buckingham Palace, 317

Caddy Top, 12, 114
    Marquetry, 55
    Two-Tier, 55
Calendar
    Dial, 118
    Church Festivals, 98
    Flyback, 222
    Gregorian, 85, 99
    Work, 124
        Setting up of, 345
    Year, 67, 76, 80, 85, 90, 105, 107, 164, 222
    Zodiac, 98
Cam, Equation, 80, 130
Campari Brothers, 70
Carrington Clock, 83
Case, 2
    Care of, 338
    Drumhead, 266, 267
    Tapering, 266, 267
Celestial Sphere, 99
Central Heating, 338
Center Seconds, 358
Center Sweep, 170
Chapter Ring, 5, 110, 170, 358
    Dutch Style, 90
    Enamelled, 149
    Skeletonized, 26, 30, 37
    Solid, 182

    Twenty-Four Hour, 84, 89
Charles I, 308
Chimes, Westminster, 166, 256
Chiming, 358
    Five Tube, 253, 256
    Movement, 166
    Nine-Tube, 169, 253
    Tube, 257, 258, 259
Chinoiserie Decoration, **131**
Chippendale
    Brickwork, 207
    Chinese, 13, 134, 177
Cleopatra's Needle, 9
Clepsydra, 9, 283, 358
Clocks
    Astrolabe, 78
    Astronomical, **79**, 83, 86, 87, 88, **90**, 95, 96, 97, 317, 320
    Early English, 23
    Electric, 278
    Equation of Time, 11, 67, 76, 79, 95, 104, 105, 114, 155, 222, **286**, 313, 321, 358
    Fire, 9
    Grandmother, 253, 257
    Mechanical, Early, 10
    Night, **70**, **71**
    Nineteeth Century, 14
    Oil, 15
    Organ, 325
    Portable, 9
    Shadow, 9
    Spring, 312
    Turret, 309
    Water, 9, 13
Clockmakers Company, 304
Cockermouth, 90, 92
Columns
    Corinthian, 24
    Double, 270
    Hood--Omission of, 157
    Quarter--Brass-strung, 148
Compensation
    Mercury, 282
    Thermal, 290
    Zinc and Steel, 252, 292
Contrate Wheel, 18
Country Clockmaking, 14, **201**
Countwheel, **19**, 358
Cornwall, 276
Correction Tables, 9
Cresting, Carved, 29, 30
Crown Wheel, 358
Crutch, 358
Cycloidal
    Curve, 283
    Cheeks, 283

Derham William, Artificial Clockmaker, 310
Devon, 276
Dial, **153**
    Astronomical, 99, 102
    Black, 238
    Brass all over engraved, 154
    Brass all over silvered, 150, **213**, 247
    Breakarch, 12, 113, **192**
    Calendar, 118

    Circular, **160**
    Convex, 238
    Dished, 238
    Early, 11
    Enamelled, 220
    Flat Brass, 14
    Fourteen-inch, 139
    Lunar Equation, 127
    Night, 70
    Nine-inch, 5
    Painted, 14, 153, 213, 219
        Restoration of, 337
    Silvering of, 337
    Ten-inch, 5
    Tidal, 94, 164, 279
    Twenty-four hour, 78, 94, 95, 207, 300, 313
    Two-part, 137
    White, **219**, **249**
Dial Center, 5
    Florally Engraving, 26
    Tudor Rose, 33, 58
Dial Feet, 219
Dial Restoration, 336
Door
    Front opening, 11
Draconic Month, 99
Dublin, 269
Duration, **65**
    Month, 54, 55, 65, 137
    Six-month, 65
    Thirty-hour, 5, 14, 65, **241**, 242, 248
    Three-month, 68, 84
    Year, **64**, 65, **66**, 67, 132, 286, 311, 312, 313, 315
Dutch Influence, 124

East Anglia, 276
Edwardian, **254**
Eight Day
    Oak, **183**
    Winding of, 344
Ellicott, Edward & Sons, 320
Ellicott, John & Son, 320
Ellicott & Taylor, 320
Elm
    Burr, 128
    Pollarded, 128
Enamels, Hard, 338
End Stones, 294
England
    Map of Counties, Towns & Villages of, 346
    North-East, 349
    North-West, 348
    South, 351
    South-East, 279, 352
    South-West, 350
Engraver Travelling, 213
Equation Clock, **79**, 95, 104, 107, 313, 315
Equation Lunar, 127
Equation of Natural Days, 105
Equation of Time, 11, 67, 76, 79, 95, 104, 105, 107, 114, 155, 222, **286**, 313, 358
Equinox, 321
    Autumnal, 103
    Vernal, 103
Escapement, 21

Anchor, **18**, **286**, 309, 358
Cross Beat Knibb, 322
Cross Beat, 283, 285
Cylinder, 314
Dead Beat, **21**, 286, 314, 358
Dead Beat--Graham, 111
Denisons, 295
Duplex, 108, 113
Gravity, 252, 294, 359
Gravity--Four Legged, 282, 294, 295
Lever, 359
Pin Wheel, 284
Recoil, 360
Sprung Pallets, 301
Tic-tac, 72
Verge, **17**, 360
Evelyn, 306
Expansion, Coefficient of, 22

Ferguson, 94, 95, 261, 319, 321
Fitzwilliam Museum, Cambridge, 6, 78
Flamstead, 286
Fly, 359
Foliot, 10
Framlingham Castle, 276
Fuzee, 359

Gama, Vasco de, 131
Galileo, 10, 16, 283
Gathering Pallet, 19
Gesso, 131
Gilding
    Fire, 337
    Mercurial, 337
Gimbals, 359
Going Ratchet--Harrison's, 20
Golden Number, 99
Gongs, 359
    Coiled, 166
    Straight, 166
Gothic, 253
Gravity, 296
Gravity Lever, 294
Gray, Lord, 301
Great Wheel, 359
Greenwich Royal Observatory, 65
Grimthorpe, Lord, 294
Guilds, 304
Guillaume, Dr., 22

Halifax, 270
Halifax Moons, 192, 270
Hammer, Vertically pivoted, 24
Hands, **327**
    Brass, 220
    Evolution from Weapons, 330
Harris, Lord, 86
Harrison, 291
Hayle, 278
High Tide, 88, 99
High Water, 94, 123, 139, 203, 217, 261, 276
Hood
    Box Top, 278
    Breakarch Top, 152
    Lift up, 11, 18
    Pillars, 114
    Removal of, 339
Hooke, Robert, 10, 310

Horizon--Perceived, 78, 102
Hull, 274, 275
Huygens, Christian, 10, 80, 283, 306
    Continous Rope, 288

Invar, 22, 290, 292, 359
Ireland, 269
Isochronicity, 286
Isochronous, 359

Japanning, 131
Jewelling, 294, 359

Key Wound, 242
Kinfauns Castle, 301

Lacquer, **13**
    Black, 138
    Dark Green, 135
    Decoration, **131**
    Japanese, 134
    Painted Scenes, 144, 145
    Red, 133, 140
    Tortoiseshell, 137
Lancashire, 270
Lenticle, 46, 137
    Glass, 29
Lincolnshire, 275
Liverpool, 270, 272
Livery, 304
London, 252
London Bridge, 164
"London" Mahogany, 147
Longcase
    Architectural, 24, 25
    Care of, 337
    Dismantling of, 339
    Setting up of, 340
Longleat, 81
Loomes, Brian, 213, 219, 220
Lunar
    Disc, 94, 113
    Equation, 127

Mahogany, 13, 242
    Cuban, 147
    Honduras, 147
Maintaining Power
    Bolt and Shutter, **20**, 51, **288**, 297
    Harrison's Going Ratchet, 288, 289
Maplewood, 128
Marquetry, **41**
    All Over, 55, 58, 61
    Arabesque, 10, 43, 45, **61**, 62
    Bird & Flower, 44, 45, 53
    Floral, 10, 41
    Floral Panelled, 50, 54
    Seaweed, 10, 42, 44, 45, **62**
Maryport, 90
Mask, 11
Matting, 23, 359
Mean Time, 79, 80, 110, 114, 286, 359
Mercury, 22
Metonic Cycle, 99
Minutes, Full 1-60 Marking of, 49
Minute Ring, Movable, 80
Moon, 276
    Arch, 138

Ball, 85
Discs, 345
Globe--Rotating, 87, 192, 202
Halifax, 270
Nodes, 99
Penny, 108
Phases, 124
Rolling, 113
Southing, 94, 98
Moore, Sir Jonas, 65, 286, 311
Mostyn, Lord, 312
Motion Work, 359
Moulding
    Crested, 31
    Ebony, 149
Movement
    Birdcage, 241
    Latched, 40
    Maintainance of, 338
    Plated, 5, 270
    Removal of, 340
    Setting in Beat, 342
    Single-Handed, 270
    Thirty-Hour, 5, 248
    Three-Month, 84
Mulberry Wood, 36, 128
Musical Clocks, 122, 125, 146, 165, 166, **170**, 240, 242, 248, 278, 280, 323
    Organ, 325

Name Plaque, 113
National Maritime Museum, 81, 82, 83
Newcastle-on-Tyne, 274
Nice Family, 99
Norfolk, 276
North East of England, 274

Oak, **183**, 242
Olivewood, 10, 48, 49
Orrery, 317
Oyster Veneers--Boxwood, 72

Pagoda, 126
    Top, **142**, 147, 194
Painted Glass, 278
Panelled, 2
    Sides, 24
Parquetry, 10, **41**, 48, 72
    Star, 41, 46, 184, 192
Pediment, Swan-neck, **196**
Pendulum
    Aluminium Rod, 282
    Compensated, **22**, 314, 320
    Effect of Gravitational Force on, 296
    Effect of Magnetic attraction on, 320
    Ellicott--Compensated, 299, 320
    Free, 294
    Gridiron, 283, 284, 290, 291
        Harrisons, 320
    Master & Slave, 294
    Mercurial, 285, 290, **291**
    One and a quarter seconds, 55, 67
    One and a half seconds, 67
    Short Bob, 25
    Rise/Fall, 74
    Royal, 286
    Thirteen-Foot, 65, 286
    Two-seconds beating, 65

Variation of Length with Latitude, 320
Wood Rod, 22, 290
Zinc/Steel Compensation, 252, 292
Penny Moon, 108
Perceived Horizon, 78, 102
Philadelphia, 99
Pillars, 359
Bracket Clocks, 163
Hood, 114
Pin barrel, 166, 172
Pinchbeck Gold, 88, 316
Pinion, 359
Pivot, 359
Plates
Dome-Shaped, 68
False, 219
Intermediate, 219
Latched, 68
Shaped, 23
Split, 69
Poker Work, 53, 55
Precision Timekeeping, **283**
Prestige, Ernest, 78
Pump Room, Bath, 107

Quaker Clockmakers, 144
Quarter Chiming, 115, 120, 158, **164, 165**, 270
Pump Action, 155, 165, 259
Quarter Striking, 165
Queen Anne, 82, 83, 327

Rack Striking, 19
Rack Tail, 19
Rating Nut, 360
Regency Period, 14
Regional Characteristics, 263
Regulation
Butterfly Nut, 68
Regulator
Arch-Topped, 299, 303
Case, **296**
Dome-Topped, 253
Remontoire, 360
Repeat Work, 315
Reifler, Sigmund, 291, 294
Relative Humidity, 338
Rocking Ship, 276
Rollers--Anti Friction, 294, 295, 301
Rolling Ball Clock, 285
Royal Clocks, 325

Sandglass
Set of four, 12
Ships, 14
Science Museum, London, 6
Scotland, 263, 347
East Coast, 264
Scottish Clockmaking, 264
Seatboard, 340, 360
Seconds in Arch, 74, 75, 76
Sectors
Sun, 207
Tides, 207
Shortt, W.H., 294
Sidereal
Day, 360
Seconds, 110
Time, 9, 80, 110

Year, 95
Single Hand, 52
Silvering, 360
Snail, 19, 360
Solar
Day, 79, 360
Hand, 223
Seconds, 110
Time, 64, 80, 108, 114, 286
Solstice
Summer, 103
Winter, 103
Spandrels, 5, **326**
Cherub, 4, 23
Chinese Style, 175
Engraved, 327
Four Seasons, 142
Lion & Unicorn, 248
Queen Anne, 85
Twin Cherub & Crown, 11, 59
Winged, 11
Spoon Lock, 18, 339
St. James Palace, 83
St. Michael Chimes, 256
Stars, Southing of, 88
Strike
Countwheel, 16, 22, 343
Double Bell, Hour, 158, 159
Double Six-Hour, 69, 322
Grande Sonnerie, **69**, 321
Rack, 19, 343, 359
Roman, 30, 68, 322
Ting Tang Quarter, 72
Suffolk, 276
Sun--Amplitude of, 90
Sun Dial
Column, 10
Garden, 9, 11
Giant, 9
Ring, 9
Tablet, 9
Sun
Fast/Slow, 105
Rise/ Set, 89
Shutters, 84
Declination of, 95, 102, 321
Sun's Time, 114
Suspension Spring, 360
Synodic Months, 99

Tables
Equation of Time, 80, 140, 286, 287, 359
Temperature Control, 292
The Temple of the Four Great, Monarchies, 325
Theelke, Anthony, 220
Thermal Compensation, 290
Thirty-Hour Clocks, 5, **241**, 242, 248
Setting up of, 344
Time
Apparent, 132, 286
Keeping, Regulation of, 342
Mean, 79, 80, 110, 114, 286, 359
Measurement of, 9
Solar
*see also* Sidereal
Timepiece, 360
Tompion, Drayton, 64
Tompion, Mostyn, 311

Trade Label, 155
Train, 360
Addition of, 166
Inverted, 102
Reversed, 286
Treffler, Johann Philipp, 10
Tubular Bells, 166, 169

Vauxhall Plate, Mirror, 136
Veneers
Oyster, 41, 46
Verre Egolomisee, 270, 272
Victorian, **254**
Vinci, Leonardo da, 10
Von Bertele, 283, 285

Wadham College Clock, 322
Walnut, 12, 29
Break arch, **113**
Warminster, 82
Wear of Movement, 294
Wells Cathedral Clock, 10, 15
West Country, 234, 236, 263, 276
Westminster Chimes, 256
Weymouth, Lord, 83
Whitehaven, 90, 92
Whittington Chimes, 256
Winding Holes, Ringed, 4, 32
Workington, 90
Worshipful Company of Clockmakers, 304
Wyke, John, 272

Yew Wood, 128

Zodiac, 89
Calendar, 99
Thirty-Degree Divisions, 99
360-Degree Divisions, 85
Scales, 85
Signs, 90